Waste Management and Life Cycle Assessment for Sustainable Business Practice

Idris Olayiwola Ganiyu
York St. John University, UK

Odunayo Magret Olarewaju
Crown College St. Bonifacius, USA

Adejoke Yesimi Ige–Olaobaju
University of Northampton, UK

Sulaiman Olusegun Atiku
Namibia University of Science and Technology, Namibia

A volume in the Advances
in Logistics, Operations,
and Management Science
(ALOMS) Book Series

Published in the United States of America by
    IGI Global
    Business Science Reference (an imprint of IGI Global)
    701 E. Chocolate Avenue
    Hershey PA, USA 17033
    Tel: 717-533-8845
    Fax: 717-533-8661
    E-mail: cust@igi-global.com
    Web site: http://www.igi-global.com

Library of Congress Cataloging-in-Publication Data

Names: Ganiyu, Idris, 1978- editor. | Olarewaju, Odunayo, 1987- editor. |
   Ige-Olaobaju, Adejoke Yemisi, 1976- editor. | Atiku, Sulaiman Olusegun,
   1980- editor.
Title: Waste management and life cycle assessment for sustainable business
   practice / edited by Idris Ganiyu, Odunayo Olarewaju, Adejoke
   Ige-Olaobaju, Sulaiman Atiku.
Description: Hershey, PA : Business Science Reference, [2024] | Includes
   bibliographical references and index. | Summary: "The book presents a
   comprehensive overview of the principles, strategies, and best practices
   for implementing green management practices by businesses globally"--
   Provided by publisher.
Identifiers: LCCN 2024005797 (print) | LCCN 2024005798 (ebook) | ISBN
   9798369325957 (hardcover) | ISBN 9798369325964 (ebook)
Subjects: LCSH: Management--Environmental aspects. | Sustainable
   development. | Social responsibility of business. | Environmental
   degradation.
Classification: LCC HD30.255 .W37 2024 (print) | LCC HD30.255 (ebook) |
   DDC 658.4/083--dc23/eng/20240312
LC record available at https://lccn.loc.gov/2024005797
LC ebook record available at https://lccn.loc.gov/2024005798

British Cataloguing in Publication Data
A Cataloguing in Publication record for this book is available from the British Library.

For electronic access to this publication, please contact: eresources@igi-global.com.

Advances in Logistics, Operations, and Management Science (ALOMS) Book Series

John Wang
Montclair State University, USA

ISSN:2327-350X
EISSN:2327-3518

MISSION

Operations research and management science continue to influence business processes, administration, and management information systems, particularly in covering the application methods for decision-making processes. New case studies and applications on management science, operations management, social sciences, and other behavioral sciences have been incorporated into business and organizations real-world objectives.

The **Advances in Logistics, Operations, and Management Science** (ALOMS) Book Series provides a collection of reference publications on the current trends, applications, theories, and practices in the management science field. Providing relevant and current research, this series and its individual publications would be useful for academics, researchers, scholars, and practitioners interested in improving decision making models and business functions.

Coverage

- Computing and information technologies
- Marketing engineering

IGI Global is currently accepting manuscripts for publication within this series. To submit a proposal for a volume in this series, please contact our Acquisition Editors at Acquisitions@igi-global.com or visit: http://www.igi-global.com/publish/.

Titles in this Series

For a list of additional titles in this series, please visit:
http://www.igi-global.com/book-series/

Transformative Roles of Women in Public and Private Sectors
Ebtihaj Al A'ali (University of Bahrain, Bahrain) Meryem Masmoudi (Applied Science University, Bahrain) and Gardenia AlSaffar (Royal Bahrain Hospital, Bahrain)
Business Science Reference • copyright 2024 • 337pp • H/C (ISBN: 9798369332085) • US $295.00 (our price)

Chaos, Complexity, and Sustainability in Management
Elif Cepni (Karabuk University, Turkey)
Business Science Reference • copyright 2024 • 288pp • H/C (ISBN: 9798369321256) • US $215.00 (our price)

Leadership Action and Intervention in Health, Business, Education, and Technology
Darrell Norman Burrell (Marymount University, USA)
Business Science Reference • copyright 2024 • 372pp • H/C (ISBN: 9798369342886) • US $345.00 (our price)

Holistic Approach to AI and Leadership
Hesham Mohamed Elsherif (Independent Researcher, USA)
Business Science Reference • copyright 2024 • 395pp • H/C (ISBN: 9798369326954) • US $295.00 (our price)

For an entire list of titles in this series, please visit:
http://www.igi-global.com/book-series/

IGI Global
PUBLISHER of TIMELY KNOWLEDGE

701 East Chocolate Avenue, Hershey, PA 17033, USA
Tel: 717-533-8845 x100 • Fax: 717-533-8661
E-Mail: cust@igi-global.com • www.igi-global.com

Table of Contents

Editorial Advisory Board xiv

Preface xv

Chapter 1
A Review of Reverse Logistics Models Based on Operations Research
Techniques 1
    Naghmeh Rabiei, Toronto Metropolitan University, Canada
    Saman Hassanzadeh Amin, Toronto Metropolitan University, Canada
    Saeed Zolfaghari, Toronto Metropolitan University, Canada

Chapter 2
Demystifying Sustainable Practices in the Hotel Industry: Evidence From
Ghana 20
    Mildred Nuong Deri, University of Energy and Natural Resources,
        Ghana
    Amrik Singh, Lovely Professional University, India
    Perpetual Zaazie, University of Energy and Natural Resources, Ghana
    David Anandene, University of Cape Coast, Ghana

Chapter 3
Eco-Friendly Solutions for Sustainable Waste Management: An Approach
Towards a Cleaner and Greener Future 43
    Subhadra Rajpoot, Amity University, Greater Noida, India

Chapter 4
Economic Implications and Scope of Wastewater Management Through Life
Cycle Cost Assessment 65
    Maria Jose Lopez-Serrano, University of Almeria, Spain
    Fida Hussain Lakho, Ghent University, Belgium
    Stijn W. H. Van Hulle, Ghent University, Belgium
    Ana Batlles-delaFuente, University of Almeria, Spain

Chapter 5
Fostering Green Product Design and Innovation for a Sustainable Future 86
    Sunil Sharma, Lovely Professional University, India

Chapter 6
Green Human Resource Management: Revealing the Route to Environmental
Sustainability... 111
    Precious T. Okunhon, University of Northampton, UK
    Adejoke Ige-Olaobaju, University of Northampton, UK

Chapter 7
Green Human Resource Management and Sustainable Performance
Management 131
    Malvern Chiboiwa, University of Hertfordshire, UK
    Elizabeth Babafemi, University of Hertfordshire, UK
    Felicia Momoh Oseghale, Alvan Ikoku University of Education, Nigeria
    Raphael Oseghale, University of Hertfordshire, UK

Chapter 8
Green Marketing Strategies for Indonesia's Micro, Small, and Medium
Enterprises 159
    Vanessa Gaffar, Universitas Pendidikan Indonesia, Indonesia
    Tika Koeswandi, Universitas Pendidikan Indonesia, Indonesia

Chapter 9
Green Organizational Culture and Sustainable Development: Nurturing
Environmental Responsibility in Businesses 189
    Omolola Ayobamidele Arise, MANCOSA, South Africa
    Meshel Muzuva, MANCOSA, South Africa

Chapter 10
Local Governments' Roles in Sustainable Waste Management for a Green
Economy 217
    Olufemi Micheal Oladejo, University of KwaZulu-Natal, South Africa
    Sybert Mutereko, Rhodes University, South Africa
    Nyikiwa Agreement Mavunda, University of KwaZulu-Natal, South
      Africa

Compilation of References 237

Related References 274

About the Contributors 298

Index 303

Detailed Table of Contents

Editorial Advisory Board xiv

Preface xv

Chapter 1
A Review of Reverse Logistics Models Based on Operations Research
Techniques 1
Naghmeh Rabiei, Toronto Metropolitan University, Canada
Saman Hassanzadeh Amin, Toronto Metropolitan University, Canada
Saeed Zolfaghari, Toronto Metropolitan University, Canada

In response to the imperative of efficient resource use, value recapture, and environmental responsibility, companies are increasingly prioritizing reverse logistics (RL) activities. In today's dynamic business environment, effectively managing returned products has become essential for companies aiming to control costs, meet customers' expectations, and align with sustainability goals. This book chapter focuses on journal papers which were published between 2020 and 2023, focusing on RL optimization models. The publications reviewed in this chapter are categorized in problem domain and techniques of operations research. The problem domain is explored in three classifications comprising literature reviews (LR), deterministic reverse logistics models (DRLM), and uncertain reverse logistics models (URLM). This book chapter also reviews the related operations research and optimization techniques. This study concludes with discussions on observations and findings, along with suggestions for future research directions.

Chapter 2
Demystifying Sustainable Practices in the Hotel Industry: Evidence From
Ghana 20
Mildred Nuong Deri, University of Energy and Natural Resources,
Ghana
Amrik Singh, Lovely Professional University, India
Perpetual Zaazie, University of Energy and Natural Resources, Ghana
David Anandene, University of Cape Coast, Ghana

Sustainability continues to be one of the most commonly discussed trends within the hotel industry. Sustainability has become a matter of concern for both individuals and corporations. Hotel chains develop environmental sustainability practices, mainly for cost-reduction purposes, accommodating the owners' demands for efficiency. Notwithstanding, there are differences according to the chain's size. This study aimed to identify sustainable hotel practices in the Sunyani Municipality of Ghana. The hospitality industry is starting to take responsibility for environmental sustainability. A strong focus on energy, waste, and water usage is directly linked with financial benefits in the operation of the hoteliers. Practices connected to the social aspect of sustainability are less developedThe hotels accepted and executed water and liquid waste management more than any other sustainable practice, and consumer demand was the element that drove these behaviours. It is advised that hotels create and execute a strategy for environmental management that will help to mitigate these consequences.

Chapter 3
Eco-Friendly Solutions for Sustainable Waste Management: An Approach
Towards a Cleaner and Greener Future.. 43
Subhadra Rajpoot, Amity University, Greater Noida, India

Solid waste management is a critical aspect of environmental stewardship that involves the collection, treatment, and disposal of various types of solid waste generated by human activities. This includes everyday items like household waste, as well as industrial, commercial, and agricultural waste. The primary goal of solid waste management is to minimize the adverse effects of waste on public health, the environment, and aesthetics. Solid waste management is a crucial aspect of maintaining environmental sustainability and public health. Proper waste management includes the collection, transportation, disposal, and recycling of solid waste materials (Nguyen-Viet et al., 2009). It is important to employ efficient strategies such as the implementation of communal bins and scheduled waste collection according to waste types (Gubernatorov, 2020). This ensures the separation of waste for recycling and reduces the amount of waste ending up in landfills.

Chapter 4
Economic Implications and Scope of Wastewater Management Through Life
Cycle Cost Assessment 65

Maria Jose Lopez-Serrano, University of Almeria, Spain
Fida Hussain Lakho, Ghent University, Belgium
Stijn W. H. Van Hulle, Ghent University, Belgium
Ana Batlles-delaFuente, University of Almeria, Spain

Global demand for freshwater is rising due to anthropic factors, exacerbating water scarcity and pollution. The economic dimensions of sustainable wastewater management offer both environmental benefits and economic promise, with studies consistently highlighting cost-effectiveness and long-term financial gains. Treated wastewater, a dependable and economically viable alternative, aligns with sustainable development goals and contributes to improving public health. An in-depth economic analysis explores the profitability and cost savings of nature-based solutions, emphasizing the importance of investing in these strategies for water sustainability within the circular economy. The study incorporates the review of a life cycle cost assessment, providing nuanced insights into the economic implications of adopting nature-based solutions in wastewater management. The findings contribute substantively to the discourse surrounding water sustainability, informing policy decisions and promoting transformative change aligned with environmental sustainability, and circular economy principles.

Chapter 5
Fostering Green Product Design and Innovation for a Sustainable Future 86

Sunil Sharma, Lovely Professional University, India

Climate change, resource depletion, and pollution pose existential threats to humanity. Traditional product design, with its focus on short-term gains and linear economy principles, has significantly contributed to these problems. Green product design offers a solution by minimizing environmental impact throughout a product's life cycle, from material sourcing to end-of-life management. In this chapter, we have discussed green products with examples and then provided the previous research for green product design. The overall advantages of green product design have been discussed to highlight their significance. Further, the barriers to the implementation of green product design in organizations have been discussed. This chapter then provides the green product design process for implementation in organizations. The support tools and methods to achieve green product design have been discussed. By understanding their strengths and limitations, designers can leverage them effectively to create more sustainable products that meet the needs of the environment, society, and the economy.

Chapter 6
Green Human Resource Management: Revealing the Route to Environmental
Sustainability.. 111

Precious T. Okunhon, University of Northampton, UK
Adejoke Ige-Olaobaju, University of Northampton, UK

Green Human Resource Management (GHRM) is an emerging field that focus on integrating HRM functions and practices to support business sustainability. Adopting Green Human Resource Management (GHRM) practices not only contributes to environmental sustainability but also amplifies employee engagement, providing organisations with a distinct competitive advantage. This chapter highlights the importance of GHRM in providing sustainable environmental solutions. This exploratory chapter will provide an introduction to GHRM and its initiatives in promoting environmental sustainability, it will further explore the concepts of GHRM, its advantages, challenges, and limitations. Additionally, the chapter will extensively examine strategies and practices integral to GHRM that contribute to environmental sustainability. Subsequently, recommendations for ensuring the successful implementation of GHRM practices and agenda for future academic research will be discussed. The conclusion of this study will highlight ways in which employees can actively foster and promote sustainable practices.

Chapter 7
Green Human Resource Management and Sustainable Performance
Management 131

Malvern Chiboiwa, University of Hertfordshire, UK
Elizabeth Babafemi, University of Hertfordshire, UK
Felicia Momoh Oseghale, Alvan Ikoku University of Education, Nigeria
Raphael Oseghale, University of Hertfordshire, UK

This study investigates how GHRM facilitates sustainable performance management. Specifically, the study investigates the role of specific GHRM activities on responsible production/consumption, climate action, clear water/sanitation, and sustainable cities. Drawing on secondary literature, the ability, motivation, and opportunity (AMO) model of HRM and the United Nations sustainability development goals (SDG) framework, the study suggests that GHRM can bolster sustainable performance management and thus lead to the attainment of responsible production/consumption, climate action, clear water/sanitation, and sustainable cities. The study extends the GHRM literature by uncovering how the SDG framework and the AMO model can interact to facilitate the development and deployment of green skills and bolster the attainment of the environmental dimension of SDGs through green leadership. The practical implications of the findings were discussed.

Chapter 8

Green Marketing Strategies for Indonesia's Micro, Small, and Medium
Enterprises 159

Vanessa Gaffar, Universitas Pendidikan Indonesia, Indonesia
Tika Koeswandi, Universitas Pendidikan Indonesia, Indonesia

Indonesia as a developing country has a long history of implementing the development transition to become a developed country. During this transition process, social and environmental phenomena also emerged as a result of this agenda. Environmental awareness campaigns for micro, small, and medium enterprises (MSMEs) in Indonesia require a different approach compared to other types of businesses. With the unique characteristics of MSMEs, a special strategy and practice is needed. This chapter explains the implementation of green marketing of MSMEs from the internal and external environment. A case study is also offered to describe how a green B2b Indonesia MSME has succeed in running sustainable business practice. This chapter offers a matrix to visualize the best positioning mapping for MSMEs to effectively and efficiently implement green marketing in Indonesia and any other developing countries. The chapters set the tone for the rest of the chapters examining the implications of the issues discussed for waste management and life cycle assessment for sustainable business practice.

Chapter 9

Green Organizational Culture and Sustainable Development: Nurturing
Environmental Responsibility in Businesses 189

    Omolola Ayobamidele Arise, MANCOSA, South Africa
    Meshel Muzuva, MANCOSA, South Africa

As society grapples with environmental challenges, businesses are increasingly compelled to align their operations with principles of sustainability. One pivotal mechanism for fostering sustainable development is the cultivation of a green organizational culture; an organizational tenet that embeds environmental responsibility within its values, norms, and behaviours. This chapter investigates the transformative role of Green Organizational Culture (GOC) within businesses in advancing Sustainable Development (SD), highlighting the imperative shift from profit-centric motives towards fostering ecological balance and social equity. Employing a qualitative synthesis of existing literature, this chapter navigates through the theoretical frameworks that underpin GOC, highlighting its significance in driving environmentally responsible behaviours within businesses. By examining case studies and scholarly works, the chapter identifies and analyzes the core components and characteristics of GOC, the strategic implementation strategies for fostering such a culture, and its profound impact on SD. The chapter systematically outlines how a commitment to environmental responsibility at all organizational levels can be achieved through aligning with external institutional norms and effectively managing internal processes for change. It provides strategic recommendations for embedding a green culture within organizational practices, thereby contributing to sustainable development goals. It concludes with strategic insights for organizations seeking to navigate the complexities of environmental sustainability, emphasizing the long-term benefits of such endeavours for ensuring corporate success and legitimacy in a globally conscious marketplace.

Chapter 10
Local Governments' Roles in Sustainable Waste Management for a Green
Economy 217

Olufemi Micheal Oladejo, University of KwaZulu-Natal, South Africa
Sybert Mutereko, Rhodes University, South Africa
Nyikiwa Agreement Mavunda, University of KwaZulu-Natal, South
Africa

Waste management has become one of the most pressing challenges facing the world today, with the increasing population and urbanization leading to an exponential increase in waste generation. Hence, sustainable waste management which includes a hygienic environment, providing waste recycling and re-use logistics, and managing the consumption pattern of the local community are key components of a green economy. The study adopts a systemic and desktop review of the literature to explore the role of Local government (as the closest government to the people) in sustainable waste management for a green economy.

Compilation of References 237

Related References 274

About the Contributors 298

Index 303

Editorial Advisory Board

Preface

In the face of the urgent and growing challenges posed by climate change, the global business community finds itself at a critical juncture. The need to achieve economic sustainability without compromising social and environmental health has never been more pressing. Businesses worldwide are under increasing pressure to adopt practices that not only bolster economic performance but also contribute positively to environmental and social outcomes (Atiku, 2019; Atiku & Fapohunda; 2021). This edited volume, waste management and life cycle assessment for sustainable business practice, is our contribution to the ongoing discourse on sustainable business practices.

The escalating effects of climate change underscore the necessity for industries to re-evaluate their operations and embrace sustainable practices. According to a report by the United Nations Environment Programme (UNEP, 2018), industries are responsible for nearly one-third of global greenhouse gas emissions. This stark reality highlights the significant role businesses play in contributing to climate change and the imperative for them to act decisively towards a more sustainable future.

Despite the increasing awareness and pressure from stakeholders, many businesses continue to prioritize short-term profitability over long-term sustainability (Porter & Kramer, 2019; Sharma & Bansal, 2017). This approach not only undermines environmental and social health but also jeopardizes the enduring success of the businesses themselves (Fields & Atiku, 2017). Effective business management practices that foster the growth of green industries are thus critical. These practices must encompass comprehensive strategies that address the entire supply chain and lifecycle of products and services, from production to disposal (Lee & Holweg, 2010). This calls for a transition from a linear economy, which emphasizes production, consumption, and disposal, to a circular economy that prioritizes the principles of reducing, reusing, and recycling (Atiku, 2020).

Transparency in reporting and communication about environmental initiatives is essential for businesses aiming to enhance their credibility and reputation as socially and environmentally responsible entities (Dzikriansyah et al., 2023; Ni et al., 2023;

Shah & Soomro, 2023). Additionally, fostering a culture of sustainability within organizations through employee education and training can lead to more motivated and engaged workforces and more efficient operations (Dzikriansyah et al., 2023; Shah & Soomro, 2023). Governments and policymakers also play a pivotal role by providing incentives and enforcing regulations that support sustainable business practices.

For businesses to truly champion green industries, sustainability must be integrated into their core strategies and decision-making processes. This involves setting environmental targets, adopting eco-friendly practices, and investing in renewable energy sources. Collaboration and partnerships with other organizations to share best practices and conduct joint research and development (Fields & Atiku, 2018) are also crucial for fostering innovation and collective responsibility towards environmental stewardship.

This comprehensive reference book delves into the multifaceted aspects of waste management and life cycle assessment for sustainable business practices. The chapters collectively explore a range of topics crucial for advancing sustainability in contemporary business. From optimization models in reverse logistics to green human resource management, each chapter offers in-depth insights and practical strategies. The discussions span diverse industries and geographic contexts, examining the latest research, case studies, and theoretical frameworks. By addressing environmental challenges, sustainable hotel practices, solid waste management, and innovative green product design, this book provides valuable knowledge for academics, practitioners, and policymakers aiming to foster a more sustainable future. Each chapter contributes to a holistic understanding of how businesses can integrate sustainability into their core operations, promoting economic, social, and environmental well-being.

In response to the imperative of efficient resource use, value recapture, and environmental responsibility, companies are increasingly prioritizing Reverse Logistics (RL) activities. In today's dynamic business environment, effectively managing returned products has become essential for companies aiming to control costs, meet customers' expectations, and align with sustainability goals. Chapter 1 focuses on journal papers published between 2020 and 2023, emphasizing RL optimization models. The reviewed publications are categorized by problem domain and operations research techniques. The problem domain is explored through three classifications: Literature Reviews (LR), Deterministic Reverse Logistics Models (DRLM), and Uncertain Reverse Logistics Models (URLM). Additionally, the chapter reviews related operations research and optimization techniques, concluding with observations, findings, and suggestions for future research directions.

Sustainability continues to be a pivotal trend within the hotel industry, with increasing concern among individuals and corporations. Hotel chains develop environmental sustainability practices mainly for cost-reduction purposes, addressing owners' demands for efficiency. Chapter 2 is aimed to identify sustainable hotel practices in the Sunyani Municipality of Ghana. The hospitality industry is taking responsibility for environmental sustainability, focusing on energy, waste, and water usage, which are directly linked to financial benefits. However, practices connected to the social aspect of sustainability are less developed. This chapter reveals that consumer demand drives sustainable practices, particularly in water and liquid waste management, and advises hotels to create and execute comprehensive environmental management strategies.

Solid waste management is a critical aspect of environmental stewardship involving the collection, treatment, and disposal of various types of solid waste generated by human activities. This includes household, industrial, commercial, and agricultural waste. The primary goal is to minimize the adverse effects of waste on public health, the environment, and aesthetics. Proper waste management strategies, such as implementing communal bins and scheduled waste collection, ensure the separation of waste for recycling and reduce landfill contributions. Chapter 3 underscores the importance of efficient waste management in maintaining environmental sustainability and public health.

Global demand for freshwater is rising due to anthropic factors, exacerbating water scarcity and pollution. Chapter 4 examines the economic dimensions of sustainable wastewater management, highlighting cost-effectiveness and long-term financial gains. Treated wastewater emerges as a dependable and economically viable alternative, aligning with Sustainable Development Goals and improving public health. The chapter includes an in-depth economic analysis of Nature-Based Solutions, emphasizing their importance for water sustainability within the circular economy. A Life Cycle Cost Assessment provides insights into the economic implications of adopting these solutions, informing policy decisions and promoting transformative change.

Climate change, resource depletion, and pollution pose existential threats to humanity. Traditional product design, focused on short-term gains, has significantly contributed to these problems. Chapter 5 explores green product design, which minimizes environmental impact throughout a product's lifecycle. It discusses previous research, the overall advantages of green product design, and the barriers to its implementation. The chapter also provides a green product design process for organizations and discusses support tools and methods for achieving sustainable design, offering strategies for leveraging these tools effectively.

Green Human Resource Management (GHRM) integrates HRM functions and practices to support business sustainability. Chapter 6 highlights GHRM's importance in providing sustainable environmental solutions and enhancing employee engagement. It explores the concepts, advantages, challenges, and limitations of GHRM, examining strategies and practices integral to GHRM that contribute to environmental sustainability. The chapter offers recommendations for successful implementation and suggests an agenda for future academic research, emphasizing the role of employees in fostering sustainable practices.

Chapter 7 investigates how GHRM facilitates sustainable performance management, focusing on activities that promote responsible production/consumption, climate action, clean water/sanitation, and sustainable cities. Drawing on secondary literature, the Ability, Motivation, and Opportunity (AMO) Model of HRM, and the United Nations Sustainability Development Goals (SDG) Framework, the study suggests that GHRM can bolster sustainable performance management. It extends the GHRM literature by showing how the SDG framework and the AMO model interact to develop green skills and support the attainment of environmental SDGs through green leadership. Practical implications are also discussed.

Indonesia, a developing country, has a long history of development transitions, leading to social and environmental challenges. Chapter 8 explains the implementation of green marketing for Micro, Small, and Medium Enterprises (MSMEs) in Indonesia. It highlights the unique characteristics of MSMEs, requiring a special strategy for environmental awareness campaigns. A case study describes how a green B2B Indonesian MSME succeeded in sustainable business practice. The chapter offers a matrix for best positioning mapping for MSMEs to implement green marketing effectively, setting the tone for further discussions on waste management and life cycle assessment for sustainable business practice.

As businesses face environmental challenges, aligning operations with sustainability principles is crucial. Chapter 9 explores how Green Organizational Culture (GOC) influences Sustainable Development (SD) in firms, prioritizing ecological and social sustainability. Through literature analysis, the chapter identifies core components and characteristics of GOC, strategic implementation strategies, and its impact on SD. It presents strategies for embedding environmental responsibility within organizations and offers recommendations to build a green culture for sustainable goals, highlighting long-term benefits for business success in a global market.

Waste management is a pressing challenge with increasing population and urbanization leading to more waste generation. Chapter 10 adopts a systemic and desktop review of the literature to explore the role of local government in sustainable waste management for a green economy. It discusses key components such as maintaining a hygienic environment, providing waste recycling and re-use logistics, and managing local consumption patterns, emphasizing the importance of local government as the closest government to the people.

Chapter 11 delves into the pivotal role of waste management, Life Cycle Assessment (LCA), and sustainable innovation in contemporary business. As global waste concerns escalate, the chapter explores the challenges businesses face, including economic, social, and regulatory aspects. It highlights the transformative potential of sustainable practices, such as energy-efficient technologies and waste reduction strategies, contributing to cost savings and brand reputation. The chapter also addresses burgeoning consumer demand for sustainable products, offering strategic recommendations for overcoming obstacles and integrating sustainable principles into operations.

Chapter 12 examines sustainable end-of-life tyres (ELT) management in India, evaluating recycling methods from origin to disposal. With the growth of the Indian vehicle sector, ELT management concerns have increased. The study identifies ELT sources and stakeholders involved in recycling, such as transport firms, tyre dealers, clients, dismantlers, and importers. The chapter evaluates thermal and mechanical shredding, with mechanical recycling emerging as the most sustainable method. The findings suggest mechanical recycling for sustainability, circular economy, and resource efficiency, providing a roadmap for sustainable ELT management in India.

The principles, strategies, and best practices discussed in this book are essential for promoting green industries and sustainable business practices globally. The shift towards environmentally friendly practices presents numerous opportunities for businesses to enhance efficiency, reduce costs, and improve their reputations among increasingly eco-conscious consumers and stakeholders. By minimizing fossil fuel use, reducing waste, and promoting eco-friendly products, businesses can make significant contributions to environmental preservation while achieving financial benefits.

We hope this book will inspire businesses to take proactive steps towards sustainability and provide practical insights and strategies for achieving economic, social, and environmental goals!

Idris Olayiwola Ganiyu
York St John University, UK

Odunayo Magret Olarewaju
Crown College St. Bonifacius, USA

Adejoke Ige-Olaobaju
University of Northampton, UK

Sulaiman Olusegun Atiku

Preface

Namibia University of Science and Technology, Namibia

REFERENCES

Atiku, S. O. (2019). Institutionalizing Social Responsibility Through Workplace Green Behavior. In Atiku, S. (Ed.), *Contemporary Multicultural Orientations and Practices for Global Leadership* (pp. 183–199). IGI Global. 10.4018/978-1-5225-6286-3.ch010

Atiku, S. O. (2020). Knowledge Management for the Circular Economy. In Baporikar, N. (Ed.), *Handbook of Research on Entrepreneurship Development and Opportunities in Circular Economy* (pp. 520–537). IGI Global. 10.4018/978-1-7998-5116-5.ch027

Atiku, S. O., & Fapohunda, T. (Eds.). (2021). *Human Resource Management Practices for Promoting Sustainability*. IGI Global. 10.4018/978-1-7998-4522-5

Dzikriansyah, M., Ni, N., Shah, S. Z., & Soomro, B. A. (2023). Transparent reporting and communication strategies for enhancing business reputation and credibility in sustainability.

Fields, Z., & Atiku, S. O. (2017). Collective Green Creativity and Eco-Innovation as Key Drivers of Sustainable Business Solutions in Organizations. In Fields, Z. (Ed.), *Collective Creativity for Responsible and Sustainable Business Practice* (pp. 1–25). IGI Global. 10.4018/978-1-5225-1823-5.ch001

Fields, Z., & Atiku, S. O. (2018). Collaborative Approaches for Communities of Practice Activities Enrichment. In Baporikar, N. (Ed.), *Knowledge Integration Strategies for Entrepreneurship and Sustainability* (pp. 304–333). IGI Global. 10.4018/978-1-5225-5115-7.ch015

Lee, H. L., & Holweg, M. (2010). Supply chain integration: The role of product and process modularity. *International Journal of Production Research*, 48(1), 169–190.

Ni, N., Shah, S. Z., & Soomro, B. A. (2023). *The impact of transparent reporting on corporate sustainability performance*.

Porter, M. E., & Kramer, M. R. (2019). Creating shared value: How to reinvent capitalism—and unleash a wave of innovation and growth. *Harvard Business Review*, 323–346. 10.1007/978-94-024-1144-7_16

Shah, S. Z., & Soomro, B. A. (2023). *Sustainable business practices: Strategies for green human resource management*.

Sharma, G., & Bansal, P. (2017). The role of sustainability in business decision-making. *Academy of Management Journal*, 60(4), 1352–1380.

UNEP (United Nations Environment Programme). (2018). *Emissions Gap Report 2018*. United Nations Environment Programme.

Chapter 1
A Review of Reverse Logistics Models Based on Operations Research Techniques

Naghmeh Rabiei

Toronto Metropolitan University, Canada

Saman Hassanzadeh Amin

Toronto Metropolitan University, Canada

Saeed Zolfaghari

Toronto Metropolitan University, Canada

ABSTRACT

In response to the imperative of efficient resource use, value recapture, and environmental responsibility, companies are increasingly prioritizing reverse logistics (RL) activities. In today's dynamic business environment, effectively managing returned products has become essential for companies aiming to control costs, meet customers' expectations, and align with sustainability goals. This book chapter focuses on journal papers which were published between 2020 and 2023, focusing on RL optimization models. The publications reviewed in this chapter are categorized in problem domain and techniques of operations research. The problem domain is explored in three classifications comprising literature reviews (LR), deterministic reverse logistics models (DRLM), and uncertain reverse logistics models (URLM). This book chapter also reviews the related operations research and optimization techniques. This study concludes with discussions on observations and findings, along with suggestions for future research directions.

DOI: 10.4018/979-8-3693-2595-7.ch001

INTRODUCTION

In recent years, Reverse Logistics (RL) has received significant attention from different aspects, and numerous researchers have delved into it. Saxena et al. (2023) stated that the growing environmental awareness among consumers and corporate responsibility have led to increased focus on RL and return policies. In response to overpopulation and insufficient environmental resources, the concept of RL has emerged as a crucial solution (Kilic et al., 2023). Lei et al. (2023) highlighted the importance of disassembly centers that optimally determine the disassembly process, manage inventory, and plan vehicle routes for the collection of products and delivery of materials. A well-developed Reverse Logistics Network (RLN) is essential for waste recycling, reducing costs, increasing profits, and enhancing efficiency (Liao and Luo, 2022).

In forward logistics, the main objectives include production, distribution, and fulfillment of customers' needs. On the other hand, RL is primarily focused on maximizing the utility of used products to improve economic efficiency and minimize irreversible environmental effects. Companies that neglect to design RL in advance may face the challenge of redesigning the RL supply chain network based on their existing forward logistics structure (Gao and Cao, 2020). In some industries, such as batteries and electric devices, it is crucial to consider RL due to the significant environmental impact of end-of-life products (Lin et al., 2023; Liao and Luo, 2022). In RL, several key decisions need to be made to efficiently manage the flows of products from the end-user back to the origin. Some of the main decisions include product disposition, RL network design, remanufacturing, transportation management, and inventory management. Fig. 1 illustrates a general RL network.

While numerous reviews on RL have been conducted in recent years, this book chapter specifically concentrates on new journal papers published between 2020 and 2024. This literature review summarizes 23 papers obtained on ScienceDirect using 'Reverse Logistics' as the primary search keyword. Our focus is on RL and the application of optimization (mathematical) models in this field. Additionally, we explore both deterministic and uncertain models in the context of RL.

The rest of this chapter is formatted as follows: Section 2 introduces the taxonomy and the classification of the literature review. In Section 3, we delve into the observations and suggestions. Finally, Section 4 covers the conclusions and outlines areas for future research.

Figure 1. A general RL network

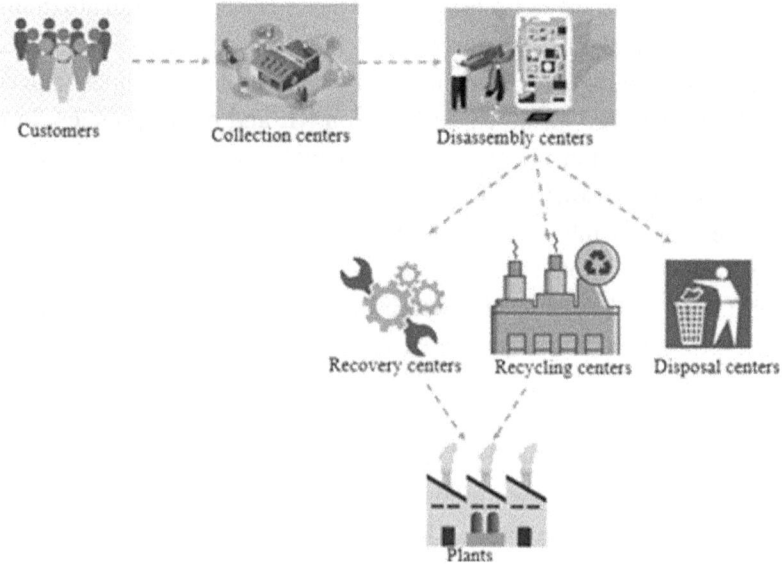

TAXONOMY

This literature review is structured around two dimensions: problem category, and operations research techniques. This structure allows for an effective analysis of RL papers.

Classification of the Problem

The problem category includes Literature Reviews (LR), Deterministic Reverse logistics Models (DRLM), and Uncertain Reverse Logistics Models (URLM). The papers are categorized in Table 1.

Literature Reviews

In this section, we examine literature reviews conducted in RL. Sar and Ghadimi (2023) presented a literature review focusing on the vehicle routing problem within the RL domain. They reviewed 109 published papers from international scientific journals, addressing five specific questions related to modeling approaches, solution techniques, variants of the vehicle routing problems, sustainability considerations, and types of waste. It was observed that Mixed-Integer Linear Programming (MILP) served as the modeling approach for 65% of single objective papers. That study also highlighted that medical waste routing problems gained significant attention during the pandemic. Until 2019, there were only a limited number of routing studies addressing medical waste generated in hospitals. Therefore, as one of the identified research gaps, it was realized that dedicating increased attention to the management of medical waste is imperative.

Table 1. Classification of references based on problem category

Problem category	References
Literature Reviews (LR) (4)	Sar and Ghadimi (2023), Guggeri et al., (2023), Ding et al., (2023), Gunasekara et al., (2023)
Deterministic Reverse Logistics Models (DRLM) (6)	Zhang et al., (2024), Kannan et al., (2023), Saxena et al., (2023), Lei et al., (2023), Nanayakkara et al., (2022), Forkan et al., (2022), Budak (2020)
Uncertain Reverse Logistics Models (URLM) (12)	Kilic et al., (2023), Lin et al., (2023), Nosrati-Abarghooee et al., (2023), Xu et al., (2023), Gholizadeh et al., (2022), Ghanbarzadeh-Shams et al., (2022), Liao and Luo (2022), Jahangiri et al., (2022), Ahmed and Zhang (2021), Roudbari et al., (2021), Hallak et al., (2021), Gao and Cao (2020)

Guggeri et al. (2023) provided a comprehensive review of recent literature examining Goal Programming (GP) and other multi-criteria methods within Closed-Loop Supply Chains (CSCs). The primary focus was on RL, highlighting remanufacturing as a frequently employed recovery option in the industry. 38 articles were examined and categorized into three groups: management and design of supply chain, strategic decisions within the circular economy, and production planning. They concluded that remanufacturing serves as a viable recovery option with numerous advantages for consumers, manufacturers, and the environment. However, solving hybrid production-remanufacturing planning problems is more complex compared to conventional production planning issues due to the incorporation of diverse item flows (used, remanufactured, and newly produced) and the consideration of distinct economic, social, and environmental goals. As a result, they recommended the utilization of Multi-Criteria Decision-Making (MCDM) approaches as a clear and direct strategy to integrate stakeholder goals and tackle the distinctive characteristics of these problems, thereby providing effective solutions.

The review paper from Ding et al. (2023) examined 81 studies of Forward Logistics (FL) and RL in construction using a Systematic Literature Review (SLR). They developed a framework for reviewing and synthesizing FL and RL operations across different stages of the construction paper life cycle. The stages included design, production, construction, and operations for FL, and deconstruction, product reuse, waste distribution, and material reprocessing for RL. They suggested that greater integration among supply chain participants is needed for RL operations to effectively close the loop in construction's Circular Economy (CE). They proposed a novel conceptual framework for Circular Logistics Integration (CLI), comprising channel creation, network integration, and inventory management to provide guidance and inspiration for future research.

The review paper from Gunasekara et al. (2023) reviewed 131 influential journal articles addressing returns Acquisition, Sorting, and Disposition (ASD) from 2012 to 2021. The aim was to evaluate the present state of ASD research for Circular Supply Chains (CSCs) and explore crucial research directions essential for facilitating the transition to a Circular Economy (CE). They included studies that examined the direct influence of one or more research considerations (such as uncertainty, legislation, industry, technology, and behavior) on ASD. They stated that the application of research was limited due to a scarcity of empirical studies, inadequate practical validation of mathematical models, a concentration on economic objectives, and restrictive modeling assumptions regarding behavior and uncertainty in return processes. In shaping the future research agenda for CSC, they recommended that concepts from CE, including collaborative decision-making between product design and returns management, cross-sector collaboration, and the integration of product-service systems should be incorporated.

Deterministic RL Models

In this section, DRL models are addressed. We delve into some publications, exploring various applications, techniques, and common themes associated with these models.

Kannan et al. (2023) introduced a Multi Objective Mixed-Integer Programming (MOMIP) model to design an RLN. They examined four reverse activities (inspection, dismantling, repair/refurbishing, and recycling), and aimed to maximize potential returns (products, components, materials). The primary objectives of that study were to minimize both overall costs and environmental impact resulting from transportation and processing. They employed a weighted goal programming technique to identify effective solutions. The validity of the mathematical model was confirmed through the application of a real-life case study involving an electronics manufacturing company in India.

Nanayakkara et al. (2022) suggested a circular RL framework consisting of three stages designed to manage e-commerce returns. In the first stage, they determined the return pattern. Then, in the second stage, they introduced a CE network among different stakeholders to actively contribute to the principles of the CE. They formulated a Mixed-Integer Linear Programming (MILP) model in the third stage based on the designed circular economy network. This model was designed to cover various aspects of e-commerce RL and optimize the overall network. The model was validated through the utilization of a case study involving an e-commerce company specializing in consumer electrical and electronics. Forkan et al. (2022) presented novel mathematical inventory management models within a reverse logistics system. The proposed models built upon the framework introduced by Nahmias and Rivera, considering the assumption that the demand for newly manufactured and remanufactured items varies. They minimized the holding cost through the development of two mathematical models and the formulation of both unconstrained and constrained optimization problems. A mathematical model with three objectives was developed, and the Pareto solution was obtained through an algorithm to address both Greenhouse Gas (GHG) emissions and energy consumption.

Budak (2020) designed a recycling network for end-of-life mobile phones. In that study, the sustainable RL network problem and disassembly line balancing were integrated for the first time to evaluate decisions using the triple bottom line approach. That study presented a multi-period Multi Objective Mixed-Integer Nonlinear Programming (MOMINLP) model. The Improved Augmented Epsilon Constraint (AUGMECON-2) technique was employed to solve it. The objectives of that study were to minimize economic and environmental impacts while maximizing social impact. Finally, it was suggested that the uncertainties related to the quantity and quality of Waste Electrical and Electronic Equipment (WEEE) can be considered to develop the model. This formulation can include recoverable materials sales prices as either fuzzy or stochastic parameters, enabling a comprehensive examination of the system's overall performance.

Uncertain RL Models

In this section, some papers that applied URLM and related techniques are considered, and the main points are highlighted. Table 2 illustrates each paper, and its source of uncertainty.

Table 2. The references and sources of uncertainty

Source of uncertainty	References
Demand	Xu et al., (2023), Gholizadeh et al., (2022), Ghanbarzadeh-Shams et al., (2022), Hallak et al., (2021), Gao and Cao (2020)
Cost	Gholizadeh et al., (2022), Ghanbarzadeh-Shams et al., (2022)
Capacity	Ghanbarzadeh-Shams et al., (2022)

continued on following page

Table 2. Continued

Source of uncertainty	References
Level of CO_2 emissions	Ghanbarzadeh-Shams et al., (2022)
Quality	Lin et al., (2023), Liao and Luo (2022), Roudbari et al., (2021)
Return	Liao and Luo (2022), Hallak et al., (2021), Roudbari et al., (2021), Gao and Cao (2020)
Waste	Xu et al., (2023), Lin et al., (2023), Kilic et al., (2023), Nosrati-Abarghooee et al., (2023), Jahangiri et al., (2022), Ahmed and Zhang (2021)
Product variety	Roudbari et al., (2021)
Bill of material	Roudbari et al., (2021)

Ghanbarzadeh-Shams et al. (2022) focused on a production planning problem that involved multiple products, sites, and periods, integrated with RL. They introduced a Possibilistic-Flexible Multi Objective Mixed-Integer Programming (PF-MOMIP) model with chance constraints to address uncertainty in the carpet industry. A novel hybrid fuzzy goal programming approach was proposed to solve the model. Fuzzy data for demand, electric energy consumption costs, capacity, and CO_2 emissions levels were considered in that study. The proposed model was evaluated by comparing it to a deterministic approach, revealing that the effectiveness of fuzzy modeling and uncertainty on sustainability performance favored the proposed fuzzy solution methodology over the deterministic approach.

In the paper published by Kilic et al. (2023), an RL network for waste batteries under a two-stage methodology was designed, in which Spherical Fuzzy Analytical Hierarchy Process (SF-AHP) was applied in the first stage to calculate the weights of objectives. In the second stage, the RL network was designed by developing a Multi Objective Mixed-Integer Linear Programming Model (MO-MILP). The model was solved for different number of waste batteries under three scenarios.

Ahmed and Zhang (2021) outlined a network-based model with multiple stages to optimize the overall cost associated with the RL management of inert construction waste throughout its entire life cycle. Two types of costing approach were considered to handle uncertainty and deficiencies in the model. The first type of cost pertained to Facility-Based Costing (FBC), while the second type of cost related to Non-Facility-Based Costing (NFBC). MILP was utilized employing the LINGO software. The developed model was validated through a case study focused on construction waste management in Hong Kong. The findings indicated a roughly 24% decrease in total costs compared to the original case. Additionally, to assess the influence of uncertainties on cost parameters, a comprehensive sensitivity analysis based on various scenarios was performed. The optimal result highlighted that a significant portion of the total cost originates from the NFBC component.

The two-stage stochastic mixed-integer programming model presented by Roudbari et al. (2021) considered a range of procedures for recovering recyclable items, comprising activities such as reuse, refurbishing, remanufacturing, recycling, and the sale of spare parts. Additionally, the model incorporated features that addressed uncertainties related to the quality and number of returned products, bill of material, and product variety. To address the intricate computational nature of this large-scale problem, a hybrid algorithm, combining a genetic algorithm and a branch-and-cut algorithm with the CPLEX solver was proposed. The examined model in that paper addressed multiple facilities and products within a stochastic environment. The authors proposed the potential extension of that study to include a multi-period scenario for an in-depth exploration of inventory control as a focus for future research. The distribution for sources of uncertainty is illustrated in Fig. 2.

Figure 2. Distribution for sources of uncertainty

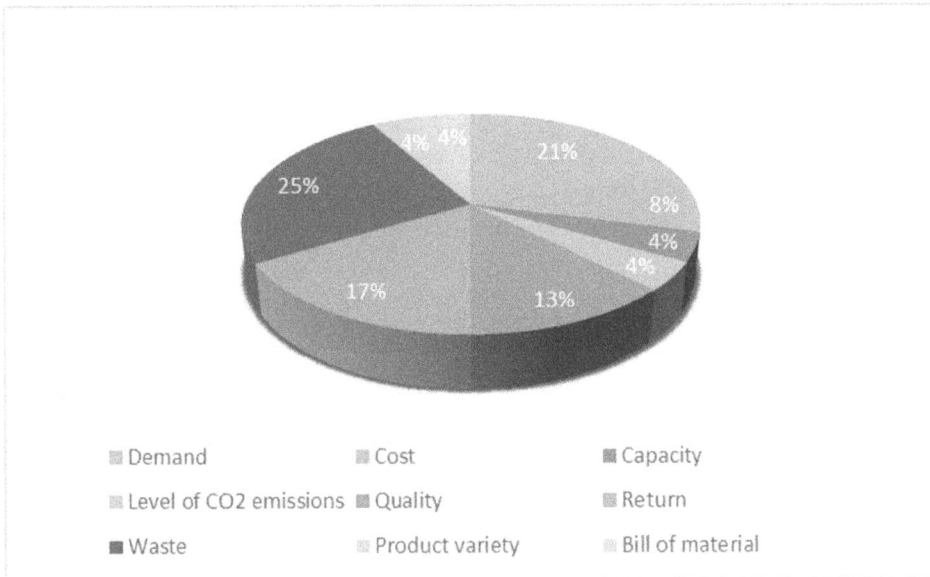

Operations Research Techniques

The publications listed in Table 3 are classified based on operations research and optimization techniques. The DRLMs and URLMs are categorized into two parts: single, and hybrid techniques. In hybrid classification, the combination of two or more techniques has been considered.

In some studies, mixed-integer linear programming has been utilized to calculate the number of end-of-life or used products in RL networks. Those models have incorporated binary variables to address the opening of some elements such as recycling facilities, landfills, and the distance matrix between recycling centers and disposal centers. Notably, due to the complexity of those problems, linear programming has not been applied in any recent studies. Consequently, MILP is gaining more attention in various research endeavors.

In this book chapter, we have categorized objective functions into two groups: single objective and multi-objective. Table 4 shows the classification of papers based on their respective objective functions. Fig. 3 illustrates the distribution of single objective and multi-objective functions of the referenced papers.

Figure 3. Distribution of single-objective and multi-objective functions

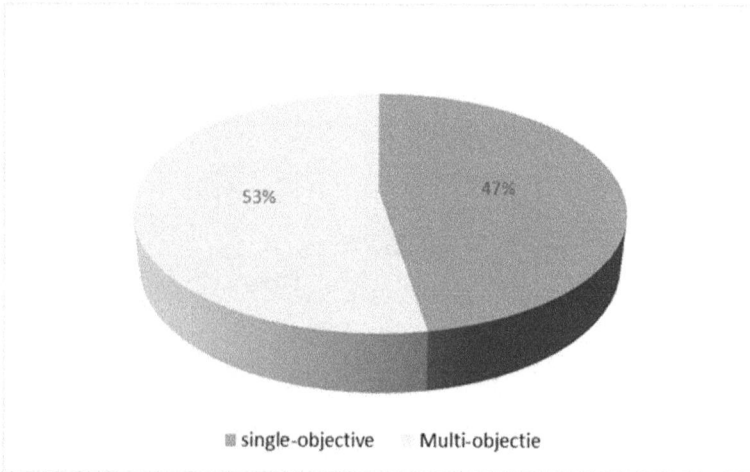

Table 3. Operations research techniques for referenced papers

	Techniques	References
Single technique	**Calculus**	**Zhang et al., (2024)**
	Weighted goal programming	Kannan et al., (2023)
	Stackelberg-Nash equilibrium	Saxena et al., (2023)

continued on following page

Table 3. Continued

	Techniques	References
Single technique	**Calculus**	**Zhang et al., (2024)**
	Mixed-Integer Linear Programming (MILP) Iterative algorithmic approach	Lin et al., (2023), Ahmed and Zhang (2021) Hallak et al., (2021)
	Fuzzy optimization	Liao and Luo (2022)
Hybrid techniques	MILP, Hybrid heuristics	Lei et al., (2023)
	MILP, Spherical Fuzzy Analytical Hierarchy Process (SF-AHP)	Kilic et al., (2023)
	K-means clustering, optimized ant colony algorithm (MACO-GKA)	Xu et al., (2023)
	MILP, Fuzzy goal programming, simulation	Nosrati-Abarghooee et al., (2023)
	Mixed-Integer Nonlinear Programming (MINLP), Heuristics, Robust optimization	Gholizadeh et al., (2022)
	MILP, Hierarchical clustering	Nanayakkara et al., (2022)
	Unconstraint optimization, Scalarization, Algorithms	Forkan et al., (2022)
	Possibilistic-Flexible Mixed-Integer Programming (PF-MIP), Hybrid fuzzy goal programming	Ghanbarzadeh-Shams et al., (2022)
	MILP, Robust optimization	Jahangiri et al., (2022)
	Stochastic MIP, Genetic algorithm, Branch and cut	Roudbari et al., (2021)
	MINLP, Augmented ε-constraint	Budak (2020)
	Integer Non-Linear Programming (INLP), Weighted-sums method, Augmented ε-constraint	Gao and Cao (2020)

Table 4. Objective functions for the referenced papers

	Objective functions	References
Single objective	**Max profit**	**Zhang et al., (2024), Liao and Luo (2022), Roudbari et al., (2021)**
	Min cost	Lei et al., (2023), Lin et al., (2023), Nanayakkara et al., (2022), Gholizadeh et al., (2022), Jahangiri et al., (2022), Ahmed and Zhang (2021), Hallak et al., (2021),
	Min cost, and emissions	Saxena et al., (2023), Kannan et al., (2023)
	Min cost, emissions, and energy consumption	Forkan et al., (2022)
	Min cost, emissions, max employment opportunity	Kilic et al., (2023), Budak (2020)
	Min cost, and risks	Nosrati-Abarghooee et al., (2023)

continued on following page

Table 4. Continued

	Objective functions	References
Single objective	Max profit	Zhang et al., (2024), Liao and Luo (2022), Roudbari et al., (2021)
	Min cost of location, vehicle usage, and transportation	Xu et al., (2023)
	Max profit, min emissions' cost, and energy consumption	Ghanbarzadeh-Shams et al., (2022)
	Max profit, and job opportunity, min emissions' cost	Gao and Cao (2020)

OBSERVATIONS AND RECOMMENDATIONS

In this section, we discuss the recommendations and observations from the review of 23 papers.

The Most Popular Category

Among the discussed domains in this chapter (LR, DRLM, URLM), URLM is the most prominent, followed by DRLM and LR. In the dominant domain (URLM), a notable observation is that most of the papers validated their mathematical models through real case studies. The rise in the popularity of URLM results from their ability to handle the complexities and uncertainties inherent in RL processes, providing businesses with more realistic, adaptable, and risk-aware solutions.

The Most Popular Source of Uncertainty

Table 2 and Fig. 2 indicate that waste is the most observed source of uncertainty in RL (25%). This uncertainty arises from both product returns and decisions related to end-of-life management. Anticipating the quantity, variety, and state of returned products poses challenges because of variables like consumer behavior, product lifespan, and technological progress. Three additional sources of uncertainty (demand, quantity, and quality of returned products) rank second, third, and fourth, contributing 21%, 17%, and 13%, respectively. Considering uncertainty in RL models leads to greater accuracy by adding complexity to the system.

The Most Popular Technique

Based on the reviewed papers, Mixed-Integer Linear Programming (MILP) emerges as the most frequently utilized technique in RL models. This preference is attributed to the effectiveness of binary and nonnegative variables in modeling fixed and variable costs. Additionally, numerous hybrid techniques, such as fuzzy programming, heuristics, ε-constraints, and weighted sum methods have been integrated with the MILP technique to solve RL models.

The Most Popular Multi-Objective Technique

As depicted in Table 4, cost minimization is the most frequently observed single objective. For multi-objective problems, the most popular objectives in the reviewed references are cost minimization and ecological impact reduction. Simple methods, such as ε-constraint and weighted-sums methods, have been employed to address multi-objective models. While it is essential to develop alternative problem-solving methods for multi-objective models, some authors have already utilized goal programming models in RL networks (e.g., Nosrati-Abarghooee et al., 2023; Ghanbarzadeh-Shams et al., 2022). Further development is expected in the future.

The Most Popular Application

The literature in Table 5 is categorized into numerical examples and case studies using real data, emphasizing practical applications. The findings reveal a notable focus among authors on waste in both case studies and numerical examples. Zhang et al. (2024) presented a case study that confirms the effectiveness of the collection strategy using an automotive engine as an example. Roudbari et al. (2021) selected medical equipment industries for a case study on designing reverse supply chain networks. The focus was on laboratory equipment, particularly the centrifuge, a crucial device capable of high-speed rotation and often refrigerated. Given that approximately 35 million metric tons of industrial waste are collected annually in Turkey, and the country's recycling industry is valued at nearly $5 billion each year, Kilic et al. (2023) implemented the proposed methodology through a case study focused on designing a waste battery network in Turkey. Kannan et al. (2023) utilized the RL model in a case study involving an Indian manufacturer of laptops and desktops to achieve sustainable advantages while dealing with various processing technologies.

Table 5. The applications of reviewed models

	Application	References
Numerical examples	Healthcare	Nosrati-Abarghooee et al., (2023)
	Waste	Saxena et al., (2023), Lei et al., (2023)
	Medical waste	Xu et al., (2023)
	Tire industry Fast/Slow-moving items	Forkan et al., (2022), Gao and Cao (2020) Hallak et al., (2021)
Case studies	Engine among auto parts	Zhang et al., (2024)
	Electronics manufacturer	Kannan et al., (2023)
	Waste batteries	Kilic et al., (2023), Lin et al., (2023), Liao and Luo (2022)
	Disposal container industry	Gholizadeh et al., (2022)
	Electrical and electronics	Nanayakkara et al., (2022)
	Carpet industry	Ghanbarzadeh-Shams et al., (2022)
	Construction waste	Jahangiri et al., (2022), Ahmed and Zhang (2021)
	Medical equipment industry	Roudbari et al., (2021)
	Mobile phone recycling	Budak (2020)

Environmental Impact and Sustainability

Reducing environmental impact and carbon emissions is crucial to 3PRL. With its focus on product movement back through the supply chain, RL helps diminish waste and lower the environmental footprint associated with logistical operations. For instance, Kilic et al. (2023) examined waste batteries in their study, recognizing that batteries contain both hazardous and valuable materials. They emphasized the importance of effective waste management through recycling due to its potential economic and environmental benefits. Additionally, they prioritized the reduction of carbon footprint as a key objective in addressing the environmental aspect of their approach. In another study by Gao et al. (2020), the environmental dimension was defined by the projected overall carbon emissions. They took carbon emissions into consideration to ensure environmental sustainability across the stages of raw material sourcing, recovery, production, and logistics operations. Ghanbarzadeh et al. (2022) considered the integration of logistics, transportation, and production planning, focusing on environmental sustainability factors such as carbon dioxide emissions and energy consumption within the carpet industry. They also stated that by considering this integration, practitioners can make informed decisions regarding their carpet production and transportation strategies, balancing cost-effectiveness with environmental considerations in their supply chain.

The List of Journals

Table 6 presents a classification of the papers, grouping them by their respective journals and domains. Many papers on RL are featured prominently in *"Journal of Cleaner Production"*.

Table 6. The list of publications

Journal	Number of articles			
	LR	**DRLM**	**URLM**	**Total**
Applied Soft Computing			2	2
Cleaner Logistics and Supply Chain		1		1
Computers & Industrial Engineering	2	1	2	5
European Journal of Operational Research		1		1
Intelligent Systems with Applications		1		1
International Journal of Production Economics	1			1
Journal of Cleaner Production	1	3	2	6
Journal of Environmental Management			1	1
Journal of Manufacturing Systems			1	1
Omega			1	1
Socio-Economic Planning Sciences			1	1
Sustainable Cities and Society			1	1
Waste Management			1	1
Total	4	7	11	23

Classification of the Articles Based on Year

Table 7 illustrates the categorization of reviewed papers based on their publication year and the relevant domain.

Table 7. Classification of reviewed papers based on year and domain

Year	Number of articles			
	LR	**DRLM**	**URLM**	**Total**
2020		1	1	2

continued on following page

Table 7. Continued

Year	Number of articles			
	LR	DRLM	URLM	Total
2021			3	3
2022		2	4	6
2023	4	3	4	11
2024		1		1
Total	4	6	10	23

CONCLUSION AND FUTURE RESEARCH

In this book chapter, three classifications have been defined for the literature: literature review, deterministic reverse logistics models, and uncertain reverse logistics models. Subsequently, we have delved into operations research techniques, conducting an analysis based on them. Finally, we have presented our recommendations, and observations.

For the future research, we recommend exploring the following directions:

I. Addressing the great challenge of a lack of data in Reverse Logistics (RL) is crucial for future studies. Machine learning and deep learning algorithms can play a crucial role in recognizing patterns, making forecasts, and enhancing various aspects of RL activities.

II. Most RL models did not consider multi-period inventory management parameters. Developing dynamic mathematical models that consider inventory management across multiple periods would enhance the models.

III. While some papers have utilized genetic algorithms, the complexity of models surpasses the capabilities of common software like GAMS and CPLEX to handle large and complex mathematical models. Hence, future models should incorporate a broader range of heuristics and metaheuristics techniques.

IV. Studying customer loyalty in RL is a promising area for future research. Understanding how loyal customers affect return rates helps businesses improve RL. Loyal customers are more likely to attend eco-friendly activities like recycling, contributing to sustainability in RL.

V. While queuing theory has been discussed before in RL models, exploring various types and models within queuing theory can enhance the prediction of return volumes. It aids businesses in anticipating the influx of customers for returns during specific time frames. Moreover, by conceptualizing the return processing system as a queuing system, researchers can evaluate the effectiveness of diverse processing strategies. It includes analyzing the influence of factors like the number of processing stations, service rates, and arrival rates of returned products.

VI. In future research, there is a need to delve into the development of hybrid models that integrate traditional optimization techniques with cutting-edge predictive analytics, aiming to optimize decision-making processes and enhance the performance of 3PRL systems.

VII. Except for Xu et al. (2023), who addressed medical waste in 3PRL, no other papers focusing on this aspect were found. Therefore, the management of medical waste within the context of 3PRL represents a potential area for future research.

REFERENCES

Ahmed, R. R., & Zhang, X. (2021). Multi-stage network-based two-type cost minimization for the reverse logistics management of inert construction waste. *Waste Management (New York, N.Y.)*, 120, 805–819. 10.1016/j.wasman.2020.11.00433279346

Budak, A. (2020). Sustainable reverse logistics optimization with triple bottom line approach: An integration of disassembly line balancing. *Journal of Cleaner Production*, 270, 122475. 10.1016/j.jclepro.2020.122475

Ding, L., Wang, T., & Chan, P. (2023). Forward and reverse logistics for circular economy in construction: A systematic literature review. *Journal of Cleaner Production*, 388, 135981. 10.1016/j.jclepro.2023.135981

Forkan, M., Rizvi, M. M., & Chowdhury, M. A. M. (2022). Multiobjective reverse logistics model for inventory management with environmental impacts: An application in industry. *Intelligent Systems with Applications*, 14, 200078. 10.1016/j.iswa.2022.200078

Gao, X., & Cao, C. (2020). A novel multi-objective scenario-based optimization model for sustainable reverse logistics supply chain network redesign considering facility reconstruction. *Journal of Cleaner Production*, 270, 122405. 10.1016/j.jclepro.2020.122405

Ghanbarzadeh-Shams, M., Yaghin, R. G., & Sadeghi, A. H. (2022). A hybrid fuzzy multi-objective model for carpet production planning with reverse logistics under uncertainty. *Socio-Economic Planning Sciences*, 83, 101344. 10.1016/j.seps.2022.101344

Gholizadeh, H., Goh, M., Fazlollahtabar, H., & Mamashli, Z. (2022). Modelling uncertainty in sustainable-green integrated reverse logistics network using meta-heuristics optimization. *Computers & Industrial Engineering*, 163, 107828. 10.1016/j.cie.2021.107828

Guggeri, E. M., Ham, C., Silveyra, P., Rossit, D. A., & Piñeyro, P. (2023). Goal programming and multi-criteria methods in remanufacturing and reverse logistics: Systematic literature review and survey. *Computers & Industrial Engineering*, 185, 109587. 10.1016/j.cie.2023.109587

Gunasekara, L., Robb, D. J., & Zhang, A. (2023). Used product acquisition, sorting and disposition for circular supply chains: Literature review and research directions. *International Journal of Production Economics*, 260, 108844. 10.1016/j.ijpe.2023.108844

Hallak, B. K., Nasr, W. W., & Jaber, M. Y. (2021). Re-ordering policies for inventory systems with recyclable items and stochastic demand–Outsourcing vs. in-house recycling. *Omega*, 105, 102514. 10.1016/j.omega.2021.102514

Jahangiri, A., Asadi-Gangraj, E., & Nemati, A. (2022). Designing a reverse logistics network to manage construction and demolition wastes: A robust bi-level approach. *Journal of Cleaner Production*, 380, 134809. 10.1016/j.jclepro.2022.134809

Kannan, D., Solanki, R., Darbari, J. D., Govindan, K., & Jha, P. C. (2023). A novel bi-objective optimization model for an eco-efficient reverse logistics network design configuration. *Journal of Cleaner Production*, 394, 136357. 10.1016/j.jclepro.2023.136357

Kilic, H. S., Kalender, Z. T., Solmaz, B., & Iseri, D. (2023). A two-stage MCDM model for reverse logistics network design of waste batteries in Turkey. *Applied Soft Computing*, 143, 110373. 10.1016/j.asoc.2023.110373

Lei, J., Che, A., & Van Woensel, T. (2023). Collection-disassembly-delivery problem of disassembly centers in a reverse logistics network. *European Journal of Operational Research*.

Liao, G. H. W., & Luo, X. (2022). Collaborative reverse logistics network for electric vehicle batteries management from sustainable perspective. *Journal of Environmental Management*, 324, 116352. 10.1016/j.jenvman.2022.11635236208516

Lin, J., Li, X., Zhao, Y., Chen, W., & Wang, M. (2023). Design a reverse logistics network for end-of-life power batteries: A case study of Chengdu in China. *Sustainable Cities and Society*, 98, 104807. 10.1016/j.scs.2023.104807

Nanayakkara, P. R., Jayalath, M. M., Thibbotuwawa, A., & Perera, H. N. (2022). A circular reverse logistics framework for handling e-commerce returns. *Cleaner Logistics and Supply Chain*, 5, 100080. 10.1016/j.clscn.2022.100080

Nosrati-Abarghooee, S., Sheikhalishahi, M., Nasiri, M. M., & Gholami-Zanjani, S. M. (2023). Designing reverse logistics network for healthcare waste management considering epidemic disruptions under uncertainty. *Applied Soft Computing*, 142, 110372. 10.1016/j.asoc.2023.110372237168874

Roudbari, E. S., Ghomi, S. F., & Sajadieh, M. S. (2021). Reverse logistics network design for product reuse, remanufacturing, recycling and refurbishing under uncertainty. *Journal of Manufacturing Systems*, 60, 473–486. 10.1016/j.jmsy.2021.06.012

Sar, K., & Ghadimi, P. (2023). A systematic literature review of the vehicle routing problem in reverse logistics operations. *Computers & Industrial Engineering*, 177, 109011. 10.1016/j.cie.2023.109011

Saxena, N., Sarkar, B., Wee, H. M., Reong, S., Singh, S. R., & Hsiao, Y. L. (2023). A reverse logistics model with eco-design under the Stackelberg-Nash equilibrium and centralized framework. *Journal of Cleaner Production*, 387, 135789. 10.1016/j.jclepro.2022.135789

Xu, X., Wang, F., Chen, Y., Yang, B., Zhang, S., Song, X., & Shen, L. (2023). Design of urban medical waste recycling network considering loading reliability under uncertain conditions. *Computers & Industrial Engineering*, 183, 109471. 10.1016/j.cie.2023.109471

Zhang, X., Zhu, S., Dai, S., Jiang, Z., Gong, Q., & Wang, Y. (2024). Optimization of third party take-back enterprise collection strategy based on blockchain and re-manufacturing reverse logistics. *Computers & Industrial Engineering*, 187, 109846. 10.1016/j.cie.2023.109846

Chapter 2
Demystifying Sustainable Practices in the Hotel Industry:
Evidence From Ghana

Mildred Nuong Deri
https://orcid.org/0000-0003-1564-4880
University of Energy and Natural Resources, Ghana

Amrik Singh
https://orcid.org/0000-0003-3598-8787
Lovely Professional University, India

Perpetual Zaazie
https://orcid.org/0000-0001-6109-2082
University of Energy and Natural Resources, Ghana

David Anandene
University of Cape Coast, Ghana

ABSTRACT

Sustainability continues to be one of the most commonly discussed trends within the hotel industry. Sustainability has become a matter of concern for both individuals and corporations. Hotel chains develop environmental sustainability practices, mainly for cost-reduction purposes, accommodating the owners' demands for efficiency. Notwithstanding, there are differences according to the chain's size. This study aimed to identify sustainable hotel practices in the Sunyani Municipality of Ghana. The hospitality industry is starting to take responsibility for environmental sustainability. A strong focus on energy, waste, and water usage is directly linked

DOI: 10.4018/979-8-3693-2595-7.ch002

with financial benefits in the operation of the hoteliers. Practices connected to the social aspect of sustainability are less developedThe hotels accepted and executed water and liquid waste management more than any other sustainable practice, and consumer demand was the element that drove these behaviours. It is advised that hotels create and execute a strategy for environmental management that will help to mitigate these consequences.

INTRODUCTION

Globally, the hotel industry is starting to embrace sustainability and becoming green has caught the interest of both consumers and businesses. Companies are being pushedto acknowledge their part in the degradation of the planet, especially hotels. Since the quality of the environment is declining, environmental issues are gaining attention on a global scale (Deri, et al., 2022). As a result of issues like global warming, air, water, and land degradation, hotels are using ecologically friendly business strategies (Leonidou, et al., 2013; Singh, et. al., 2024).

Environmental concerns in the hotel industry include waste recycling, energy and water conservation among others (Mensah, 2006). As guests become more conscious of environmental deterioration and overconsumption of resources such as commodities, energy, and water, they are becoming more interested in staying in green lodging facilities (Han et al., 2010). For hotels to become eco-friendly, lodging facilities must demonstrate responsible behaviour, such as water and energy-saving and waste reduction, to become green hotels" (Manaktola & Jauhari, 2007. Pg.17). Hotel managers have a role in motivating the implementation of green practices (Rogerson& Sims, 2012). Policies, programmes, and regulations for eco practices in hotels in Southern part of Africa arein a fragmented manner, which is cause for concern (Spenceley, 2005). In Ghana, some studies have been done by Mensah, (2006, 2008 & 2012) in Accra but few or none in the hotels sector in the Sunyani municipality. To meet the global market for eco lodgings, green management rules and regulations need to be made and put into place.The study is crucial because customers are more informed of the need for hotels to go green and hoteliers cannot afford to remain unconcerned.As a result, baseline data on eco-friendly hotel practises in the Sunyani municipality is required to create an environmental management programme for the industry.This study therefore seeks to fill in this gap. The research goals listed below served as the study's guidelines:

1. To identify sustainable practices among hotels in Sunyani;
2. To examine the level of adoption of sustainability practices among the various categories of hotels in Sunyani

3. To examine barriers in implementing sustainability practices among hotels in Sunyani

LITERATURE REVIEW

Sustainable Practices

Sustainabilityis about utilizing goods and techniques in a green way prevents pollution and the depletion of natural resources from having a detrimental influence on the environment (Perks, 2010). Environmental management can be thought of as an on-going process that is implemented through management decisions. It involves monitoring a hotel's operations and developing appropriate plans and activities to lessen any adverse environmental effects (Mensah, 2006). Hotels have a significant environmental impact, contributing to a variety of global issues, the most significant of which is climate change (Bahadanowcz et al., 2011; Singh, et. al., 2024). Hotel activities generate greenhouse gas emissions, which are harmful to the environment and human health (Singh et al., 2024). Hotels have been demonstrated to have the greatest adverse effects on the environmentof all commercial structures (Rogerson& Sims, 2012).

Liquid Waste Management

Water is extremely vital in the accommodation sector. Water management entails storing waste-water for various purposes and lowering water consumption (Tang, 2012). Water is utilised in hotels on a daily basis for cleaning, cooking, and drinking. The average water usage pattern of hotel visitors is estimated to be around 170-500 litres per day (Mungai & Urungu, 2013). According to research, green hotel activities include the deployment of water conservation techniques (Rogerson, 2012; Singh et. al., 2024; Singh et al., 2024). "The use of water-efficient devices such as low-flow or infrared-activated faucets, low-flow showerheads, low-water-volume toilets, sink aerators, regular fixing of toilet and bath leaks are examples of these measures" (Hsiesh, 2012. p.28). By collecting rainwater and using it to flush toilets, hotels can keep their operations waste-free (Moreo, 2008; Singh, et al., 2024).

Solid Waste Management

Waste management is a comprehensive approach to waste prevention that involves a wide range of solutions for waste management that are environmentally responsible, commercially viable, and ethically acceptable (McDougal, et al., 2001). A hotel that implements a solid waste minimisation programme can save money on garbage transportation fees while also being more ecologically friendly. This is becoming more and more the case as landfill costs increase and solid waste is recognised as a severe environmental problem (Moreo, 2008). Dealing with food waste is an additional component of a solid reduction and recycling approach as well as table leftovers, cooking losses, and packing mistakes that cause food waste to build up.

Green Energy Consumption and Efficiency

Adequate and efficient running of a hotel requires the utilization of power (Bohdanowicz, 2006). Therefore, energy conservation has long been regarded very important aspects of sustainability in the hotel sector, because these facilities use a lot of power for various operating purposes (Kasim, 2017; Singh, et. al., 2024). Alternative energy sources, such as renewables, are critical, resulting in a shift in energy supply to more sustainable solutions. putting in place energy-saving devices, installing smart thermostats to regulate guestroom energy usage, installing sensors that detect occupancy to turn off the lights automatically when guests leave the room, and establishing sustainable energy initiatives such as using solar and wind power are energy management practices highlighted by Gise, (2009). Green hotels use energy-saving methods such as use of energy saving bulbs to make their hotels more energy-efficient (Hsieh, 2012; Singh, et al., 2024).

Air Quality Management

As the hotel sector grows more competitive, hotel expansion is becoming the order of the day. This raises awareness of the damaging effects on air quality (Cascardo, 2007; Singh, et al., 2024). In order to improve air quality in hotel rooms, it is strongly suggested that ecologically friendly and non-toxic cleaning chemicals be used, particularly in housekeeping. Energy efficiency and environmental measures (clean air) are related and this can lower the risk of health-related liability while also improving employee and guest interactions.Chlorofluorocarbons, for example, are an ozone-depleting substance that can release toxic air pollutants through poorly maintained heating, ventilation, and air conditioning (HVAC) devices in lodging establishments (Suttell, 2005). Controlling air quality in lodging facilities is primarily motivated by concerns about human health.

Green Purchasing

Green purchasing is described as an eco-responsible method of buying that is more environmentally friendly and promotes recycling and reuse of materials while maintaining the performance standards for the products (Millar, et al., 2012). The biggest impediment to effective green purchasing is the high cost of environmental programmes. It is preferable to utilise paper created from recycled items rather than plastic-based packaging, which is very polluting (Allen, 2007). According to Moreo (2008), hotels should acquire locally cultivated food since it is fresh, indigenous, and emblematic of the place (Singh et al., 2024). It is also crucial that purchases are made from vendor who exclusively offers items that encourage environmental sustainability, social justice, and fair-trade ideals (Allen, 2007). Regarding guest rooms and food and beverage outlets, hotels can purchase regenerated sustainable and environment packaging including take-out containers (Timothy &Teye, 2009).

Current Situation in the Sunyani Municipality

Currently, increased energy demand, a greater strain on the treatment of solid waste and industrial discharges into land and air are only a few of the numerous severe environmental consequences of the hotel sector. As a result of these implications, the protracted repercussions of such negative environmental influences, notably those connected to global climate change, are quite unknown (Rogerson & Sims 2012). Hence, the need for this study is very significant. Hotels in the Sunyani municipality lacks data on their green management activities, yet a wide range of environmental effects have advised that rapid action is necessary to mitigate these impacts. Hotels in the Sunyani municipality are witnessing a lack of coordination in the development, formulation, and implementation of green management strategies and this should be a matter of concern. In order to meet the growing demand from guests for more environmentally friendly lodgings, green management rules and regulations (policies) need to be made and put into place.

Level of Adoption of Sustainable Practices

The hospitality industry plays a pivotal role in shaping sustainable practices, with hotels standing as prominent contributors to environmental impact. According to Diaz-Farina, et al. (2021), a notable shift has occurred as hotels increasingly embrace green initiatives. This transformation reflects an evolving understanding of the importance of sustainability and responsible stewardship of resources. The adoption and implementation of green practices within hotels encompass a spectrum of strategies aimed at reducing carbon footprints, conserving resources, and promoting eco-conscious operations, basic recycling efforts to comprehensive sustainability programs that permeate every aspect of hotel operations (Malik & Kumar, 2012; Bresciani & Dhir, 2021). Adoption simply refers to how far individuals go to implement innovative procedures or techniques. Adoption typically starts with the recognition of a problem, then moves on to seeking an answer, the fundamental decision to attempt adopting a solution, and finally, the sincere decision to attempt carrying out its execution (Damanpour & Schneider, 2006; Mendel et al. 2008). There has been a growing trend in hotels towards adopting eco-friendly practices due to increasing awareness of environmental issues and the benefits of sustainability (Deri et al., 2022). An overview of green practices together with the degree of acceptance and application of sustainable practices can be seen in Table 1.

Table 1. Level of adoption of sustainable practices in hotels in Sunyani Municipality

Green Practices		Level of Adoption and Implementation
Green Energy Consumption and Efficiency	i.	Hotels should embrace and implement natural light mechanism to save on energy
	ii.	Hotels should adopt and implement the use of energy-saving bulbs in the facility
	iii.	They should also embrace the use of renewable sources of energy for daily heating of water
	iv.	Employ a conventional energy source that is more dependable than renewable energy.
	v.	They should use power saving appliances to save on energy
	vi.	The hotels should also adopt and implement the use of solar which saves cost in the long run
Water and Liquid Waste Management	i.	The hotels must practice eco-friendly liquid waste-management
	ii.	Hotels must adopt easy and efficient ways to dispose liquid waste
	iii.	They should also implement water saving measures and procedures
	iv.	The hotels should also provide efficient and sufficient source of water as well as practice recycling and water re-use
Air Quality Management	i.	The hotels should embrace and implement measures to reduce carbon emissions
	ii.	Heating, Ventilation and Air Condition (HVAC) systems must be eco-friendly
	iii.	Food production activities should produce less of carbon dioxide
Solid Waste Management	i.	Solid waste which is organic should be recycled, re-used and used for other purposes like manure
	ii.	The hotel employs an environmentally friendly solid waste management technology
	iii.	The hotels must embrace and implement processes and practices which reduce solid waste production
Environmental Purchasing	i.	Use fresh products with little or no preservatives and food colouring.
	ii.	Buy in bulk rather than individually packaged items.

continued on following page

Table 1. Continued

Green Practices		Level of Adoption and Implementation
Green Energy Consumption and Efficiency	i.	Hotels should embrace and implement natural light mechanism to save on energy
	iii.	Choose and buy seasonal fruits and vegetables.
	iv.	Avoid using plastic cups or disposable table wares.

Source: Fadhil (2014)

Barriers in Implementing Sustainable Practices among Hotels in the Sunyani Municipality

In the pursuit of creating more environmentally responsible and sustainable operations, the hospitality industry faces a multitude of challenges and barriers when attempting to implement eco-friendly practices within hotels. While the commitment to sustainability is growing, various hurdles impede the seamless integration of green initiatives across the sector. From financial constraints to infrastructure limitations and regulatory challenges, these barriers pose significant hurdles for hoteliers striving to embrace eco-conscious strategies. Studies have addressed obstacles to adopting sustainable practices in hotels (Luo, et al., 2021; Singh et al., 2024). First and foremost, lack of green information and knowledge. Chan (2008) claims that without expert guidance from professionals who assist in the execution and development of sustainability standards, many managers are ignorant of regulations pertaining to environmental management and struggle to comprehend what is required of them in the early stages. Secondly is the uncertainty of sustainable outcome.Hotel management may have concerns about the continuous efficacy of green measures in achieving goals and may find it challenging to evaluate the environmental impacts and weigh the importance of implementing sustainable practices. Therefore, the outcome can hardly be guaranteed. Several managers are hesitant to establish environmentally conscious hotels because they are unsure about how to manage green operations and what the operation's outcome will be. They think that operating in a green manner entails a complex process that requires additional time, money, and procedures (Chang & Ho, 2006). This uncertainty can deter hotel owners or management from committing to such projects.

Furthermore is about the difficulty in managing and training staff.According to Mandarin Oriental KL manager, it takes a lot of efforts and time to train staffs to become green and understand green practices (Yusof & Jamaludin, 2013). In addition, there is a lack of government regulation and policy. Rivera et al. (2009) define government or regulation as a coercive instrument that puts pressure on organizations to become greener.Because of the negative environmental consequences of hotels, increasing pressure is being put on the government to encourage more eco-friendly firms and green consumerism in hotels (Moreo, 2008; Singh.A et. al., 2024). Meanwhile, in some countries, the absence of supportive policies for sustainable practices can hinder hotels from prioritizing eco-friendly initiatives. Last but not the least is lack of networking with green suppliers. Supplier constraints are a major impediment to the effective adoption of sustainable practices. These limitations encompass various challenges that hoteliers encounter when seeking eco-friendly products, services, or materials from their suppliers (Saud, 2013).Limited availability or higher costs of eco-friendly supplies may deter hotels from integrating them into their operations. Suppliers might have a limited range of eco-friendly products or services compared to conventional options. Finally, is lack of customer support. Perceived customer demand—or the perception that guests might not actively seek or prioritize eco-friendly accommodations—can indeed serve as a barrier to implementing sustainable practices in hotels. While consumer preferences are shifting towards sustainability, there might still be uncertainties among hoteliers regarding the actual demand for and willingness to pay for eco-friendly initiatives (Mohan, 2017).While guest preferences for sustainable accommodations are increasing, there may still be a perception gap regarding the actual demand for eco-friendly services. Hotels might hesitate to invest in sustainable practices if they believe customers are not actively seeking or willing to pay for these features.

Theories Underpinning the Study

"Sustainability, institutional, the planned behaviour and resource-based theories" were all considered in this study.The idea of sustainable tourism serves as the foundation for green practices in the travel and hospitality sectors, which have received a lot of attention. According to Daft (2008), a company that is sustainable is one that creates money, satisfies present demands, and protects the planet for coming generations. For the benefit of society, the travel and hospitality sectors must assume accountability for their effects on the environment and promote sustainable growth (Gössling, 2009; Singh, et. al., 2024).

Several empirical studies show that the hospitality industry benefits economically from adopting environmental measures (Álvarez, 2001). According to Hendrix and Vasilind (2005), businesses place a high value on sustainable growth due to regulatory measures, economic opportunities presented by pollution prevention, or their recognition of the strategic significance of environmental challenges.It focuses on strategic decision-making, entrusting the leadership of the company with the critical duties of determining, developing, and utilizing vital resources to optimize profitability (Singh et al., 2024)

The resource-based approach states that companies with strategic resource own-ership have significant market advantages over those without. In addition to reducing opportunities and risks from rivals, a strategic resource helps the company become more successful and efficient (Muhittin & Reha, 1990). A wide spectrum of human behaviours can be accurately predicted by the "Theory of Planned Behavior (TPB)", utilizing high predictive value (Aragon-Correa & Sharma, 2003).The motivating factors that affect behaviour are expected to be captured by intentions, which are measures in relations to the energy"people are willing" to put forth to carry out the intended behaviour. The "Ajen's Theory of Planned Behaviour (TPB)" explains the ability of guest' intentions to visit a green hotel (Hsu & Sheu, 2009). A growing number of consumers are beginning to search for eco lodging facilities and are ready to pay more as they take into account numerous environmental challenges (Laroche, et al., 2001).

Conceptual Framework of the Study

As can be seen in the image below, the research employed a conceptual framework to identify its independent and dependent variables. Sustainable practises are the independent variable in this study, while obstacles to adopting sustainable practises are the dependent variable. The underlying framework's list of sustainable practices includes managing liquid and solid waste, managing air quality, managing green energy consumption and efficiency, and managing solid waste. It also includes environmental purchasing. And the barriers to the adoption of the sustainable practices in this study are cost, information and knowledge, managing and training staff, green outcome, customer and employee support, and green suppliers. Figure 1 is the conceptual framework.

Figure 1. conceptual framework

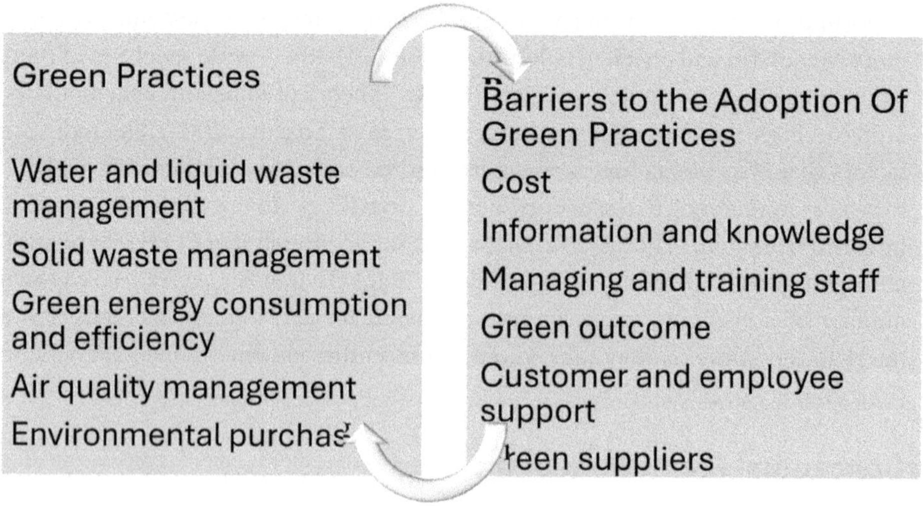

Green Practices	Barriers to the Adoption Of Green Practices
Water and liquid waste management	Cost
Solid waste management	Information and knowledge
Green energy consumption and efficiency	Managing and training staff
Air quality management	Green outcome
Environmental purchas̆	Customer and employee support
	ᵗeen suppliers

Source: Authors construct

METHODOLOGY

Study Area

The Bono region is one of the sixteen (16) administrative regions of Ghana and Sunyani as its regional capital also known as green city of Ghana. Sunyani can pride itself with a major conference destination with hotels springing up. Sunyani is surrounded by tourist destinations, making the region a hub for tourist. Their distinctive tourist attractions draw visitors from far and near. The region is a stopover for visitors from the south and the north. The region's tourism commercialization has increased the demand for accommodation. As a result, many hotels, guesthouses, and budget hotels have sprung up to fulfil the increased demand for lodging. The study location was chosen because there is a serious demand for accommodation by travellers, both leisure and business within the region. Figure 2 represents the map of the study area.

Figure 2. Map of the Sunyani Municipality

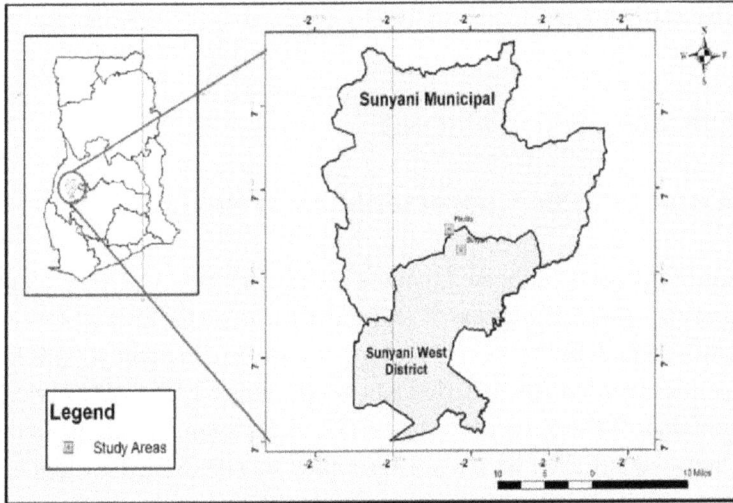

Source: Foli, Awuni & Amponsah (2020)

Research Design and Paradigm

This study is based on a quantitative method-based post-positive paradigm. The target population of the study encompassed hotels managers within the Sunyani municipality. Data from GTA (2019) shows that 74 registered hotels are situated within the Sunyani municipality, consisting of 1-star, 2-star, 3-star, and budget hotels. The study used simple random sampling and stratified technique. This sampling approach was used for the investigation because the study population and sample frame were known. Thus, all units within the sample population had an equal probability of being chosen. The hotel managers (n =40) were chosen using stratified random sampling.

The questionnaire was the primary tool used to collect data from respondents. The data was gathered by distributing surveys directly to hotel managers. The questionnaires are divided into three modules. The first section described the background characteristics of respondents such as gender, age, educational background etc. Section two documented hotel characteristics such as ownership type, number of rooms, years of operation etc. Section three, highlighted hotels' green practices.

Out of the total of 50 hotels sampled based on stratified sampling technique, 40 hotels returned their questionnaires for analysis. For data analysis purpose Version 26 of the Statistical Package and Service Solution (SPSS) program was used to analyse the field data. The SPSS program was used to decode and input this for additional analysis and interpretation. This data was however, meticulously adjusted in order to exclude any outliers that could have jeopardised the validity of the findings. Descriptive statistics and factor analysis were used to present results.

RESULTS AND DISCUSSIONS

Socio-Demographic Characteristics of Respondent

Results of this study indicated that the highest age range of respondents is 31-40 years representing 42.0% whereas 51 and above recorded the least representing 17.5% of the total respondent. More males were recorded 65% against 35% for females. Marital status recorded 70% married and work status 62.5% fulltime workers of the total respondent. With respect to the level of education of the respondents, more (72.5%) had their tertiary education. This implies that there are more males' managers in the hotel industry in Sunyani as compared to their female counterparts, who are between the ages of 31-40 and majority of them are married and also are employed on full time basis. This information is represented in Table 2Socio-Demographic Characteristics of Respondent (N=40).

Table 2. Socio-Demographic Characteristics of Respondent (N=40)

Demographics	Categories	Frequency	Percentage (%)
Age	Below 30	8	20.0
	31-40	17	42.5
	41-50	8	20.0
	51 and above	7	17.5
Gender	Male	26	65
	Female	14	35
Marital Status	Single	12	30.0
	Married	28	70.0
Work Status	Fulltime	25	62.5
	Casual	15	37.5
Status	Proprietor	7	17.5
	General Manager	17	42.5
	Manager	16	40.0
Level of education	SHS	11	27.5
	Tertiary	29	72.5
Years of service in this hotel	Less than 1 year	4	10.0
	1-3 years	12	30.0
	4-6 years	13	32.5
	More than 6 years	11	27.5

continued on following page

Table 2. Continued

Demographics	Categories	Frequency	Percentage (%)
Years of service in the hotel/ hospitality industry	Less than 1 year	4	10.0
	1-3 years	6	15.0
	4-6 years	14	35.0
	More than 6 years	16	40.0

Source: field survey (2020)

Hotel Characteristics

The study sought some information about the hotels in the Sunyani municipality. From table 2, Hotels that have been operating between 6-10 years, represents 67.5% of the total sample size whiles hotels operating above 20 years representing 5% of the total sample. The highest number of rooms ranged from 11-20 which constitute 45% compared to 50 rooms representing 5%. From the survey, majority (72.5%) of the respondents were from the budget hotels and the least hotelrated 3-star represents 2.5% of the total sample size. Most hotels were proprietorship owned representing 55.0% whilst limited liability hotels were made up of 45.0% of the total sample size. This means the municipal has more budget hotels with rooms ranging 11-20 and these hotels are individually owned and have been in operation between 6 – 10 years. The municipality is lacking 5 -4 star rated hotels and also multinational hotels as compared to other regions like Kumasi and Accra. Table 3represent Hotel Characteristics (N=40).

Table 3. Hotel Characteristics (N=40)

Hotel Characteristics	Categories	Frequency	Percentage (%)
Number of years in operation	Under 5 years	3	7.5
	6-10	27	67.5
	11-15	8	20.0
	Over 21 years	2	5.0
Number of rooms available	0-10 rooms	13	32.5
	11-20 rooms	18	45.0
	21-30 rooms	5	12.5
	31-50 rooms	2	5.0
	Over 51 rooms	2	5.0
Classification of the hotel	3-star	1	2.5
	2-star	4	10.0
	1-star	6	15.0
	Budget	29	72.5
Type of ownership	Sole proprietorship	22	55.0
	Limited Liability	18	45.0

Source: field survey (2020)

Sustainable Practices Among Hotels
in the Sunyani Municipality

The first objective of the study was to identify green practices adopted among hotels in the Sunyani Municipality. The information is represented in percentage in figure 3. The pie chart showed 22% indicated they have an environmental management policy whiles majority (78%) of the hotels do not have any environmental management policy. It is observed that the high graded hotels are the ones having policies as compared to the budget hotels but the question here is, is it really working effectively with those hotels having these policies? This is in line with the research by (Mbasera, Plessis, Saayman& Kruger, 2016) highlighting that without the direction of a policy, the execution of green projects is done haphazardly. As stated by Hsieh, (2012), a sustainable policy for leadership is a statement from top management about the organization's environmental commitments.

Figure 3. Pie chart showing an environmental management policy

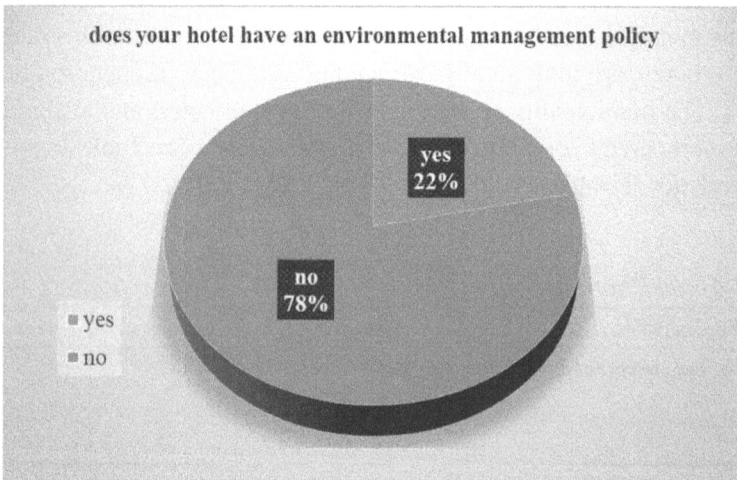

Source: field survey, (2020)

It is indicated that, 57.5% of hotels in the municipality employ liquid waste management largely, whiles 20% employ it to a moderate extent and 22.5% employ water and liquid waste management to a small extent. Most of the hotels had water collecting ducts that stored rain water and was used for irrigation and other purposes. The study demonstrates that while hotels make significant efforts to conserve water, they have very subpar practises for recycling and water reuse. Previous research, Harju, (2012), conform to this present study that, waste water treatment necessitated a high level of technical expertise. This might be the primary reason why some hotels in the municipality are performing poorly in water recycling and re-use without the necessary information and expertise.

The extent to which the hotels in the Sunyani municipality employ solid waste management is highlighted in this study. Results indicates that, 65% of the hotels employ solid waste management most, whiles 5% employ it to a moderate extent and 30% employ solid waste management to a small extent. Hotels in the Sunyani Municipality had much knowledge on solid waste management as compared to the other green practices. This ascertains to the fact that solid waste management is of great concern to hotels in the municipality.However, the reusing and recycling of solid garbage in their hotels is quite unsatisfactory. Some earlier studies have agreed with this, particularly on solid waste recycling (Cohen, 2006). Some operators are uninterested in decreasing and recovering waste.

The extent to which the hotels in the Sunyani municipality employ green energy consumption and efficiency. From the study, 32% of the hotels employ green energy consumption and efficiency to a large extent whiles 28% employ it to a moderate extent and 40% employ green energy consumption and efficiency to a small extent. Some of the star rated hotels had solar panels installed whiles others used energy saving bulbs and used power saving appliances. The study reveals that some of the hotels in Sunyani are trying to save energy which is in line with Dutta (2008) study on measures in saving energy. Odeku (2018) stressed in his study that "switching to the useof renewable energy as an energy source is the perfect solution to save the earth's biosphere and isessential to the stabilization of the climate world-wide".

It is shown that 25% of the hotels employ air quality management mostly whiles 22.5% employ it to a moderate extent and 52.5% employ air quality management to a small extent. One of the least adopted variables is the air quality management. In this category, Sunyani hotels have been slow to implement this green practise. This indicates a low-level knowledge on this particular green practise adoption. Previous research, particularly by Emblem and Hewett in 2001, demonstrated that "air quality" has received attention in the hotel business.

From the study, it is indicated that 17.5% of the hotels largely employ environmental purchasing whiles 32.5% employ it to a moderate extent and 50% employ environmental purchasing to a small extent. Environmental purchasing was the worst adopted known green practice of the hotels studied. "Hotels can acquire recycled eco-friendly packaging such as take-out boxes and bags, stationery, toilet paper, and other items created from previously recycled biodegradable packaging for their guest rooms, administrative offices, and kitchens" (Timothy &Teye, 2009 p. 67). There is the need for more training to strengthen hotel manager's knowledge on environmental purchasing in the municipality.

Level of Adoption of Sustainable Practices Among the Hotels in Sunyani

Based on the study, it is evident that the hotels use the following strategies for energy conservation: "energy-efficient light bulbs" (mean of 1.70), adopting natural light mechanisms (mean of 2.00) and with a mean of 3.92 for hot water use, this demonstrates unequivocally how widely hotels have embraced green energy efficiency and usage. Table 4 shows the level of adoption and implementation of the sustainable practices in the hotels in Sunyani.

Table 4. Level of adopting sustainable practices factors N rank (f) Mean St deviation

	N rank (f)	Mean	St deviation
Green energy consumption and efficiency			
In order to preserve energy, our hotel has welcomed and integrated natural light mechanisms	19		
Our lodging facility uses energy-efficient light bulbs.	26		
The hot water in our hotel is produced by solar energy.	7		
In our hotel, green energy is easily accessible.	9		
Compared to green energy, conventional energy sources are more dependable.	6		
The hotel employs a lot of energy-efficient equipment	15		
The hotel is able to employ more renewable energy	9		
The hotel uses solar energy to save money.	6		
Overall score	**9.7**		
Water and liquid waste management			
We make use of environmentally friendly liquid waste handling	9		
The hotel finds it quite simple to get rid of liquid waste.	15		
The hotel has implemented water-saving measures.	13		
There is plenty water from the hotel's supply	21		
This hotel has incorporated recycling and water conservation	6		
Overall score			
Air quality management			
The hotel engages in carbon-emitting activities.	3		
Heating, ventilation and air condition (HVAC) systems are eco-friendly	10		
Carbon dioxide emissions from food manufacturing operations are high	6		
This hotel has adopted strategies to lower its carbon footprint	4		
Overall score	**5.7**		
Solid waste management			
A substantial amount of biodegradable solid waste is present.	7		

continued on following page

Table 4. Continued

Green energy consumption and efficiency			
Solid trash is disposed of by the hotel using environmentally acceptable methods.	7		
The majority of the hotel's non-organic garbage is recycled and used again.	2		
Our hotel uses its organic waste for various things including compost.	4		
Our facility has adopted procedures and methods that lower solid waste	4		
Overall score	**4.8**		
Environmental purchasing			
The hotel uses fresh products with little or no preservatives and food colouring	6		
The hotel buys in bulk rather than individually packaged items	12		
We choose and buy seasonal fruits and vegetables	10		
The hotel cuts down the use of plastic cups or disposable tableware	9		
Overall score	9.2		

Source: field survey, (2020)

According to the findings of the study, "plenty water from the hotel's supply"recorded a mean of 2.00, "We make use of environmentally friendly liquid waste handling", with a mean of 2.55,"the hotel has implemented water-saving measures" had a mean of 2.55, the statement "The hotel finds it quite simple to get rid of liquid waste" with a mean of 2.27 and the statement"this hotel has incorporated recycling and water conservation" had a mean of 3.35.

In the category of air quality management, "Heating, Ventilation and Air Condition (HVAC) systems are eco-friendly"recorded a mean of 2.72. Results show that hotels are adopting environmentally friendly air and water policies. Additionally, food preparation operations generate a significant amount of carbon dioxide. Hotel engages in carbon-emitting activities have a mean of 3.52, Carbon dioxide emissions from food manufacturing operations are high as a mean of 3.42, this hotel has adopted strategies to lower its carbon footprintwith a mean of 3.80 and standard deviation of 1.364.

The hotel has implemented prudent solid waste management practices, as evidenced by the following: a large portion of solid waste is organic, scoring 3.41, Our facility has adopted procedures and methods thatlower solid wasterecorded a mean of 3.64,Our hotel uses its organic waste for various thingsincluding compost., with a mean of 4.33 and the majority of the hotel's non-organic garbage is recycled and used again, with a mean of 4.27 standard deviation 1.109.

The results further indicate that the hotels have implemented environmentally friendly purchasing practices. Specifically, the hotel purchases goods in bulk rather than individually packaged items (mean of 2.70, standard deviation 1.505); selects and purchases seasonal fruits and vegetables (mean of 3.02, standard deviation

1.544); reduces the use of plastic cups or disposable tableware (mean of 3.17, standard deviation 1.550); and uses fresh products (mean of 3.27, standard deviation 1.568) that have little to no preservatives or food colouring.

Table 5. Ranking of sustainable practices in their level of adoption

Sustainable practice	mean	Rank
Water and liquid waste management	2.72	13
Green energy consumption and efficiency	2.54	10
Environmental purchasing	3.36	9
Air quality management	3.50	6
Solid waste management	3.04	5

Source: field survey, (2020)

The ranking in table 6 indicates that, water and liquid waste management was adopted most among the various sustainable practices identified in the conceptual framework, followed by green energy consumption and efficiency and environmental purchasing. Air quality management and solid waste management were the less ranked in terms of their level of adoption and implementation respectively. This contradicts the research by Fadhil (2014) whose research found out that the most adopted and implemented green practice is green energy consumption and efficiency.

Barriers in Implementing Sustainable Practices Among Hotels in the Sunyani Municipality

The third and final objective of the study sought to examine the barriers in implementing sustainable practices among hotels in the Sunyani municipality. It is realised, 57.5% of the hotels have a challenge, 22.5% moderately and 20% little extend. Lack of green information and knowledge constitute a major barrier in implementing green practices among hotels. Findings from the study shows that lack of networking with green suppliers constitute a major barrier in implementing green practices among hotels. It is seen that, 72.5% of the hotels see it to be a challenge largely whiles 17.5% to a moderate extent and 10% to a small extent. The result from the study indicates that the extent to which uncertainty of green outcome constitute a major barrier in implementing green practices among hotels. It is seen that, 70% of the hotels see it to be a challenge mostly whiles 15% to a moderate extent and 15% to a small extent.

The extent to which difficulty in managing and training staff constitute a major barrier in implementing green practices among hotels. It is observed that, 57.5% of the hotels see it to be a challenge to a large extent, 25% to a moderate extent and 17.5% to a small extent. The study also considered customer support as a barrier.

The extent to which lack of customer support constitute a major barrier in implementing green practices among hotels. 40% of the hotels have challenge mostly, 42.5% moderate extent and 17.5% small extent.

Lack of implementation and maintenance cost constitute a major barrier in implementing green practices among hotels. 77% of the hotels largely have challenges, 10% moderately and 13% small extent. Finally, lack of resources such as manpower and equipment also constitute a major barrier in implementing green practices among hotels. It is seen that, 80% of the hotels considers this as a challenge mostly, 7.5% moderate extent and 12.5% to a small extent.

CONCLUSION

It is concluded that solid waste management, followed by liquid waste management, green energy consumption, and efficiency, were the three green practises most frequently used by hotels in Sunyani Municipality. From the study, it came to light that water and liquid waste management is the most widely adopted sustainable practice in hotels, with solid waste management being the least. The study revealed that the major barrier that influences the adoption and implementation of green practices were lack of resources, followed by lack of implementation and maintenance cost and lack of networking with green suppliers. Government regulation and policy, Customer demand, level of competition have impact on waste management practices according to this study. Implications are that sustainability should be encouraged through recognising, promoting, and honouring green hotel mentors for current and prospective hotel managers and green practise promotion.

MANAGERIAL IMPLICATION

Hotel operators are recommended to apply sustainable business practices in their facilities. These regulations offer standards, recommendations and also specify how it must be accomplished to lessen the influence on the environment of hotel operations and also hotels should adopt and track the success of green policies in their facilities. Hotel managers should send their staff for training on green management and environmentally friendly practises developed by green initiative agencies, institutions, and universities (Singh et. al., 2024; Singh et al., 2024). These trainings would help strengthen sustainability in hotel operation which depends heavily on how accessible and supportive green training is to employees. This would also help workers understand how their operations relate to the current generation's environmental issues. They will also recognise that they have a responsibility to reduce environmental consequences by implementing environmentally sustainable projects.

LIMITATIONS AND FUTURE RESEARCH

There are some limitations as far as this paper is concerned. First of all, our research is not generalizable. Only hotels in the Sunyani metropolis were of interest to us. In order to compare our findings with those from the other region in Ghana, we would like to look into further, hotels in the entire Bono Region. We presume that it would be beneficial to compare different countries and areas. In addition, we would like to explore in the near future research on relationship between the hotel's financial success and the adoption of environmental policies.

REFERENCES

Allen, Y. (2007). *Innovation pushes Edmonton to the leading edge of waste management*. FCM. https://www.fcm.ca/Documents/presentations/2007/mission/Innovation _pushes_Edmonto to the leading edge of waste management EN.pdf

Aragon-Correa, J. A., & Sharma, S. (2003). A contingent resource based view of proactive corporate environmental strategy. *Academy of Management Review*, 28(1), 71–88. 10.5465/amr.2003.8925233

Cascardo, A. (2007, October 9-13). Indoor air pollution: An ever-growing threat to our society. *Executive Housekeeping Today*.

Deri, M. N. (2022). Green practices among hotels in the Sunyani Municipality of Ghana. *Journal of Business and Environmental Management*, 1(1), 1–22. 10.59075/ jbem.v1i1.147

Diaz-Farina, E., Díaz-Hernández, J. J., & Padrón-Fumero, N. (2021). Analysis of hospitality waste generation: Impacts of services and mitigation strategies. *Annals of Tourism Research Empirical Insights*, 4(1), 12–24.

Gise, (2009). Improving Operations Performance in a small Company: A Case Study. *International Journal of Operations & Production Management*, 20(3).

Han, H., Hsu, J., & Sheu, C. (2010). Application of the theory of planned behavior to green hotel choice: Testing the effect of environmental friendly activities. *Tourism Management*, 31(3), 325–334. 10.1016/j.tourman.2009.03.013

Hsieh, Y. (2012). Hotel companies' environmental policies & practices: A content analysis of web pages. *International Journal of Contemporary Hospitality Management*, 24(1), 97–121. 10.1108/095961112

Kasim, A. (2017). Corporate environmentalism in the hotel sector: Evidence of drivers and barriers in Penang, Malaysia. *Journal of Sustainable Tourism*, 15(6), 680–699. 10.2167/jost575.0

Laroche, M., Bergeron, J., & Barbaro-Forleo, G. (2001). Targeting consumers who are willing to pay more for environmentally friendly products. *Journal of Consumer Marketing*, 18(6), 503–520. 10.1108/EUM0000000006155

Leonidou, L. C., Leonidou, C. N., Fotiadis, T. A., & Zeriti, A. (2013). Resources and Capabilities as Drivers of Hotel Environmental Marketing Strategy: Implications for Competitive Advantage and Performance. *Tourism Management*, 35, 94–110. 10.1016/j.tourman.2012.06.003

Luo, J. M., Chau, K. Y., Fan, Y., & Chen, H. (2021). Barriers to the implementation of Green Practices in the Integrated Resort Sector. *SAGE Open*, 11(3), 1–15. 10.1177/21582440211030277

Malik, S., Kumar, S. (2012). Management of Hotel Waste: A Case Study of Small Hotels of Haryana State. *ArthPrabandh: A Journal of Economics and Management*, *1*(09) 43-55.

Manaktola, K., & Jauhari, V. (2007). Exploring consumer attitude and behaviour towards green practices in the lodging industry in India. *International Journal of Contemporary Hospitality Management*, 19(5), 364–377. 10.1108/09596110710757534

Mbasera, M., Du Plessis, E., Saayman, M. & Kruger, M. (2016). Environmentally-friendly practices in hotels. *ActaCommercii* 16(1)

McDougal, F., White, P., Franke, M., & Hindle, P. (2001). *Integrated Solid Waste Management: A Life Cycle Inventory* (2nd ed.). Blackwell Science. 10.1002/9780470999677

Mensah, I. (2006). Environmental management practices among hotels in the greater Accra region. *International Journal of Hospitality Management*, *25*(3), 0–431.

Millar, M., Mayer, K. J., & Baloglu, S. (2012). Importance of green hotel attributes to business and leisure travellers. *Journal of Hospitality Marketing & Management*, 21(4), 395–413. 10.1080/19368623.2012.624294

Mohan, V., Deepak, B., & Mona, S. (2017). Reduction and Management of Waste in Hotel Industries. *International Journal of Engineering Research and Applications*, 7(7), 34–37. 10.9790/9622-0707103437

Moreo, A. (2008). Green Consumption in hotel Industry an examination of consumer attitudes. Google scholar accessed 27 March 2019.

Mungai, M. &Urungu, R. (2013).An assessment of management commitment to application of green practices in 4-5-star hotels in Mombasa, Kenya.*Information and knowledge management*, *3*(6), 40-47.

Odeku, K. O. (2018). Proactive responses to mitigate climate change impacts by the hospitality sector in South Africa. *African Journal of Hospitality, Tourism and Leisure*, 7, 1–13.

Rogerson, J. M. (2012). The Boutique hotel industry in South Africa: Definition, scope and organisation. *Symposium on Motivation*, 27, 65-116.

Rogerson, J. M., & Sims, S. R. (2012). The greening of urban hotels in South Africa: Evidence from Gauteng. *Urban Forum23*(3), 391–407.

Saud, J. S. (2013). Solid waste management utilizing microbial consortia and its comparative effectiveness study with vermicomposting. *International Journal of Engineering Research & Technology (Ahmedabad)*, 2(10), 2870–2885.

Singh, A., Tyagi, P. K., & Garg, A. (Eds.). (2024). *Sustainable Disposal Methods of Food Wastes in Hospitality Operations*. IGI Global. 10.4018/979-8-3693-2181-2

Singh, V., Archana, T., Singh, A., & Tyagi, P. K. (2024). Utilizing Technology for Food Waste Management in the Hospitality Industry Hotels and Restaurants. In Singh, A., Tyagi, P., & Garg, A. (Eds.), *Sustainable Disposal Methods of Food Wastes in Hospitality Operations* (pp. 287–295). IGI Global. 10.4018/979-8-3693-2181-2.ch019

Spenceley, A. (2005). Tourism certification initiatives in Africa. The International Ecotourism Society (TIES), Washington, DC.

Suttell, R. (2005). Hospitality and IAQ. *Buildings*, (November), 62–74.

Tang, F. E. (2012). A study of water consumption in two Malaysian resorts. *International journal of environmental, Ecological and geophysical engineering*, 6(8), 88-93.

Timothy, D. J. &Teye, V. B. (2009). *Tourism & Lodging sector*. UK-Oxford Elsevier INC.

Chapter 3
Eco–Friendly Solutions for Sustainable Waste Management:
An Approach Towards a Cleaner and Greener Future

Subhadra Rajpoot
https://orcid.org/0000-0003-3177-1871
Amity University, Greater Noida, India

ABSTRACT

Solid waste management is a critical aspect of environmental stewardship that involves the collection, treatment, and disposal of various types of solid waste generated by human activities. This includes everyday items like household waste, as well as industrial, commercial, and agricultural waste. The primary goal of solid waste management is to minimize the adverse effects of waste on public health, the environment, and aesthetics. Solid waste management is a crucial aspect of maintaining environmental sustainability and public health. Proper waste management includes the collection, transportation, disposal, and recycling of solid waste materials (Nguyen-Viet et al., 2009). It is important to employ efficient strategies such as the implementation of communal bins and scheduled waste collection according to waste types (Gubernatorov, 2020). This ensures the separation of waste for recycling and reduces the amount of waste ending up in landfills.

DOI: 10.4018/979-8-3693-2595-7.ch003

INTRODUCTION

Waste control is taken into consideration to be intently associated with sustainable development. This paper highlights that Traditional structures for waste disposal and recycling are now not appropriate. Many developing and rising nations are Going through primary assignment in enhancing their inadequate and unsustainable waste management systems. Soil, air and water pollutants Are continuously posing risk to sustainable development. The paper emphasized that waste ought to no longer be deposited in Residential regions and out of control landfills. Waste management hierarchy has been described within the present look at to address Waste disposal troubles. Similarly, advantages of opting sustainable waste management strategies as well as demanding situations in the region of waste disposal also are protected by way of the authors. It's far recognized that sustainable waste control provides a suitable selection in adopting a technique for lowering waste with the involvement of all stakeholders in a network (Guerrero et al., 2012).

High rates of resource consumption and an improving standard of living have had an unanticipated and detrimental effect on the urban environment, generating garbage that is far more than what local governments and agencies can handle. Large waste quantities, associated costs, disposal methods and procedures, and the effects of wastes on the local and global environment are currently issues that cities are dealing with. However, these issues have also given cities a window of opportunity to create solutions, including combining the public and private sectors, utilizing cutting-edge technologies and disposal techniques, and involving behavioural adjustments and awareness-raising. The excellent practices from numerous cities across the world provide ample evidence of these difficulties.

A thorough re-evaluation of "waste" is required to determine whether it truly qualifies as such. Reconsideration that demands that waste become wealth, refuse become resources, and stash become cash. It is obvious that the current waste disposal strategy, which is centred on municipalities and uses high energy/high technology, needs to shift more in the direction of waste processing and recycling, which entails public-private partnerships, community-driven waste minimization as the goal, and low energy/low technology resources. Future waste minimization programs will be defined by several factors, such as increased community involvement, comprehension of the financial advantages and waste recovery, an emphasis on life cycles rather than end-of-pipe solutions, decentralized waste administration, reduction of environmental effects, and balancing investment costs with long-term benefits.

A waste management system is a streamlined procedure used by businesses to avoid, minimize, and repurpose garbage. Alternatively referred to as waste disposal, this method involves businesses putting in place all-encompassing plans to effectively handle wastes from the point of origin to the point of disposal. Recycling, composting, incineration, landfills, bioremediation, waste-to-energy, and waste minimization are examples of potential waste disposal technique (Hettiarachchi et al., 2018). Organizations can eliminate, minimize, reuse, and prevent trash by using a waste management system, which is a simplified procedure. This method, which is often referred to as waste disposal, involves businesses putting thorough plans into place to effectively manage wastes from the point of origin to the point of disposal. Recycling, composting, burning, landfills, bioremediation, waste-to-energy, and waste minimization are some possible techniques for disposing of trash.

Figure 1. Waste Management cycle

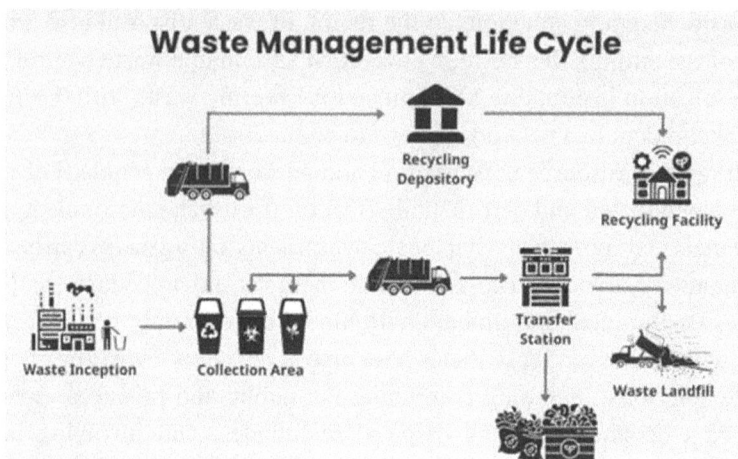

Source: safety culture (2014, p.1)

SUSTAINABLE WASTE MANAGEMENT

Since its inception, the idea of sustainable development has gone through several stages of growth. Several institutions and organizations that now focus heavily on putting the concept's tenets and goals into practice were involved in its historical development. Over time, the notion of sustainable development has been subjected to various criticisms and interpretations while maintaining acceptance in various domains of human endeavour. As a result, the definition has garnered significant citations within academic literature. The idea has evolved to meet the demands of a complex, global world, but its fundamental ideas and objectives—as well as the challenges associated with putting them into practice—have largely not changed (Scarlat, 2015). Nevertheless, new objectives were established, and some old ones were revised. These objectives are shared within the scope of the 2015 Millennium Development Goals, which list the obstacles that mankind must overcome in order to both achieve sustainable development and continue to exist on Earth.

A definition of sustainable development is "development that meets the needs of the present without compromising the ability of future generations to meet their own needs," as stated in the Brundtland Report of the World Commission on Environment and Development. (Brundtland Report 1987) According to this notion, current activities shouldn't endanger a society's culture or level of living. Countries vary in their ability to achieve sustainable development because of differences in their size, income, standards of living, cultures, and political and administrative structures. While industrialized nations may find it simpler to pursue sustainable development when they are wealthy and have access to cutting-edge technology, this is not always the case.

Sustainable waste management aims to minimize waste generation, maximize resource recovery, and reduce environmental impact. There are numerous approaches and techniques for managing garbage. An enterprise can customize a waste management system by combining or rearranging these tactics. Reducing waste production, making effective use of material resources, and acting in a way that actively advances the social, economic, and environmental objectives of sustainable development are some ways to implement sustainable waste management (Abubakar,2017). The following are some proactive ways that process planning can address sustainable waste management:

1. Making the best use of the materials needed for the construction.
2. Cutting back on the quantity of garbage produced.
3. Waste management for building and demolition projects.
4. Specifications for the materials (such as the usage of recycled and repurposed materials).
5. Providing facilities and space for recycling.

SUSTAINABLE WASTE MANAGEMENT PRINCIPLES

The Waste Management Hierarchy, which represents the relative sustainability of the various waste management alternatives, can be used to arrange them. The goal of contemporary waste management techniques is sustainability. Waste reduction, reuse, and recycling are additional options for getting rid of waste. Our society's material usage is crucial to both the environment and the economy in the future. Growth in the world's population and economies will increase competition for limited resources on a global scale (Abul.2010). When resources become scarcer in the future, a more efficient and less harmful use of materials keeps our civilization prosperous, competitive in the market, and environmentally safe. Over the past century, the United States and the world have seen a sharp rise in material consumption. The worldwide raw material use increased during the 20th century at a rate roughly twice that of population growth, according to the Annex to the June 8, 2015, Declaration of the G7 Leaders (Levente. 2021)The usage of raw materials has increased by 0.4 percent for every percent increase in the gross domestic product. The consequences of this growing demand have been felt by the environment, including desertification, overfishing stress, habitat damage, and biodiversity loss. 140 million acres of agricultural land, 5.9 trillion gallons of blue water, 778 million pounds of pesticides, 14 billion pounds of fertilizer, 664 billion kWh of energy, and 170 million MTCO2e GHG are "wasted" by food loss and waste in the United States alone (Yang Q, 2018). Here are some key objectives:

- Waste Reduction: Implement strategies to minimize waste generation at the source through practices such as waste prevention, product redesign, and encouraging the use of reusable products.
- Recycling and Recovery: Increase recycling rates by promoting the separation and collection of recyclable materials such as paper, plastics, glass, and metals. Implement programs for composting organic waste to recover valuable nutrients.
- Waste-to-Energy Conversion: Explore technologies for converting non-recyclable waste into energy through processes like incineration with energy recovery or anaerobic digestion.
- Resource Conservation: Encourage the reuse of materials and products to extend their lifespan and reduce the consumption of virgin resources.
- Education and Awareness: Raise public awareness about the importance of sustainable waste management practices through education campaigns, community outreach, and involvement in waste reduction initiatives.
- Regulatory Compliance: Develop and enforce regulations and policies that promote sustainable waste management practices, including waste diversion targets, recycling mandates, and landfill diversion fees.
- Innovation and Research: Invest in research and development of innovative technologies and approaches for waste management, such as advanced recycling methods, circular economy models, and waste-to-value processes.
- Community Engagement: Involve local communities, businesses, and stakeholders in waste management planning and decision-making processes to ensure solutions are tailored to local needs and preferences.
- Environmental Protection: Minimize the environmental impact of waste management activities by preventing pollution, reducing greenhouse gas emissions, and protecting ecosystems and natural resources.

- Economic Viability: Promote economically viable solutions by considering the cost-effectiveness of waste management practices, including the potential for generating revenue from recycled materials or energy recovery.

By pursuing these objectives, sustainable waste management can contribute to environmental protection, resource conservation, and the transition to a more circular economy. Not only this number of new concepts are being used day by day for sustainable waste management for example the concept of circular economy **The concept of a circular economy** is gaining traction as a sustainable alternative to the traditional linear economy model. In a linear economy, resources are extracted, used to make products, and then discarded as waste. In contrast, a circular economy aims to minimize waste and make the most of resources. This is achieved through various strategies that focus on closing the loop of product lifecycles through greater resource efficiency. Key Components of Circular Economy is to Design the products for Longevity and Durability, to last longer, reducing the frequency of replacement. Products can be easily disassembled, and parts replaced or upgraded, and the Products can be updated with new features without replacing the entire item. No. of Companies offer repair services to extend the life of their products. Consumers are also provided with tools and instructions to repair products themselves. Product as a Service (PaaS) concept has been introduced so that instead of selling products, companies lease them, retaining ownership and responsibility for their lifecycle. Used products are disassembled, cleaned, repaired, and reassembled to "like-new" condition. End-of-life products are processed to recover raw materials.

Examples of Circular Economy Models

* Clothing and Textiles:
* Patagonia: Offers a repair and reuse program, encouraging customers to return worn-out garments for refurbishment.
* H&M's Garment Collection Program: Collects used clothing for recycling or resale.

Benefits of Circular Economy Models

* **Environmental Impact:** Reduces waste, lowers greenhouse gas emissions, and conserves natural resources.

Economic Opportunities: Creates new business models and revenue streams, such as leasing and refurbishment services.

Consumer Benefits: Often results in cost savings and longer-lasting products.

Resilience: Builds more resilient supply chains by reducing dependency on raw material extraction and improving resource efficiency.

• **Challenges and Considerations**

Initial Costs: Higher upfront investment for designing and implementing circular systems.

Behavioural Change: Requires significant shifts in consumer and business behaviour.

Regulatory and Policy Support: Needs supportive policies and regulations to incentivize circular practices.

By embracing circular economy models, businesses and societies can move towards more sustainable and resilient systems, ensuring resources are used efficiently and responsibly.

Artificial intelligence in waste management

The utilization of artificial intelligence has the potential to bring about a revolution in municipal waste management by enhancing the effectiveness of waste collection, processing, and classification. Artificial intelligence-based technologies like intelligent garbage bins, classification robots, predictive models, and wireless detection enable the monitoring of waste bins, predict waste collection, and optimize the performance of waste processing facilities.

1. **Smart Waste Bins**

AI-powered smart bins can identify and sort waste automatically. These bins use sensors and image recognition to detect different types of waste, such as plastics, metals, and organic materials, and sort them into appropriate categories. This reduces contamination in recycling streams and increases recycling rates.

2. Automated Sorting Systems

AI-driven robotic arms and conveyor systems in recycling facilities can quickly and accurately sort waste materials. Machine learning algorithms enable these systems to recognize various types of waste and sort them more efficiently than human workers. This not only speeds up the recycling process but also improves the quality of the sorted materials.

3. Predictive Maintenance

AI can predict when waste management equipment, such as collection trucks and recycling machinery, will need maintenance. By analysing data from sensors on these machines, AI systems can schedule maintenance before a breakdown occurs, reducing downtime and repair costs.

4. Route Optimization

AI algorithms can optimize waste collection routes for garbage trucks. By analysing factors such as traffic patterns, waste generation rates, and truck capacity, AI can create the most efficient routes, reducing fuel consumption and emissions while ensuring timely waste collection.

5. Waste Generation Prediction

Machine learning models can predict waste generation patterns in different areas. This helps municipalities and waste management companies plan for adequate resources and infrastructure to handle future waste volumes, ensuring efficient waste management and reducing the risk of overflow.

6. Recycling Process Optimization

AI can optimize the recycling process by analysing data on the types and quantities of waste being processed. This information can be used to adjust sorting and processing techniques in real-time, maximizing the recovery of valuable materials and reducing the amount of waste sent to landfills.

7. Public Awareness and Engagement

AI can enhance public awareness and engagement in waste management. Smart apps and chatbots can provide information on recycling guidelines, collection schedules, and the environmental impact of waste. AI-driven campaigns can also personalize messages to encourage more sustainable waste disposal behaviours among individuals and businesses.

8. Waste-to-Energy Conversion

AI can improve waste-to-energy (WTE) processes by optimizing the conversion of waste into electricity or fuel. Machine learning models can analyse the composition of waste and adjust the WTE process parameters to maximize energy output and minimize emissions.

9. Circular Economy Support

AI can support the development of a circular economy by tracking the lifecycle of products and materials. This helps in identifying opportunities for reuse, refurbishment, and recycling, ensuring materials stay in use for as long as possible and reducing the need for virgin resources.

10. Environmental Impact Monitoring

AI systems can monitor the environmental impact of waste management practices. By analysing data from sensors, satellites, and other sources, AI can provide insights into pollution levels, greenhouse gas emissions, and other environmental indicators. This information can be used to improve waste management strategies and reduce their ecological footprint.

Case Studies and Examples

AMP Robotics: This company uses AI-powered robots to sort recyclables at high speed and accuracy, significantly improving the efficiency of recycling facilities (Abdul,2007).

Bigbelly: Smart waste bins from Bigbellb use sensors and AI to monitor waste levels and optimize collection schedules, reducing the frequency of collections and associated costs.

Rubicon: An AI-driven platform that helps cities and businesses optimize their waste management operations, including route optimization and predictive analytics for equipment maintenance. The details are shown in Fig.2 given below. By leveraging artificial intelligence, municipalities bin reduces costs, improve safety, and reduce environmental impacts associated with waste management.

Figure 2. Application of artificial intelligence in waste management. Source: Bing-bing Fang (2023, p.2)

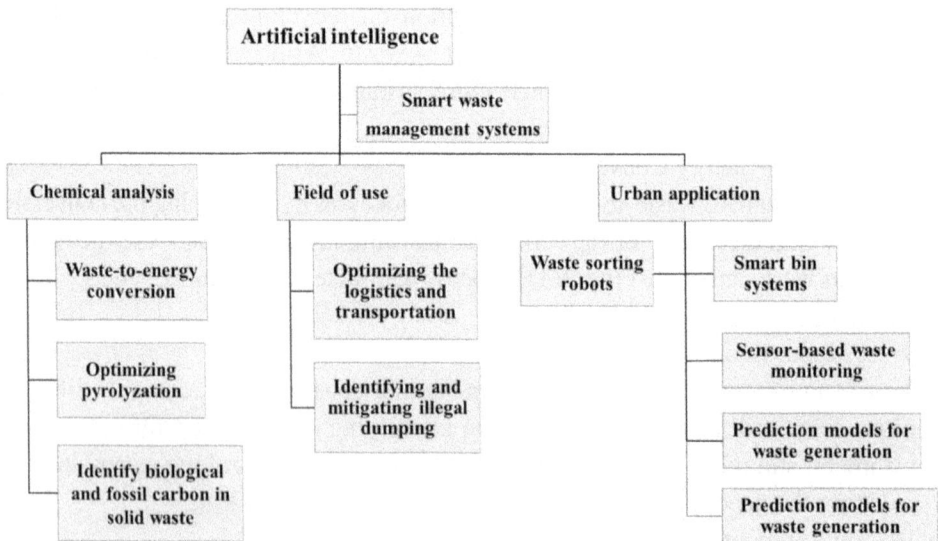

THE LIFE-CYCLE VIEWPOINT OF SUSTAINABLE MATERIAL MANAGEMENT

A comprehensive examination of a product's life cycle, from the extraction of materials to the management of its end of life, can reveal new avenues for cost-cutting, resource conservation, and environmental impact reduction. For instance, a product might be redesigned to be produced with fewer, alternative, less hazardous, and longer-lasting materials. Its design facilitates easy disassembly at the end of its useful life. The maker of the product keeps in touch with its clients to guarantee optimal usage, upkeep, and return of the product at the end of its useful life (Kabera,2019). This lowers the risk associated with the material supply, fosters customer loyalty, and helps the manufacturer recognize the evolving needs of their clients. Additionally, the producer's connection with its suppliers is comparable, which enables the company to react to shifting requirements—such as lowering environmental impacts along the supply chain.

WASTE MANAGEMENT HIERACHY

Waste management hierarchy, also known as the waste management pyramid, is a framework that prioritizes actions to minimize waste generation and maximize resource recovery. At its core are the principles of reduce, reuse, recycle, and dispose. The hierarchy is structured to guide decision-making towards the most environmentally and economically beneficial methods of waste management (Abubakar, 2016).

Understanding that no single waste management strategy is appropriate for handling all materials and waste streams in all situations, the EPA created the non-hazardous materials and waste management hierarchy. The hierarchy presents the different management options in order of environmental preference. The least preferred choice in the hierarchy is disposal, while the most environmentally beneficial is waste prevention. Most waste streams are covered by the hierarchy. Reducing, reusing, recycling, and composting are emphasized in the hierarchy as being essential to sustainable materials management. These tactics lessen the emissions of greenhouse gases, which fuel climate change (Khan,2013).

IMPORTANCE OF WASTE MANAGEMENT HIERACHY:

The waste management hierarchy is crucial for effective waste management and environmental conservation. It prioritizes actions according to their environmental impact, aiming to reduce waste generation and promote sustainable practices. Here's why each level of the hierarchy is important:

*   **Prevention:** The top priority is to prevent waste generation in the first place. This involves reducing consumption, using products with minimal packaging, and implementing efficient production processes. Prevention minimizes resource depletion, energy consumption, and pollution associated with waste production.

- **Minimization:** When waste generation cannot be avoided entirely, the next step is to minimize it. This includes strategies like reuse, recycling, and composting. Minimization reduces the volume of waste sent to landfills or incinerators, conserving resources and reducing environmental pollution.
- **Reuse:** Reusing products or materials extends their lifespan and reduces the need for new resources. It encompasses activities such as repairing, refurbishing, and repurposing items instead of disposing of them. Reuse reduces waste generation, conserves resources, and saves energy compared to manufacturing new products.
- **Recycling:** Recycling involves converting waste materials into new products, thereby conserving raw materials and reducing energy consumption and pollution associated with extraction and manufacturing processes.
- **Recovery:** Waste recovery refers to processes like energy recovery, where energy is generated from waste through methods like incineration or anaerobic digestion. While not as environmentally preferable as prevention, minimization, reuse, or recycling, recovery can still extract value from waste that cannot be avoided or recycled.
- **Disposal:** Disposal is the least preferred option and should only be considered for waste that cannot be prevented, minimized, reused, recycled, or recovered. Landfilling and incineration are common disposal methods, but they can have significant environmental impacts .

Figure 3. waste management hierarchy

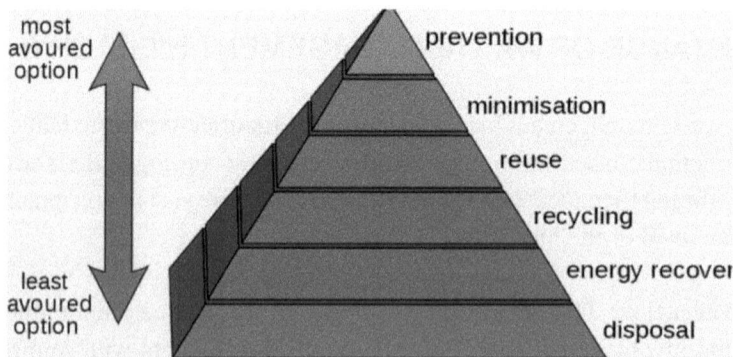

Source: Daniel O.(2018,P-3)

By following the waste management hierarchy, communities and industries can minimize environmental pollution, conserve natural resources, reduce greenhouse gas emissions, and move towards a more sustainable and circular economy. Overall, the waste management hierarchy provides a structured approach to managing waste that prioritizes prevention, reuse, and recycling over disposal. By following this hierarchy, communities and businesses can minimize their environmental footprint, conserve resources, and move towards a more sustainable future (Zafar 2012).

CURRENT STATE OF WASTE MANAGEMENT

The current state of waste management varies greatly depending on the region and country. However, globally, there has been a growing recognition of the need for more sustainable waste management practices due to concerns about environmental pollution, resource depletion, and climate change. Many countries are striving to implement more comprehensive waste management strategies that prioritize waste reduction, reuse, recycling, and proper disposal. This includes initiatives such as promoting composting, developing advanced recycling technologies, and implementing stricter regulations on waste disposal and pollution. However, challenges remain, including inadequate infrastructure in some areas, lack of public awareness and participation, and economic barriers to implementing more sustainable practices. Additionally, the rise of single-use plastics and electronic waste presents new challenges that require innovative solutions. Overall, while progress is being made in many places, there is still much work to be done to achieve truly sustainable waste management on a global scale (Balogun, 2022).

The following succinctly describes the condition of municipal solid waste management as of right now: -Roughly two-thirds of municipal solid trash are composted or recycled. - The Central Pollution Control Board (CPCB) estimates that approximately 160,000 metric tons of solid trash are produced nationwide each day (TPD). Around 153,000 TPD of waste is collected with an efficiency of about 96%. Roughly 30,000 TPD (18.4%) of garbage is landfilled, while 80,000 TPD (50%) of waste is processed. Of the entire garbage created, around 50,000 TPD (31.2%) is still unaccounted for. Between 2015–16 and 2020–21, the per-capita generation of solid waste increased slightly from 118.7 gm/day to 119.1 gm/day. Delhi produces the highest amount of solid trash produced per person. Solid waste processing has improved dramatically, rising from 19% in 2015–16 to around 50% in 2020–21. The percentage of solid trash that is landfilled has decreased from 54% to 18.4% over the same time. Chhattisgarh has a 100% solid waste treatment rate; 89% of Daman & Diu and Dadra & Nagar Haveli (DDDNH) and 87% of Goa have similar rates. A World Bank report states that every year, 2.01 billion tonnes of municipal solid garbage are produced worldwide, at least thirty-three percent of which are not managed in a way that is safe for the environment. The average daily production of garbage per person worldwide is 740 gm, while country-to-country variations range from 110 gm to 4.54 kg/day. Even though they only account for 16% of the world's population, high income countries generate 34% of the trash produced worldwide.

LIMITATIONS OF WASTE MANAGEMENT HIERACHY:

The waste management hierarchy is a prioritized approach to managing waste, with the goal of reducing waste generation and promoting sustainable practices (Gazzeh,2022)). However, it does have its limitations:

* **Implementation Challenges:** While the waste hierarchy provides a clear framework for waste management, implementing it effectively can be challenging due to various factors such as lack of infrastructure, resources, and enforcement mechanisms.
* **Resource Intensive:** Some waste management options higher in the hierarchy, such as recycling and composting, can be resource-intensive and require significant investments in technology, infrastructure, and human resources.
* **Economic Constraints:** Certain waste management options, such as recycling and energy recovery, may not always be economically viable, especially in regions with limited market demand for recycled materials or where energy prices are low.
* **Technological Limitations:** The effectiveness of certain waste management options, such as recycling and composting, can be limited by technological constraints, such as the availability of suitable recycling facilities or the capacity to process organic waste.
* **Behavioural and Cultural Factors**: Changing consumer behaviour and cultural attitudes towards waste generation and disposal can be difficult and may hinder the effective implementation of the waste hierarchy.
* **Environmental Concerns:** While the waste hierarchy aims to minimize the environmental impact of waste management, some waste management options, such as incineration and landfilling, can still have negative environmental consequences, such as air and water pollution, greenhouse gas emissions, and habitat destruction.
* **Inadequate Monitoring and Enforcement:** In many regions, there may be inadequate monitoring and enforcement of waste management regulations, leading to illegal dumping, improper disposal practices, and failure to comply with the waste hierarchy (Addo,2015).

Addressing these limitations requires a holistic approach that involves government regulation, public education and awareness campaigns, investment in infrastructure and technology, and collaboration between various stakeholders, including governments, businesses, communities, and non-profit organizations.

WHAT IS INDIA'S SOLID WASTE MANAGEMENT CHALLENGES?

- **Growing Trash Generation:** Growing economies produce more trash, which in turn causes consumption to rise. The production of e-waste will multiply as the digital economy grows. Furthermore, the population's rapid growth will increase trash. According to a 2014 Planning Commission Report, India is expected to produce 165 million tonnes by 2030, with an annual production of about 60 million tonnes in 2020.
- **Inadequate Garbage Management:** (a) Inadequate Processing: Merely 50% of the garbage undergoes processing. Approximately thirty percent of the garbage is not accounted for, and twenty percent ends up in landfills—a very bad way of disposal. (b) Inadequate and Incorrect Segregation Techniques: There is inadequate segregation at the source; hazardous waste is not tagged or sealed, which results in incorrect disposal. Moreover, improper disposal of e-waste happens; (c) waste is reused and recycled by scavengers in the unofficial sector; (d) recyclables are not regularly collected by the government; and (e) trash is frequently thrown into improper receptacles, creating unclean (Ojok,2013).
- **Littering And Illegal Dumping:** There are no sanitary landfills with leachate collection and gas recovery, therefore about half of waste is disposed of in uncontrolled dumps. This has negative effects on the ecosystem.
- **Lack Of Funds:** Cleaning and sanitation departments that are understaffed and underpaid are the result of a lack of funding with municipal bodies. Vehicles and other collection infrastructure are not well maintained. Having insufficient money makes it impossible to buy new cars and equipment.
- **Inconsistent Collection:** Inconsistent rubbish collection is caused by understaffing and under compensation. Not all localities are served by sanitation workers.
- **Inadequate Infrastructure:** Waste collection vehicles are not built for this function. This frequently results in overloading, which causes spills to occur while being transported. Since vehicles lack lifting equipment, loading is done manually, which is dangerous and unclean.

- **Lack Of Civic Responsibilities:** Low motivation and a lack of environmental knowledge have stifled innovation and the adoption of innovative technology that may revolutionize garbage management in India.

Negative Impacts Of Inadequate Waste Management

Health Concerns: (a) Poor and careless waste collection and handling exposes sanitation workers to several diseases; (b) municipal waste is frequently combined with hazardous and medical wastes, exacerbating health risks; (c) open burning of waste creates dangerous particles that can cause lung diseases; (d) poor collection results in garbage dumps that serve as havens for rats, mosquitoes, and other vermin. Mosquitoes are vectors of diseases like dengue and malaria (Zhang, 2010).

Environmental Concerns: (a) Careless disposal in landfills creates dangerous compounds that seep into the groundwater and soil. This makes groundwater unsafe to drink and can result in several illnesses; (b) garbage in landfills creates toxic gasses that pollute the air. The makeup of landfill gases varies depending on the kind of garbage, but 90–98% of the gases are made up of carbon dioxide and methane. Nitrogen, oxygen, ammonia, sulphide's, hydrogen, and other gases make up the remaining 2 to 10%. They also contribute to global warming; (c) A large amount of garbage from land eventually finds its way into the ocean, causing marine pollution (Hoorn Weg, 2017).

Economic Impacts: (a) Growing landfills take up valuable area and result in the wasteful use of a resource; (b) Recycling waste can save costs and even bring in money. Ineffective waste management loses out on this beneficial chance; (c) Ineffective waste collection causes drain obstruction, which contributes to urban flooding and causes financial losses; (d) Ineffective waste management causes overall city squalor, which reduces the potential for tourism (Manaf, 2008).

Steps Taken by the Government Regarding Waste Management

Institutional Framework: In India, waste management is overseen by the ULBs (12th Schedule of the Constitution), the Central Pollution Control Board (CPCB), the Ministry of Urban Development (MoUD), the Ministry of Environment, Forest, and Climate Change (MoEFCC),(SPCBs).Along with the government initiatives taken at central and state level for proper handling of waste it is the responsibility of every individual also. **Community engagement** plays a crucial role in effective waste management. By involving residents in waste reduction initiatives, recycling programs, and proper disposal practices, communities can significantly reduce the volume of waste sent to landfills and incinerators. Engaged communities are more likely to adopt sustainable habits, such as composting organic waste and participating in local clean-up events, which help to maintain cleaner and healthier environments. Furthermore, when community members are informed and involved, they can provide valuable feedback and innovative ideas, enhancing the overall efficiency and effectiveness of waste management systems. Collaborative efforts between local governments, organizations, and residents foster a sense of shared responsibility, leading to more robust and long-lasting solutions for waste management challenges. Through education, participation, and a collective commitment to sustainability, community engagement ensures that waste management practices are not only implemented but embraced, leading to significant environmental and societal benefits (Saeed,2009).

Fostering sustainable consumption patterns is critical for addressing pressing environmental challenges such as climate change, resource depletion, and biodiversity loss. Behavioural change and education play pivotal roles in this process. Below given are few of the ways of promoting community engagement for sustainable consumption:

1. Raising Awareness

Education increases awareness about the environmental impact of consumption patterns. Through educational programs, people can learn about issues like carbon footprints, waste generation, and the depletion of natural resources. Knowledge is the first step toward fostering a sense of responsibility and urgency in individuals and communities.

2. Changing Attitudes and Beliefs

Behavioural change often stems from shifts in attitudes and beliefs. Education can challenge existing consumption norms and promote values aligned with sustainability. By emphasizing the long-term benefits of sustainable practices and highlighting the environmental costs of unsustainable behaviours, educational initiatives can reshape how people think about their consumption choices.

3. Providing Skills and Tools

Education equips individuals with the skills and tools needed to make sustainable choices. This includes teaching practical skills such as recycling, energy conservation, and sustainable food choices. It also involves educating consumers about sustainable products and services, enabling them to make informed decisions.

4. Encouraging Behavioural Change

Behavioural change strategies, such as nudging and social marketing, are essential for encouraging sustainable consumption. Nudging involves subtly guiding choices without restricting options, for example, by placing sustainable products at eye level in stores. Social marketing uses commercial marketing techniques to promote social good, such as campaigns that highlight the benefits of reducing single-use plastics.

5. Building Community and Social Norms

Education and behavioural change initiatives can help build community around sustainable practices. When sustainable behaviours are adopted by a critical mass within a community, they can become social norms, making it easier for others to follow suit. Programs that encourage collective action, such as community gardens or local recycling schemes, foster a sense of shared responsibility and commitment to sustainability (Menikpura, 2012).

6. Influencing Policy and Corporate Practices

Equity considerations in waste management planning are essential to ensure that all communities, especially marginalized and vulnerable populations, benefit from waste management services and are not disproportionately burdened by waste-related environmental and health issues.

India's Policy and Legal Environment for Waste Management: Regarding solid waste management, the Government of India (GOI) has created several rules and regulations (SWM). These consist of rules for the management of plastic waste, e-waste, and solid waste, among others. These regulations were created in accordance with the Environment Protection Act of 1986 and are reviewed on a regular basis. Outside of municipal areas, the Solid Waste Management Rules, 2016 also apply to urban agglomerations, census towns, notified industrial townships, airports, areas under Indian Railways control, special economic zones, places of pilgrimage, sites of historical and religious significance, and State and Central Government Organizations within their purview.

Governmental Programs: (a) **Swachh Bharat Mission** – Urban (SBM-U): Door-to-door collection, source segregation, etc., has been started with the adoption of new regulations; (b) **Swaccha Surekha:** An yearly assessment of sanitation, hygiene, and cleanliness is conducted in Indian cities and towns. It was introduced by the Ministry of Housing and Urban Affairs (MoHUA) as a component of the SBM-U. It serves as a motivator for cities by assigning stars based on several criteria to garbage-free cities and villages. (d) **Compost Banao, Compost Apnao Campaign** the MoHUA initiated Campaign, a multi-media initiative promoting waste-to-composting under SBM-(U). The intention is to lessen the quantity of garbage that ends up in landfills by encouraging people to turn their kitchen waste into compost, which may be used as fertilizer; (e) **Waste to Energy Promotion:** To encourage the establishment of waste-to-energy projects and to offer central financial assistance, the Ministry of New and Renewable Energy (MNRE) launched the Program on Energy from Urban, Industrial, Agricultural waste/residues and Municipal Solid Waste(vidanaarachchi, 2006).

What further actions can be done to enhance waste management?

Scientific Waste Management: Robust scientific and engineering research ought to serve as the foundation for waste management strategy. garbage composition, initial and ongoing operational expenses, transportation distances, and the locations of facilities for processing and disposing of garbage should all be considered. Planning for solid waste management requires precise data, which can only be obtained through thorough waste characterisation investigations (Miezah. 2015).

Waste-to-energy: Anaerobic digestion, or bio-methanation, employs microorganisms to break down organic waste into fuel-grade methane. Plants for bio methanation ought to be expanded.

Strict Enforcement of the Laws: The "Polluter Pays Principle" has been included in the Waste Management Rules. Strict enforcement of the regulations is required to penalize noncompliance (Hoornweg 2012).

India's fast urbanization combined with a multiplication in garbage creation will present new governance challenges. The government needs to put more energy into improving garbage management. However, without public participation, government initiatives may remain useless. Therefore, public participation is crucial. Therefore, it is intended to give information, planning, funding, unified waste management, and community education greater weight. It is important to actively promote the 4 R's idea of reducing, reusing, recycling, and recovering resources (Batool, 2009).

CASE STUDIES: SUCCESSFUL AND UNSUCCESSFUL WASTE MANAGEMENT INITIATIVES

* Indore, India: Case Study

Indore has become a model for effective waste management in India. The city faced severe challenges before 2016, including inadequate waste segregation, inefficient collection systems, and open dumping. With the implementation of the Swachh Bharat (Clean India) campaign, Indore transformed its waste management system. Key measures included mandatory waste segregation, door-to-door collection with GPS-enabled vehicles, and the establishment of advanced waste processing facilities like a 15 MW waste-to-energy plant. Public awareness campaigns and strict enforcement significantly improved compliance. As a result, Indore has consistently ranked as the cleanest city in India since 2017, with a 95% waste recovery rate and improved public health outcomes (Earth5R).

India has faced significant challenges with several waste management initiatives that have proven to be unsuccessful. Here are some notable examples:

* **Delhi's Waste-to-Energy (WTE) Plants:** Despite the significant investment in WTE plants, these projects have been plagued by inefficiencies, high operational costs, and environmental concerns. Many plants have failed to process waste effectively, leading to continuous reliance on landfills. The plants have also faced strong opposition from local communities due to pollution and health risks associated with incineration (Frontiers).

* **Bangalore's Solid Waste Management: Bangalore** has struggled with waste management due to rapid urbanization and inadequate infrastructure. Efforts to implement new technologies and decentralized waste management systems have been hampered by bureaucratic delays, poor planning, and lack of public awareness. This has resulted in persistent garbage heaps and environmental hazards across the city (Frontiers).

* **Ahmedabad's Waste Segregation Initiatives:** Initiatives to promote waste segregation at source in Ahmedabad have largely been ineffective. Despite campaigns and policies to encourage segregation, there has been a lack of sustained enforcement and community engagement, leading to low compliance rates and continued mixed waste collection (Frontiers).

* **Mumbai's Landfill Crisis:** Mumbai's reliance on massive landfills like Deonar has led to severe environmental and health issues. Efforts to reduce the load on these landfills through recycling and waste processing plants have largely failed due to corruption, mismanagement, and inadequate technological solutions. Frequent fires at these landfills have further exacerbated the situation (Frontiers).

These case studies highlight the complexities and challenges in implementing effective waste management solutions in India, often due to systemic issues, technological shortcomings, and insufficient community involvement.

REFERENCES

Abdul-Aziz H.M.et.al A. *Study of Baseline Data Regarding Solid Waste Management in the Holy City of Makkah during Hajj.* The Custodian of the Two Holy Mosques Institute of the Hajj Research; Medina, Saudi Arabia: 2007. Unpunished Report.

Abubakar, I. R. (2017). Household response to inadequate sewerage and garbage collection services in Abuja, Nigeria. *Journal of Environmental and Public Health*, 2017, 5314840. 10.1155/2017/531484028634496

Abubakar, I. R., & Aina, Y. A. Population Growth and Rapid Urbanization in the Developing World. IGI Global; Hershey, PA, USA: 2016. Achieving sustainable cities in Saudi Arabia: Juggling the competing urbanization challenges; pp. 234–255.

Abul, S. (2010). Environmental and health impact of solid waste disposal at Manganin dumpsite in Manzini: Swaziland. *Journal of Sustainable Development in Africa*, 12, 64–78.

Addo, I. B., Adei, D., & Acheampong, E. O. (2015). Solid Waste Management and Its Health Implications on the Dwellers of Kumasi Metropolis, Ghana. *Curr. Res. J. Soc. Sci.*, 7(3), 81–93. 10.19026/crjss.7.5225

Akmal T., Jamil F. Health impact of Solid Waste Management Practices on Household: The case of Metropolitans of Islamabad-Rawalpindi, Pakistan. *Heliyon*. 2021. doi: . 2021.e07327.10.1016/j.heliyon

Balogun, A. L., Adebisi, N., Abubakar, I. R., Dano, U. L., & Tella, A. (2022). Digitalization for transformative urbanization, climate change adaptation, and sustainable farming in Africa: Trend, opportunities, and challenges. *Journal of Integrative Environmental Sciences*, 19(1), 17–37. 10.1080/1943815X.2022.2033791

Batool, S. A., & Chaudhry, M. N. (2009). The impact of municipal solid waste treatment methods on greenhouse gas emissions in Lahore, Pakistan. *Waste Management (New York, N.Y.)*, 29(1), 63–69. 10.1016/j.wasman.2008.01.01318387288

Daniel, O. Okanigbe et al IOP Conf. Series: Materials Science and Engineering 391 (2018) 012006 10.1088/1757-899X/391/1/012006

Fang, B., Yu, J., Chen, Z., Osman, A. I., Farghali, M., Ihara, I., Hamza, E. H., Rooney, D. W., & Yap, P.-S. (2023). Artificial intelligence for waste management in smart cities: A review. *Environmental Chemistry Letters*, 21(4), 1959–1989. 10.1007/s10311-023-01604-3

Gazzeh, K., Abubakar, I. R., & Hammad, E. (2022). Impacts of COVID-19 Pandemic on the Global Flows of People and Goods: Implications on the Dynamics of Urban Systems. *Land (Basel)*, 11(3), 429. 10.3390/land11030429

Giusti L. A review of waste management practices and their impact on human health. *Waste Manag.* 2009; **29:2227**–2239.n .10.1016/j.wasman.2009.03.028

Guerrero, L. A., Maas, G., & Hogland, W. (2012). Solid waste management challenges for cities in developing countries. *Waste Management (New York, N.Y.)*, 33(1), 220–232. 10.1016/j.wasman.2012.09.00823098815

Hettiarachchi, H., Meegoda, J., & Ryu, S. (2018). Organic Waste Buyback as a Viable Method to Enhance Sustainable Municipal Solid Waste Management in Developing Countries. *International Journal of Environmental Research and Public Health*, 15(11), 2483. 10.3390/ijerph1511248330405058

Hoorn Weg, D., & Giannelli, N. (2007). *Managing Municipal Solid Waste in Latin America and the Caribbean: Integrating the Private Sector, Harnessing Incentives*. World Bank.

Hoornweg, D., & Bhada-Tata, P. What a Waste: A Global Review of Solid Waste Management. Urban Development Series. World Bank; Washington, DC, USA: 2012. Knowledge Papers No. 15.

Kabera, T., & Nishimwe, H. (2019). *Systems Analysis of Municipal Solid Waste Management and Recycling System in East Africa: Benchmarking Performance in Kigali City, Rwanda*. EDP Sciences.

Khan, M. S. M., & Kaneesamkandi, Z. (2013). Biodegradable waste to biogas: Renewable energy option for the Kingdom of Saudi Arabia. *International Journal of Innovation and Applied Studies*, 4, 101–113.

Levente Szász, Ottó Csíki, Béla-Gergely Rácz, Sustainability management in the global automotive industry: A theoretical model and survey study, International Journal of Production Economics, Volume 235,2021

Manaf, L. A., Samah, M. A. A., & Zukki, N. I. M. (2009). Municipal solid waste management in Malaysia: Practices and challenges. *Waste Management (New York, N.Y.)*, 29(11), 2902–2906. 10.1016/j.wasman.2008.07.01519540745

McAllister, J. Factors Influencing Solid-Waste Management in the Developing World. All Graduate Plan B and Other Reports. 528. 2015. ((accessed on 9 November2021)).https://digitalcommons.usu.edu/grad reports/528

Menikpura, S. N. M., Gheewala, S. H., & Bonnet, S. (2012). Sustainability assessment of municipal solid waste management in Sri Lanka: Problems and prospects. *Journal of Material Cycles and Waste Management*, 14(3), 181–192. 10.1007/s10163-012-0055-z

Miezah, K., Obiri-Danso, K., Kádár, Z., Fei-Baffoe, B., & Mensah, M. Y. (2015). Obiri-DansoK.,KádárZ.,Fei-Baffoe B.,Mensah M.Y. Municipal solid waste characterization and quantification as a measure towards effective waste management Ghana. *Waste Management (New York, N.Y.)*, 46, 15–27. 10.1016/j.wasman.2015.09.00926421480

Ojok, J. (2013). Rate and quantities of household solid waste generated in Kampala City, Uganda. *Sci.J.Environ.Eng. Res.*, 2013. Advance online publication. 10.7237/sjeer/237

Pokhrel, D., & Viraraghavan, T. (2005). Municipal solid waste management in Nepal: Practices and challenges. *Waste Management (New York, N.Y.)*, 25(5), 555–562. 10.1016/j.wasman.2005.01.02015925764

Ramachandra, T. V., Bharath, H. A., Kulkarni, G., & Han, S. S. (2018). Municipal solid waste: Generation, composition and GHG emissions in Bangalore, India. *Renewable & Sustainable Energy Reviews*, 82, 1122–1136. 10.1016/j.rser.2017.09.085

Saeed, M. O., Hassan, M. N., & Mujeebu, M. A. (2009). Assessment of municipal solid waste generation and recyclable materials potential in Kuala Lumpur, Malaysia. *Waste Management (New York, N.Y.)*, 29(7), 2209–2213. 10.1016/j.wasman.2009.02.01719369061

Scarlat, N., Motola, V., Dallemand, J. F., Monforti-Ferrario, F., & Mofor, L. (2015). Evaluation of energy potential of municipal solid waste from African urban areas. *Renewable & Sustainable Energy Reviews*, 50, 1269–1286. 10.1016/j.rser.2015.05.067

Yang, Q., Fu, L., Liu, X., & Cheng, M. (2018). Evaluating the Efficiency of Municipal Solid Waste Management in China. *International Journal of Environmental Research and Public Health*, 15(11), 2448. 10.3390/ijerph1511244830400237

Yousif, D. F., & Scott, S. (2007). Governing solid waste management in Mazatenango, Guatemala: Problems and prospects. *International Development Planning Review*, 29(4), 433–450. 10.3828/idpr.29.4.2

Zafar, S. Solid Waste Management in Saudi Arabia. EcoMENA. 2015. ((accessed on 24 February 2021)). Available online: https://www.ecomena.org/tag/dammam/

Zhang, D. Q., Tan, S. K., & Gersberg, R. M. (2010). Municipal solid waste management in China: Status, problems and challenges. *Journal of Environmental Management*, 91(8), 1623–1633. 10.1016/j.jenvman.2010.03.01220413209

Chapter 4
Economic Implications and Scope of Wastewater Management Through Life Cycle Cost Assessment

Maria Jose Lopez-Serrano
University of Almeria, Spain

Fida Hussain Lakho
Ghent University, Belgium

Stijn W. H. Van Hulle
Ghent University, Belgium

Ana Batlles-delaFuente
https://orcid.org/0000-0003-2108-0516
University of Almeria, Spain

ABSTRACT

Global demand for freshwater is rising due to anthropic factors, exacerbating water scarcity and pollution. The economic dimensions of sustainable wastewater management offer both environmental benefits and economic promise, with studies consistently highlighting cost-effectiveness and long-term financial gains. Treated wastewater, a dependable and economically viable alternative, aligns with sustainable development goals and contributes to improving public health. An in-depth economic analysis explores the profitability and cost savings of nature-based solutions,

DOI: 10.4018/979-8-3693-2595-7.ch004

emphasizing the importance of investing in these strategies for water sustainability within the circular economy. The study incorporates the review of a life cycle cost assessment, providing nuanced insights into the economic implications of adopting nature-based solutions in wastewater management. The findings contribute substantively to the discourse surrounding water sustainability, informing policy decisions and promoting transformative change aligned with environmental sustainability, and circular economy principles.

INTRODUCTION

The escalating demand for freshwater, understood as water with low salt content crucial for drinking, agriculture, and ecosystems, has witnessed a consistent rise in recent decades, and the trajectory portends a further aggravation due to a myriad of anthropic factors, instigating global apprehension (Bunn, 2016; Tociu et al., 2019; Ashu & Lee., 2021; Baggio et al., 2021; Truchado et al., 2021). The multifaceted challenges posed by climate change, encompassing global warming and erratic weather patterns, coupled with the escalating global population, anticipated to surge in the coming years, have precipitated heightened water scarcity and pollution (Castillo-Díaz et al., 2023). This hazardous convergence confluence jeopardizes the equitable distribution of finite freshwater resources, placing unprecedented stress on water and food security (Melián-Navarro & Ruiz-Canales, 2020; Al-agele et al., 2021; Wang et al., 2021; Arellano-Gonzalez et al., 2021; Weerasooriya et al., 2021; Maniam et al., 2022). The intensification of this already dire situation, wherein more than two billion people grapple with water stress and an additional four billion contend with severe water scarcity for at least a month annually, underscores the imperative to explore alternative and sustainable water resources (UNESCO, 2021; UNESCO, 2022). As freshwater resources become scarcer, the economic implications of wastewater management emerge as a pivotal facet in the pursuit of a sustainable solution to this burgeoning crisis.

In contemplating the economic dimensions, it becomes evident that the sustainable management of wastewater not only addresses environmental concerns but also holds significant economic promise. Studies have consistently highlighted the cost-effectiveness and long-term financial benefits of adopting sustainable wastewater management practices (Abdelhay & Abumaser., 2021; Basi et al., 2022). Efficient wastewater treatment not only mitigates pollution but also unlocks the potential for reclaimed water, thereby reducing the strain on conventional freshwater resources. Furthermore, from a broader economic standpoint, sustainable wastewater management contributes to the establishment of a circular economy. The recuperation and reuse of valuable resources from wastewater, such as nutrients and energy, can serve as a catalyst for economic growth (Declercq et al., 2020). Moreover, the establishment of circular economy principles within wastewater management aligns with broader sustainability goals. . The integration of circular economy principles in wastewater management closely aligns with key Sustainable Development Goals (SDGs). By promoting efficient water resource utilization, minimizing waste generation, and fostering innovation in resource recovery, these principles contribute significantly to achieving specific SDGs. Notably, they support universal access to clean water (SDG 6), advocate for sustainable consumption and production (SDG 12), mitigate climate impact (SDG 13), and contribute to the preservation of terrestrial ecosystems and biodiversity (SDG 15). Additionally, the establishment of circular economy practices encourages collaborative efforts among stakeholders, reinforcing SDG 17 on partnerships for collective sustainability objectives. This connection between circular economy and SDG foster resilience in the face of water scarcity and promoting responsible resource utilization.

In this dynamic landscape, the utilization of treated wastewater emerges as a dependable and economically viable water alternative, providing a steadfast guarantee of water supply while simultaneously addressing multifaceted challenges associated with water constraints (Hristov et al., 2021). By integrating circular economy principles into wastewater management strategies from the inception of the research, the study not only addresses the specific challenges related to water treatment but also underscores a commitment to sustainable and resource-efficient solutions. This proactive approach ensures that the study's objectives, methodologies, and outcomes are inherently linked to the overarching theme of circular economy principles, fostering a more comprehensive and meaningful exploration of sustainable wastewater management practices. Reclaimed water, as a resource, holds pivotal significance in the global pursuit of the United Nations' 17 Sustainable Development Goals (SDGs) outlined in 2015. Its direct alignment with Goal 6, focusing on ensuring clean water and sanitation for all, amplifies its impact and contributes to the comprehensive attainment of interconnected SDGs (Brennan et al., 2021; Di Vaio et al., 2021; Rodríguez de Sá Silva et al., 2022). The critical role of treated wastewater becomes particularly evident in developing countries, where empirical data underscores the dire consequences of inadequate water supply and wastewater management. The repercussions extend beyond mere environmental concerns, profoundly impacting aspects of hygiene, health, economic stability, and societal well-being (Lutterbeck et al., 2017; Sánchez Pérez et al., 2020). The transformative potential of this alternative water source lies in its ability to convert waste into a valuable resource, aligning seamlessly with circular economy principles and bolstering sustainable development without compromising environmental integrity.

The tangible impact of treated wastewater extends beyond theoretical consid-erations, as exemplified by real-world statistics. In regions grappling with water scarcity and deficient sanitation infrastructure, the implementation of sustainable wastewater management has demonstrated a substantial reduction in waterborne dis-eases, contributing to improved public health and well-being (López-Serrano et al., 2021). The convergence of treated wastewater with circular economy principles not only mitigates the strain on traditional water resources but also catalyzes economic growth. Reclaiming valuable nutrients and energy from wastewater contributes to a regenerative economic model, fostering resilience and resource efficiency (Omole et al., 2019; Otter et al., 2020). This symbiotic relationship between sustainable wastewater management, economic prosperity, and environmental stewardship underscores its pivotal role in shaping a resilient and water-secure future.

Within this context, the primary aim of the investigation analysed in this chapter is to analyse and explore an in-depth economic analysis previously carried out by López-Serrano et al., 2023. This thorough economic assessment not only delves into the financial profitability of Nature-Based Solutions (NBS) coupled with advanced technologies but also scrutinizes the operational cost savings realized by integrating NBS into wastewater management practices. The objective of this chapter offers paramount significance as it seeks to quantifiably demonstrate to legislators, stakeholders, and researchers the imperative nature of investing in NBS as a strategic approach toward achieving water sustainability within the overarching framework of the circular economy. Nature-Based Solutions (NBS) are approaches that utilize natural processes and ecosystems to address environmental challenges and provide sustainable solutions. These strategies leverage the inherent resilience of ecosystems to enhance biodiversity, mitigate climate change impacts, and promote overall environmental and social well-being. By attaining the outlined objective, this research here studied and detailed endeavors to elucidate the pivotal role that both proper wastewater management and the transformation of waste into a resource play in advancing the United Nations' SDGs. Wastewater management is directly linked to several Sustainable Development Goals (SDGs), with a primary connection to SDG 6, "Clean Water and Sanitation." The effective treatment and safe disposal of wastewater are pivotal elements in ensuring access to clean water and preventing waterborne diseases. According to the World Health Organization (WHO), inade-quate sanitation, including poor wastewater management, contributes to the spread of waterborne diseases, affecting 2.4 billion people globally. Addressing this issue aligns with SDG 6's target to achieve universal access to adequate and equitable sanitation and hygiene for all. Furthermore, proper wastewater management has implications for SDG 3, "Good Health and Well-being," as untreated wastewater poses health risks. According to the United Nations, over 80% of global wastewater is discharged without proper treatment, impacting water quality and human health.

Indirectly, wastewater management intersects with various other SDGs. SDG 1, "No Poverty," can benefit from wastewater reuse in agriculture, promoting sustainable livelihoods. According to the United Nations Development Programme (UNDP), wastewater reuse can enhance agricultural productivity, contributing to poverty reduction. SDG 2, "Zero Hunger," is also influenced, as treated wastewater supports agricultural irrigation, positively impacting food security. A study by the International Water Management Institute indicates that wastewater irrigation contributes significantly to global food production, benefiting over 50 million hectares of farmland. In essence, effective wastewater management emerges as a cross-cutting solution, addressing water-related challenges and contributing to broader sustainable development objectives.The ensuing quantifiable insights are anticipated to underscore the tangible contributions of sustainable wastewater management and circular economy practices to the broader agenda of achieving the SDGs. To lay the groundwork for this research, a comprehensive Life Cycle Assessment (LCA) was previously conducted, adhering to ISO 14044 standards. This antecedent assessment meticulously evaluated and quantified environmental impacts within the same scenario under consideration in the present study.

This undertaking is especially pertinent in the current global landscape, where the escalating demand for water resources intersects with the imperative of environmental conservation. The economic ramifications of adopting NBS are underscored by real-world statistics and empirical evidence. Studies reveal that integrating NBS into water management strategies not only enhances ecological resilience but also presents a compelling economic case. For instance, in regions where NBS has been implemented, there has been a notable reduction in water treatment costs and a concurrent increase in the cost-effectiveness of water supply infrastructure. These empirical observations substantiate the argument for prioritizing NBS as a viable and economically advantageous solution in the pursuit of sustainable water management.

As the global community grapples with the imperative to balance economic growth with environmental stewardship, the findings of this economic analysis are poised to contribute substantively to the discourse surrounding water sustainability and the circular economy. By quantifying the economic advantages of NBS over conventional alternatives, this research aspires to offer actionable insights that can inform policy decisions, drive sustainable investments, and catalyze transformative change in water resource management practices.

Employing a Life Cycle Cost Assessment (LCCA), this research here studied endeavors not only to gauge the financial viability of the investment through well-established economic indicators but also to conduct a comprehensive monetary evaluation of the cleaning costs associated with the implementation of Nature-Based Solutions (NBS) vis-à-vis their traditional counterparts. The innovative approach adopted in this research involves the amalgamation of two distinct methodologies, rendering it comprehensive and pioneering.

In the realm of traditional financial metrics, the analysis delves into a Life Cycle Cost Analysis (LCCA) through Cost-Benefit Analysis (CBA). This approach employs established decision criteria such as Net Present Value (NPV), Benefit-Cost Ratio (BCR), Internal Rate of Return (IRR), and Discounted Payback Period (DPP). These metrics serve as the bedrock for evaluating the economic viability and desirability of investing in NBS compared to traditional alternatives. Real-world statistics and empirical evidence from analogous studies emphasize the importance of considering these financial indicators in decision-making processes. Simultaneously, an innovative dimension of this research involves a monetary assessment of the cleaning costs associated with environmental indicators stemming from both scenarios under consideration—the traditional approach and the Constructed Wetland (CW). This dual-pronged approach facilitates a comprehensive and nuanced understanding of the real costs inherent in both alternatives. Incorporating real-world data, this aspect of the analysis considers not only the immediate financial implications but also the long-term economic and environmental consequences associated with the chosen wastewater management strategies. This holistic research design, encompassing both traditional financial metrics and an innovative assessment of cleaning costs, is pivotal in providing a nuanced perspective on the economic implications of adopting NBS within the context of sustainable wastewater management. The quantitative nature of this analysis, grounded in real-world data, positions it as a valuable resource for policymakers, stakeholders, and researchers seeking to make informed decisions aligned with economic prudence, environmental sustainability, and circular economy principles.

Recognizing the interconnectedness of economic, environmental, and societal dimensions, this chapter aims to contribute substantively to the discourse surrounding sustainable water management within the framework of circular economy principles. Therefore, the incorporation of diverse perspectives through a comparative lens enhances the robustness of this research, offering the author a panoramic view of the economic landscape associated with NBS in wastewater management. The resulting analysis is poised to stand as a benchmark, guiding policymakers, researchers, and stakeholders toward informed decisions aligned with economic prudence, environmental sustainability, and the broader goals of a circular economy.

In conclusion, this pivotal chapter not only offers a comprehensive understanding of economic indicators but also contextualizes them within the urgent global imperative for sustainable wastewater management. By delving into real-world statistics and empirical evidence, the article establishes a persuasive narrative on the economic viability, environmental sustainability, and societal benefits of NBS. This chapter stands as an indispensable contribution to the literature, providing the author with a compelling rationale for recognizing the paramount importance of adopting sustainable practices in wastewater management within the realms of economic prudence and environmental stewardship.

METHODOLOGY

In light of the considerations previously stated in the introduction, this chapter crucially presents an in-depth exploration of the article of Lopez-Serrano et al. (2023) that undertakes a meticulous examination of Life Cycle Cost Assessment (LCCA) after a comprehensive LCA carried out in Lakho et al., 2022. This ground-breaking study goes beyond traditional economic evaluations and delves into the nuanced realm of life cycle assessments, where indicators of economic sustainability are rigorously examined. The chosen approach seeks to unravel the intricate layers of economic implications associated with adopting Nature-Based Solutions (NBS) in wastewater management compared to conventional methods. By employing a combination of established financial metrics and cutting-edge life cycle assessment techniques, the study offers a holistic perspective on the long-term economic viability and environmental ramifications of these contrasting approaches. In conjunction with the primary investigation by Lopez-Serrano et al. (2023), this chapter endeavours to broaden its scope by elaborating some visual graphs that enhance the visual results of the research. By incorporating a diverse range of graphs, this research seeks to enrich the understanding of sustainable wastewater management practices and their economic implications in terms of both, costs and environmental implications.

All the calculations in the study here analyzed were conducted with a focus on a Vertical Flow Constructed Wetland (VFCW) coupled with a membrane-based potable water production system, proposed as an investment. This specific scenario, previously subjected to a Life Cycle Assessment (LCA), operates at a restaurant in Belgium, managing a water flow of 4 m^3 per day. This system was juxtaposed with its conventional counterpart, namely a public water supply and sewerage system located approximately 300 meters away from the restaurant in the same geographical area of Belgium, as detailed in the research by Lakho et al. (2022).

This research is grounded in the findings of the study conducted by Lakho et al. in 2022, which focused on the sustainability of Decentralized Wastewater Treatment Technologies (DWTS) in rural areas lacking connection to conventional municipal treatment systems due to extended transport distances. The primary objective of Lakho et al.'s investigation was to employ the Life Cycle Assessment tool for a comprehensive life cycle impact assessment of two distinct decentralized water treatment systems implemented in Belgium. The first scenario involved a mobile constructed wetland treating grey water at music festivals, coupled with a membrane-based drinking water production system. This system produced 100 m^3 of potable water from 400 m^3 of wastewater generated per festival. The second scenario featured a vertical flow constructed wetland treating black water at a restaurant with 135 visitors per day, also coupled with a membrane system. Lakho et al.'s research conducted a rigorous comparison with conventional alternatives, specifically PET bottled water supply and a public drinking water supply for Scenarios 1 and 2, respectively.

Across various impact categories, Scenarios 1 and 2 exhibited approximately an order of magnitude lower environmental impact compared to their conventional alternatives. Lakho et al. further conducted a sensitivity analysis, varying parameters such as the distance traveled for both the mobile constructed wetland and PET bottles in Scenario 1, and the distance of the restaurant from a drinking water supply and a sewerage system in Scenario 2. Encouragingly, the results demonstrated that DWTS remained environmentally feasible compared to their conventional alternatives, even at the shortest distances studied (Scenario 1: 175 km and Scenario 2: 75 m). In conclusion, Lakho et al.'s research provides compelling evidence that under specific conditions, DWTS can be considered environmentally beneficial, offering a sustainable alternative to conventional water treatment systems in rural areas with logistical challenges. Furthermore, results derived from (López-Serrano et. al., 2023) have been included and analysed.

Within the context of this book chapter, the analysis draws upon data derived from the referred prior investigations, where a Life Cycle Cost Analysis (LCCA) was conducted to assess the economic viability and efficiency of utilizing a constructed wetland as an alternative to a traditional sewerage system in a restaurant setting. LCCA, an analytical method rooted in cost-benefit analysis (CBA), is widely employed for appraising the profitability of investment projects, particularly in the realm of alternative water supply systems like Vertical Flow Constructed Wetlands (VFCWs) (Diaz-Elsayed et al., 2020; Cao et al., 2021). While CBA is a pivotal tool in project evaluation, its application within methodologies pertaining to wastewater reuse projects has been limited, creating a notable research gap (Declercq et al., 2020). The research underscores the significance of introducing a new factor, specifically the calculation of interest rates, into the economic methodology of LCCA. Unlike previous literature where a fixed interest rate for the entire period was considered, this study acknowledges the potential impact of interest rate fluctuations over extended periods, addressing a critical aspect often overlooked in assessments lasting up to 20 years (Abdulfatah et al., 2019; Otter et al., 2020). Against the backdrop of uncertainty in international financial markets, the study places special emphasis on incorporating varying rates within a yield curve for the next two decades. The Euro area yield curve, considered in this research, integrates Euribor rates based on Euro area interbank loans and longer maturities of interest rates from significant European bonds, particularly the German bond. Projections of the yield curve are sourced from September 27th and are based on Bloomberg analyses. Additionally, the financial indicators calculated in this research consider cash inflows as monetary benefits or savings, with gross income and revenues based on savings resulting from the utilization of the alternative under study compared to the traditional option (Zadeh et al., 2013; Abdelhay et al., 2021). Prior to calculating traditional indicators, a comprehensive LCCA necessitates the development of an initial cost structure based on start-up costs required for the constructed wetland. Construction and operational costs have been meticulously evaluated based on primary data. The initial investment is computed based on construction materials provided by local supply companies, cross-validated with international counterparts to ensure data reliability and applicability to areas grappling with water scarcity.

In response to the notable inflation rates observed in recent months, a thorough price update has been undertaken to align the database with current economic realities. This adjustment aims to enhance and adapt results based on deviations in the annual growth rate of the Gross Domestic Product (GDP) implicit deflator (Agiakloglou & Gkouvakis, 2022). The Eurozone Consumer Price Index (CPI) has been employed to measure price changes, being an expenditure-weighted index that incorporates the most relevant goods and services in the consumer market basket. Widely acknowledged as the primary index for calculating price variations, the CPI holds particular importance as it reflects the most significant impacts of changes in prices on consumers (Krimpas et al., 2021).

Key assumptions guiding this analysis include a project lifetime of 20 years, consistent with prior VFCW assessments (Abdelhay et al., 2021; Lakho et al., 2022), varying discount rates each year based on the Euro area yield curve, and costs derived from constructing and installing a VFCW system tailored to a medium-level restaurant with local knowledge and resources, as ascertained through a comprehensive market survey. Environmental repercussions stemming from the use of traditional alternatives for water supply and sewerage often elude thorough examination due to inherent measurement challenges.

Environmental repercussions stemming from the utilization of conventional water supply and sewerage alternatives often go unnoticed due to challenges in their quantification. In pursuit of a comprehensive investigation encompassing not only traditional methodologies evaluating new water supply alternatives from a financial perspective but also innovative ways of assessing Decentralized Water Treatment Systems (DWTS), the removal costs for each indicator previously assessed in a Life Cycle Assessment (LCA) have been meticulously calculated. These costs, derived from data published in the "Environmental Prices Handbook-EU28," signify the social marginal value of averting emissions from an average source in Europe (De Bruyn et al., 2018). Notably, these prices are tailored for impact indicators obtained through the ReCiPe midpoint method 2016 (hierarchical approach) during the LCA, consistent with a previous study on decentralized water treatment systems and their conventional counterparts (Lakho et al., 2022).

Furthermore, a price update has been conducted to align the database with current economic conditions, addressing deviations in the annual growth rate of the Gross Domestic Product (GDP) implicit deflator owing to recent inflation rates. The Eurozone Consumer Price Index (CPI), an expenditure-weighted index incorporating essential goods and services, has been utilized to measure price changes. This index is particularly significant as it accurately reflects how alterations in prices impact consumers and is widely employed for calculating price variations (Krimpas et al., 2021; Karaduić & Đalović, 2021).

Moving to the results, the initial investment and fixed costs associated with constructing a Vertical Flow Constructed Wetland (VFCW) and a membrane-based potable water production system have been evaluated. The initial investment considers a VFCW designed to treat wastewater from a restaurant operating five days a week for twenty years. Total costs (TC) include expenses for materials, labor, transportation, and control tasks, with labor costs based on construction worker agreements in Belgium. Transportation costs, although negligible, were calculated for an average distance of 250km. Control tasks encompass site visits, sample collections, and follow-ups, validated by a minimum of two specialized companies per entry. Fixed costs, accounting for membrane materials and their lifespan, were computed, with energy costs factored in due to fluctuations in energy prices during the study year. The inclusion of energy costs as fixed, despite its variable nature, is justified by its significant impact on the overall project investment.

RESULTS

Table 1 presents the diverse environmental impacts stemming from a comprehensive Life Cycle Assessment (LCA) for each variable under examination, comparing the traditional water supply and sewerage alternative to DWTS. The depicted impacts represent those to which an economic valuation has been assigned, reflecting the cost implications associated with their reversal or cleaning. These quantified environmental costs have been integrated into Graphs 1, 2, and 3 to afford a visually insightful representation of their economic dimensions.

The LCA framework employed in the study here analysed serves as a robust analytical tool for evaluating the holistic environmental footprint of both the conventional water supply and sewerage system and the DWTS. Each variable within the scope of analysis contributes distinctive environmental consequences, and the subsequent economic valuation underscores the financial implications entailed in addressing or mitigating these impacts. In the pursuit of a comprehensive understanding of the environmental ramifications, the LCA extends beyond a mere enumeration of impacts, delving into the economic quantification of the associated costs. The allocation of economic values to environmental impacts is instrumental in providing a tangible basis for comparative analysis, facilitating the identification of economically favorable alternatives. Moreover, the integration of these quantified environmental costs into graphical representations, as delineated in Figures 1, 2, and 3, not only enhances the visual interpretability of the data but also affords a nuanced perspective on the economic dimensions of each variable. The graphs serve as a visual narrative, elucidating the financial implications of adopting the DWTS over the traditional water supply and sewerage system. In essence, Table 1 acts as a pivotal reference point, encapsulating the quantified environmental impacts and their corresponding economic values, thereby establishing a foundation for the subsequent graphical representations. This meticulous integration of scientific rigor and economic considerations contributes to a comprehensive understanding of the environmental and financial dynamics inherent in the comparison between traditional and innovative water treatment alternatives.

Table 1. Napierian logarithm of removal cost

	IMPACTS	
	Conventional water supply and sewerage	**DWTS**
Fine Particulate Matter Formation	$1.27 \times 10^{+02}$	$4.95 \times 10^{+01}$
Freshwater Ecotoxicity	$3.91 \times 10^{+03}$	$3.63 \times 10^{+03}$
Freshwater Eutrophication	$3.20 \times 10^{+01}$	$1.14 \times 10^{+01}$
Global Warming	$9.37 \times 10^{+04}$	$2.42 \times 10^{+04}$
Human Carcinogenic Toxicity	$1.86 \times 10^{+04}$	$5.33 \times 10^{+03}$
Human Non-Carcinogenic Toxicity	$5.44 \times 10^{+04}$	$3.05 \times 10^{+04}$
Ionizing Radiation	$5.69 \times 10^{+03}$	$2.26 \times 10^{+03}$
Land Use	$2.49 \times 10^{+03}$	$7.73 \times 10^{+02}$
Marine Ecotoxicity	$5.28 \times 10^{+03}$	$4.55 \times 10^{+03}$
Marine Eutrophication	$2.62 \times 10^{+00}$	8.45×10^{-01}
Ozone Formation, Human Health	$2.39 \times 10^{+02}$	$5.63 \times 10^{+01}$
Ozone Formation, Terrestrial Ecosystems	$2.46 \times 10^{+02}$	$5.92 \times 10^{+01}$
Stratospheric Ozone Depletion	3.85×10^{-02}	3.24×10^{-02}
Terrestrial Acidification	$2.80 \times 10^{+02}$	$1.01 \times 10^{+02}$
Terrestrial Ecotoxicity	$2.68 \times 10^{+05}$	$4.16 \times 10^{+04}$

Figures elaborated in this study that display the results of the data derived from LCA and LCCA show how cleaning costs of DWTS are always preferred from an economic point of view rather than the conventional water supply and sewerage.

In pursuit of presenting data in the most visually effective manner, four distinct graphs have been meticulously crafted for this research, given the inherent diversity of the dataset necessitating such a nuanced approach. Primarily, the removal costs, denominated in euros, have been segregated into three distinct figures due to the considerable variations in numerical scales among different variables. This segregation into three graphs was imperative owing to the magnitude differences observed; some variables were expressed in the hundreds, others in the thousands, and one considerably higher, denoted in hundreds of thousands. The substantial discrepancies inherent in these three groups of variables rendered a singular graph impractical, as the visually imperceptible lower figures, when amalgamated, would compromise clarity and comprehension.

Consequently, Figure 1 encapsulates variables wherein cleaning costs are expressed in tens or hundreds, Figure 2 portrays those articulated in thousands, and Figure 3 singularly represents the variable expressed in hundreds of thousands. Notably, across all three figures, a discernible pattern emerges: for each variable, the cleaning cost associated with the traditional alternative consistently presents a less favorable scenario compared to the implementation of a Distributed Water Treatment System (DWTS).

In a strategic effort to facilitate seamless and visual comparison among all variables, Figure 4 has been meticulously constructed. In this figure, Napierian or natural logarithms have been employed on the cleaning cost data, unifying all variables into a cohesive visual representation. Notably, the logarithmic transformation serves to compress the wide range of values, thereby enabling a more coherent visualization and interpretation of the comparative trends. The overarching observation gleaned from Figure 4 unequivocally supports the assertion that, across all instances, the traditional water treatment alternative is consistently surpassed by the DWTS. This visual analysis serves as a compelling testament to the robust superiority of the latter approach. Figure 4 unequivocally demonstrates that the cleaning costs of a decentralized wastewater treatment system outperform those of the conventional water supply and sewerage system across all fifteen variables considered. Beyond the environmental desirability of this outcome, it proves to be economically more viable, as the calculated costs often go unnoticed, ultimately being absorbed by humans, animals, and the environment through negative externalities. This underscores the dual benefit of the VFCW – not only as an environmentally favorable alternative but also as a financially sound and cost-effective solution.

In conclusion, the deliberate structuring of the graphical representation, coupled with the judicious use of logarithmic transformations, not only enhances the visual appeal but also serves as a robust analytical tool. The comprehensive and meticulous nature of this approach ensures that nuances within the dataset are not only preserved but also presented in a manner conducive to insightful interpretation, thereby fortifying the argument in favor of the DWTS over the conventional water treatment alternative. The removal costs, as presented consistently prove to be more economical for the VFCW compared to the conventional system. While certain variables exhibit comparable cleaning costs for both scenarios, resulting in marginal differences, the cumulative disparity between the two potential investments amounts to €1,984,335. Notably, when considering that these costs are based on a project investment of €63,021.03, the financial implications become even more substantial. Emphasizing the significance of the findings, it's crucial to underscore that 80% of the variables, or 4 out of 5, exhibit removal costs for the traditional water supply and sewerage system that are at least double those of the VFCW, with almost half of them being three times as much. These results show how this chapter has examined a research that transcends conventional financial metrics typically employed in economic assessments of water-related analyses.

Figure 1. Cleaning costs of the variables with lower figures.

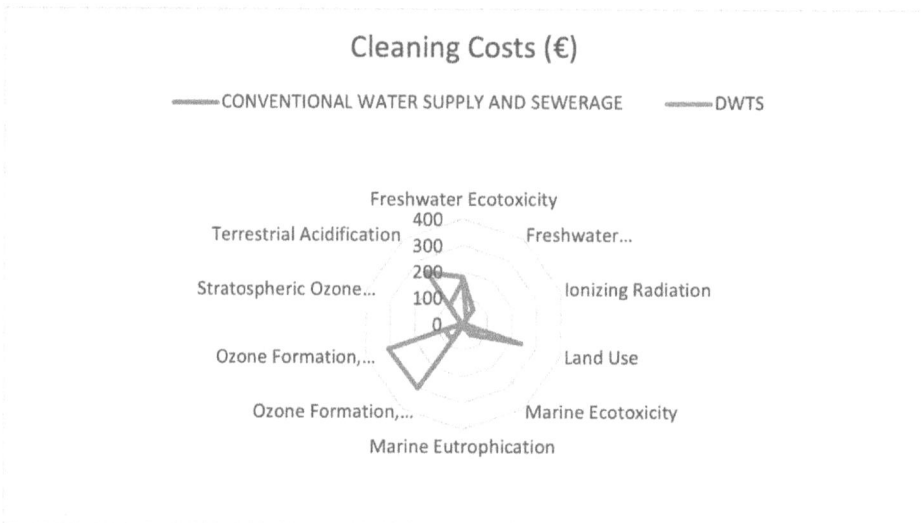

Figure 2. Cleaning costs of the variables with higher figures.

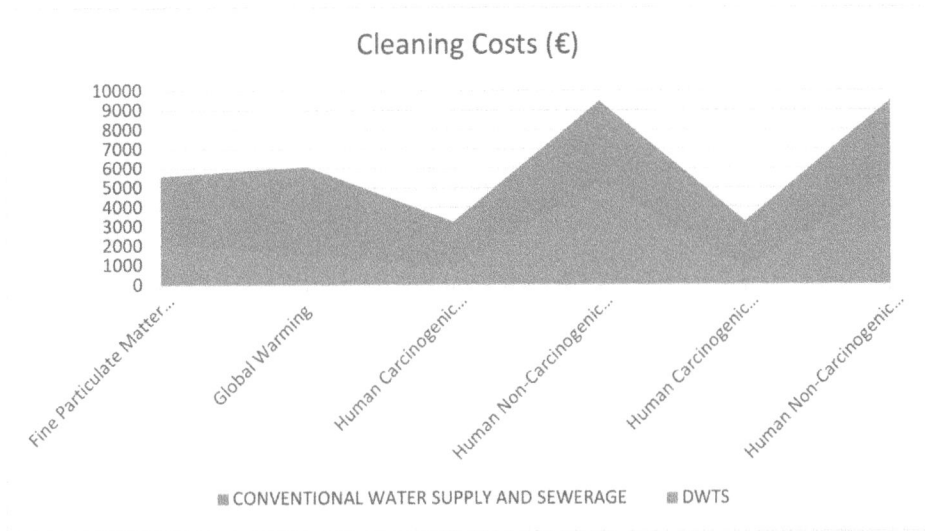

Figure 3. Cleaning costs of Terrestrial Ecotoxicity.

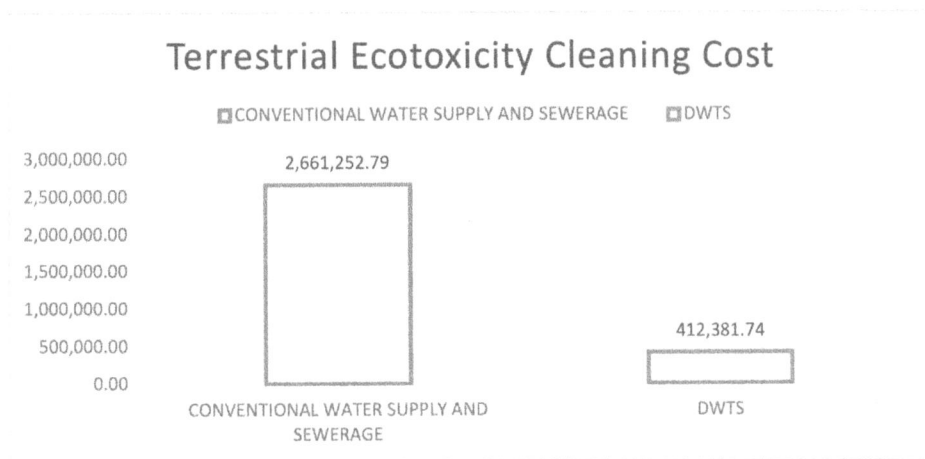

Figure 4. Napierian logarithm of removal cost

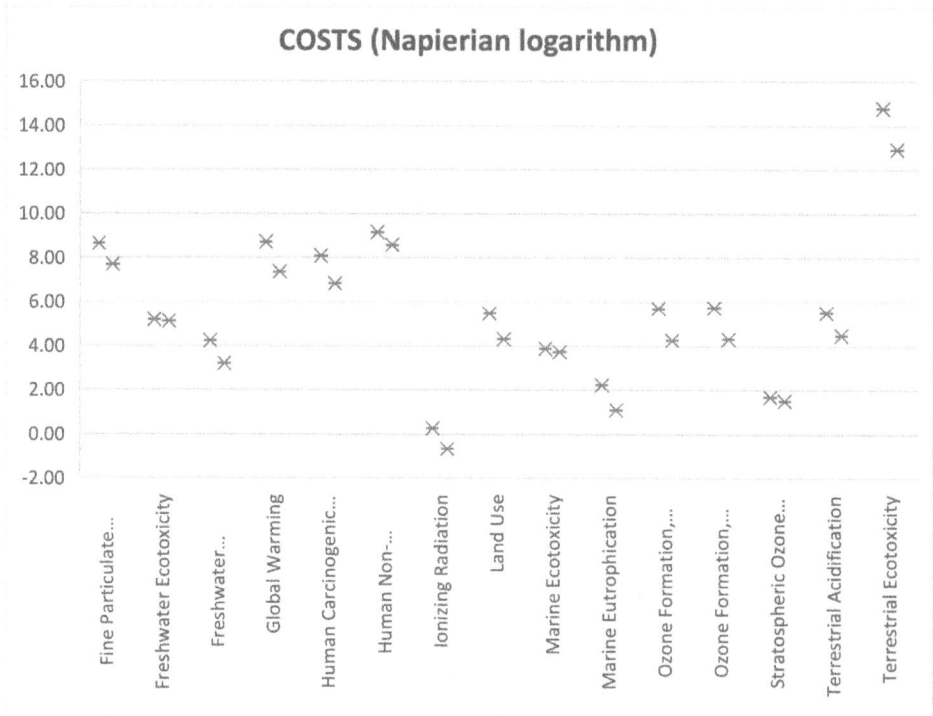

CONCLUSION

In summary, the insights gleaned from this analysis, expounded upon within the confines of this study, posit the implementation of Vertical Flow Constructed Wetlands (VFCWs) in conjunction with a membrane-based potable water unit as a compelling and sustainable alternative to conventional sewage treatment and water supply systems. This assertion is grounded in a meticulous examination, encompassing traditional financial metrics, comprehensive ratios, and a nuanced exploration of the economic implications associated with VFCW implementation in the environmental context.

It is noteworthy that these conclusions are not arbitrary but are intricately linked to the research findings published by López-Serrano et al. in 2023 and Lakho et al. in 2022. The empirical evidence presented in these research forms the bedrock upon which the financial and economic viability of VFCWs is substantiated. Rigorous assessments and calculations, echoing the recommendations of López-Serrano et al., affirm a commendable return on investment, solidifying the economic soundness of VFCWs as an efficacious wastewater treatment solution. Delving beyond traditional financial evaluations, a robust Life Cycle Cost Analysis (LCCA) underscores not only the profitability of the VFCW project but also its far-reaching positive impact on sanitation costs. This dual validation, both in fiscal terms and within the broader economic landscape, fortifies the argument for widespread adoption of VFCWs as a sustainable alternative to conventional water treatment and supply methodologies.

A granular examination of calculations, scrutinizing the adverse effects stemming from both traditional and VFCW alternatives, unveils a monetary evaluation of environmental consequences. This meticulous monetization, rooted in López-Serrano et al.'s research, accentuates the imperative of embracing VFCWs to alleviate negative externalities and accentuate positive outcomes. The economic justification, therefore, extends beyond the immediate fiscal returns, encompassing a broader environmental and social economic perspective. Furthermore, the global significance of this research comes to the forefront when considering the intrinsic value of water worldwide. As elucidated by López-Serrano et al., VFCWs emerge as a practical solution to enhance water supply, particularly in regions where traditional systems prove impractical or are entirely absent. The efficacy of VFCWs is especially pronounced in rural areas lacking connections to public sewerage systems, whether due to geographical distances or a limited population. In these contexts, VFCWs, as underscored by López-Serrano et al., not only address acute water supply challenges but also play a pivotal role in curbing rural exodus by providing a cost-effective and environmentally friendly wastewater treatment solution. The multifaceted benefits of VFCWs as extend beyond direct economic considerations. Governments, as key stakeholders, stand to accrue substantial savings when compared to traditional alternatives. Citizens, on the other hand, gain access to an environmentally sound wastewater treatment solution, contributing to enhanced public health and well-being. At the societal level, the development of sustainable communities aligns seamlessly with the broader objectives encapsulated in the Sustainable Development Goals (SDGs).

The significance of this research is underscored when considering the global value of water. This study verifies Vertical Flow Constructed Wetlands (VFCWs) as a viable alternative capable of enhancing water supply in regions where traditional systems are either absent or economically unviable. The traditional financial indicators employed here not only affirm the economic desirability of the project but also emphasize the importance of fostering VFCW construction for the benefit of both the environment and society. Particularly, VFCWs prove especially useful in rural areas without access to public sewerage systems due to long distances or a small population, potentially mitigating rural exodus by offering a cost-effective wastewater treatment solution.

The outcomes of this research validate the cost-effectiveness of constructing VFCWs, benefiting investors, the environment, and society at large. VFCW implementation not only contributes indirectly to the economy by supporting economic development through establishments like restaurants but also plays a vital role in safeguarding water bodies, enhancing their availability and regeneration. This, in turn, contributes to the achievement of Sustainable Development Goals (SDGs) by recognizing the direct and indirect impact of water supply on their fulfillment.

Despite the positive outcomes, certain limitations are evident. Challenges related to the availability of construction supplies in underdeveloped regions, particularly in rural areas of great importance, pose a hurdle. The "yuck factor," rooted in misconceptions about treated wastewater, could impede its reuse. Additionally, the relatively high initial investment, coupled with prevailing high-interest rates, may pose barriers to VFCW implementation. Public authorities, through grants or loans, could play a pivotal role in overcoming these barriers, fostering the adoption of VFCWs. Further, international institutions and public initiatives could provide subsidies or tax incentives to incentivize investment in this environmentally friendly technology within the circular economy framework.

Looking ahead, future research should explore readily available alternatives and materials for VFCW construction, addressing potential supply challenges. Additionally, an extension of this research to include Life Cycle Cost Assessment (LCCA) and monetary evaluations of treatment costs for other applications, such as the analysis of a mobile CW for greywater treatment at music festivals, could provide a more comprehensive understanding of the economic implications.

REFERENCES

Abdelhay, A., & Abunaser, S. G. (2021). Modeling and Economic Analysis of Greywater Treatment in Rural Areas in Jordan Using a Novel Vertical-Flow Constructed Wetland. *Environmental Management*, 67(3), 477–488. 10.1007/ s00267-020-01349-732856093

Abdulfatah, H. K., Stanley, O. I., Nzerem, P., & Jakada, K. (2019). Defining the optimal development strategy to maximize recovery and production rate from an integrated offshore water-flood project. *Society of Petroleum Engineers - SPE Nigeria Annual International Conference and Exhibition 2019, NAIC 2019*. ACM. 10.2118/198843-MS

Agiakloglou, C., & Gkouvakis, M. (2022). Policy implications and welfare analysis under the possibility of default for the Euro zone area. *Journal of Economic Asymmetries, 25*(November 2020), e00246. 10.1016/j.jeca.2022.e00246

AL-agele, H. A., Nackley, L., & Higgins, C. W.AL-agele. (2021). A Pathway for Sustainable Agriculture. *Sustainability (Basel)*, 13(8), 4328. 10.3390/su13084328

Ashu, A. B., & Lee, S. (2021). The Effects of Climate Change on the Reuse of Agricultural Drainage Water in Irrigation. *KSCE Journal of Civil Engineering*, 25(3), 1116–1129. 10.1007/s12205-021-0004-2

Baggio, G., Qadir, M., & Smakhtin, V. (2021). Freshwater availability status across countries for human and ecosystem needs. *The Science of the Total Environment*, 792, 148230. 10.1016/j.scitotenv.2021.14823034147805

Brennan, M., Rondón-Sulbarán, J., Sabogal-Paz, L. P., Fernandez-Ibañez, P., & Galdos-Balzategui, A. (2021). Conceptualising global water challenges: A transdisciplinary approach for understanding different discourses in sustainable development. *Journal of Environmental Management, 298*. https://doi.org/10.1016/j .jenvman.2021.113361

Bunn, S. E. (2016). Grand challenge for the future of freshwater ecosystems. *Frontiers in Environmental Science*, 4(MAR), 1–4. 10.3389/fenvs.2016.00021

Cao, Z., Zhou, L., Gao, Z., Huang, Z., Jiao, X., Zhang, Z., Ma, K., Di, Z., & Bai, Y. (2021). Comprehensive benefits assessment of using recycled concrete aggregates as the substrate in constructed wetland polishing effluent from wastewater treatment plant. *Journal of Cleaner Production*, 288, 125551. 10.1016/j.jclepro.2020.125551

Castillo-Díaz, F. J., Belmonte-Ureña, L. J., Batlles-delaFuente, A., & Camacho-Ferre, F. (2023). Strategic evaluation of the sustainability of the Spanish primary sector within the framework of the circular economy. *Sustainable Development (Bradford)*, 1–16. 10.1002/sd.2837

Declercq, R., Loubier, S., Condom, N., & Molle, B. (2020). Socio-Economic Interest of Treated Wastewater Reuse in Agricultural Irrigation and Indirect Potable Water Reuse: Clermont-Ferrand and Cannes Case Studies' Cost–Benefit Analysis. *Irrigation and Drainage*, 69(S1), 194–208. 10.1002/ird.2205

Di Vaio, A., Trujillo, L., D'Amore, G., & Palladino, R. (2021). Water governance models for meeting sustainable development Goals: A structured literature review. *Utilities Policy, 72*, 101255. 10.1016/j.jup.2021.101255

Diaz-Elsayed, N., Rezaei, N., Ndiaye, A., & Zhang, Q. (2020). Trends in the environmental and economic sustainability of wastewater-based resource recovery: A review. *Journal of Cleaner Production*, 265, 121598. 10.1016/j.jclepro.2020.121598

Goedkoop, M., Heijungs, R., Huijbregts, M., De Schryver, A., Struijs, J., & Van Zelm, R. (2009). ReCiPe 2008. *Potentials, May 2014*, 1–44. https://www.pre-sustainability.com/download/misc/ReCiPe_main_report_final_27-02-2009_web.pdf

Hristov, J., Barreiro-Hurle, J., Salputra, G., Blanco, M., & Witzke, P. (2021). Reuse of treated water in European agriculture: Potential to address water scarcity under climate change. *Agricultural Water Management*, 251(March), 106872. 10.1016/j.agwat.2021.10687234079159

Krimpas, N. A., Salamaliki, P. K., & Venetis, I. A. (2021). Factor decomposition of disaggregate inflation: The case of Greece. *International Journal of Computational Economics and Econometrics*, 11(1), 84–104. 10.1504/IJCEE.2021.111713

Lakho, F. H., Qureshi, A., Igodt, W., Le, H. Q., Depuydt, V., Rousseau, D. P. L., & Van Hulle, S. W. H. (2022). Life cycle assessment of two decentralized water treatment systems combining a constructed wetland and a membrane based drinking water production system. *Resources, Conservation and Recycling, 178*, 106104. 10.1016/j.resconrec.2021.106104

López-Serrano, M. J., Lakho, F. H., Van Hulle, S. W. H., & Batlles-delaFuente, A. (2023). Life cycle cost assessment and economic analysis of a decentralized wastewater treatment to achieve water sustainability within the framework of circular economy. *Oeconomia Copernicana*, 14(1), 103–133. 10.24136/oc.2023.003

López-Serrano, M. J., Velasco-Muñoz, J. F., Aznar-Sánchez, J. A., & Román-Sánchez, I. M. (2021). Financial evaluation of the use of reclaimed water in agriculture in Southeastern Spain, a mediterranean region. *Agronomy (Basel)*, 11(11), 2218. 10.3390/agronomy11112218

Lutterbeck, C. A., Kist, L. T., Lopez, D. R., Zerwes, F. V., & Machado, E. L. (2017). Life cycle assessment of integrated wastewater treatment systems with constructed wetlands in rural areas. *Journal of Cleaner Production*, 148, 527–536. 10.1016/j. jclepro.2017.02.024

Maniam, G., Zakaria, N. A., Leo, C. P., Vassilev, V., Blay, K. B., Behzadian, K., & Poh, P. E. (2022). An assessment of technological development and applications of decentralized water reuse: A critical review and conceptual framework. *WIREs. Water*, 9(3), 1–31. 10.1002/wat2.1588

Melián-Navarro, A., & Ruiz-Canales, A. (2020). Evaluation in carbon dioxide equivalent and chg emissions for water and energy management in water users associations. A case study in the southeast of spain. *Water (Basel)*, 12(12), 3536. 10.3390/w12123536

Omole, D. O., Jim-George, T., & Akpan, V. E. (2019). Economic Analysis of Wastewater Reuse in Covenant University. *Journal of Physics: Conference Series*, 1299(1), 012125. 10.1088/1742-6596/1299/1/012125

Otter, P., Sattler, W., Grischek, T., Jaskolski, M., Mey, E., Ulmer, N., Grossmann, P., Matthias, F., Malakar, P., Goldmaier, A., Benz, F., & Ndumwa, C. (2020). Economic evaluation of water supply systems operated with solar-driven electro-chlorination in rural regions in Nepal, Egypt and Tanzania. *Water Research*, 187, 116384. 10.1016/j. watres.2020.11638432980605

Rodríguez de Sá Silva, A. C.., Bimbato, A. M., Balestieri, J. A. P., & Vilanova, M. R. N. (2022). Exploring environmental, economic and social aspects of rainwater harvesting systems: A review. *Sustainable Cities and Society, 76.* https://doi.org/ 10.1016/j.scs.2021.103475

Sánchez Pérez, J. A., Arzate, S., Soriano-Molina, P., García Sánchez, J. L., Casas López, J. L., & Plaza-Bolaños, P. (2020). Neutral or acidic pH for the removal of contaminants of emerging concern in wastewater by solar photo-Fenton? A techno-economic assessment of continuous raceway pond reactors. *The Science of the Total Environment*, 736(May), 139681. 10.1016/j.scitotenv.2020.13968132479960

Tociu, C., Ciobotaru, I. E., Maria, C., Déak, G., Ivanov, A. A., Marcu, E., Marinescu, F., Savin, I., & Noor, N. M. (2019). Exhaustive approach to livestock wastewater treatment in irrigation purposes for a better acceptability by the public. *AIP Conference Proceedings*, 2129(July), 020066. Advance online publication. 10.1063/1.5118074

Truchado, P., Gil, M. I., López, C., Garre, A., López-Aragón, R. F., Böhme, K., & Allende, A. (2021). New standards at European Union level on water reuse for agricultural irrigation: Are the Spanish wastewater treatment plants ready to produce and distribute reclaimed water within the minimum quality requirements? *International Journal of Food Microbiology*, 356(June), 109352. 10.1016/j.ijfoodmicro.2021.10935234385095

United Nations. (2021). The United Nations World Water Development Report: Va*luing Water*. UNESCO, Paris.

Weerasooriya, R. R., Liyanage, L. P. K., Rathnappriya, R. H. K., Bandara, W. B. M. A. C., Perera, T. A. N. T., Gunarathna, M. H. J. P., & Jayasinghe, G. Y. (2021). Industrial water conservation by water footprint and sustainable development goals: a review. In *Environment, Development and Sustainability, 23*(9). 10.1007/s10668-020-01184-0

Zadeh, S. M., Hunt, D. V. L., Lombardi, D. R., & Rogers, C. D. F. (2013). Shared urban greywater recycling systems: Water resource savings and economic investment. *Sustainability (Basel)*, 5(7), 2887–2912. 10.3390/su5072887

Chapter 5
Fostering Green Product Design and Innovation for a Sustainable Future

Sunil Sharma
http://orcid.org/0000-0001-9936-5103
Lovely Professional University, India

ABSTRACT

Climate change, resource depletion, and pollution pose existential threats to humanity. Traditional product design, with its focus on short-term gains and linear economy principles, has significantly contributed to these problems. Green product design offers a solution by minimizing environmental impact throughout a product's life cycle, from material sourcing to end-of-life management. In this chapter, we have discussed green products with examples and then provided the previous research for green product design. The overall advantages of green product design have been discussed to highlight their significance. Further, the barriers to the implementation of green product design in organizations have been discussed. This chapter then provides the green product design process for implementation in organizations. The support tools and methods to achieve green product design have been discussed. By understanding their strengths and limitations, designers can leverage them effectively to create more sustainable products that meet the needs of the environment, society, and the economy.

DOI: 10.4018/979-8-3693-2595-7.ch005

INTRODUCTION

A green product or a sustainable product is made with the purpose of minimizing its environmental effect over the course of its whole life cycle, even after it has outlived its usefulness. These products minimize waste and maximize resource efficiency. They are made using non-toxic materials and ecologically friendly production techniques. These products are produced in hygienic circumstances and without the use of dangerous chemicals. They can be recycled, reused, and are biodegradable, i.e. produced with environmental consideration. They use the least resources and have less carbon footprint (Durif et al., 2010). Green Design is about designing products that are environmentally and ecologically friendly and have no social or moral issues. Green product design, in turn, might mean adding a new "green" variant to a product family, thus increasing product variety and stimulating the adoption of flexible manufacturing (Ciccullo et al., 2018). Green design has two characteristics. First, the design should consider the entire life cycle. Second, in order to ensure the normal function of the product, designers should give priority to the object of environmental properties, from the fundamental prevention and control of environmental pollution and energy (He & Li, 2017). Green product design is a proactive business approach that addresses environmental considerations in the early stages of the product development process in order to minimize negative environmental impacts throughout the product's life cycle. It is also referred to as design for environment (DfE), design for eco-efficiency, or sustainable product design (Z. Li et al., 2015; Sun et al., 2003). Green product design includes planning for a product's ultimate disposal (recycling, reuse, or disposal), resource utilization, material selection, and production needs. It is not a stand-alone process; rather, it is used with an organization's current methods for designing products in order to balance environmental considerations with more conventional features like usefulness, quality, and producibility. Compared to traditional products, green products are easier to update, disassemble, recycle, and reuse. They also use less material and may be broken down into modular pieces that can be replaced. By concentrating on resource efficiency, which may lower costs and frequently cut production times, as well as by bringing varied functional groups to the design table and encouraging product and process innovation, a firm can reap several benefits from implementing green product design. Green product design may help reconcile the historically incompatible goals of sustained development and environmental excellence, which can be the first step in closing the loop on a company's industrial operations. Because of this, an increasing number of businesses are emphasizing the importance of green product design in their plans for sustainable business. Making the most profit possible for a company while reducing its environmental effect is mostly dependent on green product design. For businesses, there are several clear advantages to using green design. These advantages include the potential for more sales, more customer happiness, increased social responsibility and environmental performance, compliance with present and future requirements, and enhanced brand recognition. These days, more and more buyers are drawn to purchasing green items that are pollution-free and favorable to the environment (G. Li et al., 2020; Xue et al., 2021).

EXAMPLES OF GREEN PRODUCT DESIGN

Many businesses are creating greener products and promoting their environmentally friendly features. Companies brag about huge sales of these environmentally friendly products. In an effort to "green" its goods, PepsiCo creates recyclable PET plastic soft drink bottles rather than corrugated ones (Shoda, 2013). Additionally, Coca-Cola created the first PET plastic bottle that is completely recyclable (Hong et al., 2019). Table 1 shows the list of green product design examples (Dell, 2022; Fairphone, 2023; Iannuzzi, 2016; IKEA, 2023; Lush Retail, 2024; Notpla, 2023; Patagonia, 2023, 2024; Philips, 2024; Tesla, 2024; The Body Shop, 2023).

Table 1. Examples of green product design with description, key sustainability features, and benefit

Product	Description	Key Sustainability Features	Benefits
Patagonia ReTool & Worn Wear program	The company repairs, refurbishes, and resells used Patagonia items, extending their lifespan and reducing waste.	Circular economy, reduced resource consumption, extended product life.	Reduces waste, lowers environmental impact, and saves money for consumers.
IKEA Kungsbacka Kitchen	The kitchen uses recycled plastic bottles and reclaimed industrial wood, showcasing the use of recycled materials.	Recycled content, reduced reliance on virgin resources, and affordable, sustainable options.	Lessens dependence on new materials, lowers environmental impact, offers accessible sustainability.
Tesla Electric Vehicles	Electric cars emit no tailpipe emissions, promoting cleaner transportation and renewable energy use.	Zero tailpipe emissions, reduced greenhouse gas emissions, promotes renewable energy transition.	Improves air quality, combats climate change, fosters cleaner transportation alternatives.
Fairphone	Smartphone prioritizes ethical sourcing, fair labor practices, and modular design for easy repair and upgrades.	Ethical sourcing, fair labor practices, modular design, extended product lifespan.	Supports ethical production, empowers users, reduces waste and resource consumption.

continued on following page

Table 1. Continued

Product	Description	Key Sustainability Features	Benefits
The Body Shop Refillable Packaging	Products come in refillable containers, reducing need for single-use plastic packaging.	Reduced packaging waste, promotes responsible consumption, circular economy approach.	Lessens plastic pollution, empowers consumers to make sustainable choices, minimizes waste.
Lush Solid Shampoos and Bars	Concentrated and waste-free alternative to bottled hair care, often using natural and biodegradable ingredients.	Water-free, reduced packaging, often biodegradable ingredients, concentrated formula.	Conserves water, minimizes packaging waste, potentially uses eco-friendly materials.
Dell Luna Laptop	Laptop incorporates ocean-bound plastics and recycled materials in its chassis.	Recycled content uses waste materials and reduces reliance on virgin resources.	Diverts waste from landfills, promotes resource efficiency, demonstrates innovative use of recycled materials.
Patagonia Black Hole Bags	Made from recycled nylon and features a lifetime warranty, minimizing need for replacements and material consumption.	Durable, repairable, recycled content, minimizes need for replacements.	Reduces need for new materials, minimizes waste, promotes product longevity.
Notpla Edible Packaging	Company creates packaging materials from seaweed and other organic materials that are completely edible and compostable.	Biodegradable, edible, eliminates plastic waste, reduces reliance on traditional packaging.	Minimizes waste and pollution, offers innovative and sustainable alternative, promotes circularity.
Philips Hue Smart Lighting	Energy-efficient LED bulbs can be controlled remotely and programmed, allowing for reduced energy consumption.	Energy-efficient, smart controls, reduced energy use, personalized lighting management.	Lowers energy consumption, empowers users to manage lighting efficiently, saves money on energy bills.

SIGNIFICANCE AND ADVANTAGES OF GREEN PRODUCT DESIGN

To prevent the worst effects of climate change, global greenhouse gas (GHG) emissions must be cut in half by 2030 and reach net zero by 2050. Organizations should closely monitor product design decisions as they map out their path toward net zero due to the influence these decisions will have on reducing these emissions. Using green product design can assist businesses in achieving their overall environmental objectives. Businesses in all sectors are susceptible to interruptions in the raw material supply chain brought on by a lack of resources. The design of modular and easily disassemblable items can enhance the recoverability of materials. Products designed for easy disassembly and component replacement extend their lifespan and simplify material recovery. Survey reports indicate that customer satisfaction levels have increased as a result of green product design initiatives, highlighting customers' interest in products' ethical standing. Demand for products is no longer confined to utilitarian convenience but increasingly includes environmentally friendly items (Rau et al., 2023). Green product design practices minimize environmental pollution and resource waste throughout the life cycle of the product. Consequently, one of the most important steps in helping businesses adopt green growth models is green product design. Green product design is significant due to its multifaceted positive impacts on the environment, economy, society, and overall well-being. It represents a proactive and responsible approach to creating products that align with the principles of sustainability and contribute to a more resilient and regenerative future. Green product design plays a crucial role in achieving sustainability by minimizing the environmental impact of products throughout their lifecycle. It goes beyond just creating "eco-friendly" products; it encompasses a holistic approach that considers the entire system, from conception and design to manufacturing, use, and end-of-life. Green product design, which focuses on creating products with minimal environmental impact and maximum sustainability, offers numerous advantages across various dimensions. Following are some key advantages of green product design.

- **Environmental Conservation**
  o   Reduced Resource Consumption: Green product design aims to minimize the use of raw materials, energy, and water, contributing to the conservation of natural resources (A. Sharma & Iyer, 2012; Tseng et al., 2013).
  o   Emission Reduction: By incorporating sustainable practices, green products often result in lower emissions of pollutants and greenhouse gases during manufacturing and use.
- **Waste Reduction**
  o   Promotes Circular Economy: Green product design emphasizes the principles of a circular economy, encouraging product reuse, recycling, and remanufacturing to reduce overall waste generation.
  o   Minimized Landfill Impact: Products designed with eco-friendly materials and end-of-life considerations result in reduced landfill waste.
- **Energy Efficiency**
  o   Optimized Energy Consumption: Green products are often designed to be more energy-efficient, reducing the overall energy consumption during manufacturing, transportation, and use.
  o   Renewable Energy Integration: Many green designs incorporate the use of renewable energy sources, contributing to a cleaner and more sustainable energy mix.
- **Cost Savings**

o   Long-Term Economic Benefits: While initial costs for green product design may be higher, the long-term economic benefits often include reduced energy and resource costs, as well as potential savings through improved efficiency.

o   Risk Mitigation: Businesses adopting green practices may experience reduced exposure to regulatory risks and potential costs associated with environmental non-compliance.

- **Market Differentiation**

o   Enhanced Brand Reputation: Green products contribute to a positive brand image, attracting environmentally conscious consumers who prioritize sustainability.

o   Market Competitiveness: Businesses that invest in green product design can gain a competitive edge in the market by meeting the growing demand for sustainable options.

- **Regulatory Compliance**

o   Adherence to Environmental Standards: Green product design ensures compliance with environmental regulations and standards, reducing the risk of legal issues and penalties.

o   Future-Proofing: Companies adopting green practices are better positioned to navigate evolving regulations and anticipate future environmental requirements.

- **Innovation and Creativity**

o   Encourages Innovation: Green product design often necessitates creative solutions and innovative thinking to achieve sustainability goals.

o   Fosters Design Excellence: Sustainability challenges inspire designers to explore new materials, technologies, and approaches, leading to enhanced design excellence.

- **Employee Engagement**

o   Attracts and Retains Talent: Companies committed to sustainability, including green product design, are often more attractive to environmentally conscious employees, contributing to talent attraction and retention.

o   Employee Satisfaction: Engaging employees in sustainability initiatives can foster a sense of purpose and satisfaction among the workforce.

- **Improved Health and Safety**

o   Reduced Exposure to Harmful Substances: Green products typically avoid or minimize the use of toxic materials, contributing to a healthier and safer environment for both consumers and workers.

- o Healthier Indoor Environments: Sustainable design considerations extend to factors like indoor air quality, promoting healthier living and working spaces.
- **Longer Product Lifespan**
  - o Durability and Quality: Green products are often designed for longevity, emphasizing durability and quality to extend their useful life.
  - o Reduced Replacement Frequency: Products that withstand wear and tear contribute to lower overall consumption and waste by requiring less frequent replacement.
- **Positive Social Impact**
  - o Community and Stakeholder Relations: Adopting green product design practices can enhance relationships with local communities and stakeholders who prioritize social and environmental responsibility.
  - o Job Creation: The growth of sustainable industries and practices can contribute to the creation of green jobs, supporting local economies.
- **Global Environmental Impact**
  - o Cumulative Positive Effect: As more businesses adopt green product design, the cumulative positive impact on the environment can be substantial, contributing to global sustainability goals.

Green product design, therefore, not only benefits the environment but also brings about economic, social, and reputational advantages for businesses and society as a whole.

BACKGROUND

The references to green product design go back to the 1980s when companies and organizations started discussing green product development and clean technologies (Waldegrave & Davis, 1987). Green product design is not a brand-new concept. It was created during the boom of environmental consciousness in the late 1960s and early 1970s, coinciding with the establishment of the Environmental Protection Agency and the enactment of legislation like the Resource Conservation and Recovery Act, the Clean Water Act, and the Clean Air Act. Though concepts like design for recycling and remanufacturing were explored in technical publications and conferences during the 1980s, politicians and the general public paid little attention to the notion. The topic has had a resurgence in the last several years, maybe as a result of recent worrying headlines about overflowing landfills, ozone depletion, and global climate change (Eyring, 1992). AT&T presented a seminal paper on green product design in 1995 where they described the design for environment (DFE) as a method to achieve the same. They stated that green product design is an activity that can be implemented by product designers, manufacturing engineers, and supply chain manufacturers (Graedel et al., 1995). One of the most important steps in helping businesses adopt green growth models is green product design (Hu et al., 2022). Yi-Fei discussed the principles of green product design such as the principle of complete utilization of resources, energy conservation, no pollution, color coding green design products, and best ecological and economic benefit (Yi-Fei, 2017). Research for innovation of green product design has become significant over the years. Batwara et al. compile various barriers in the implementation of green product design and development and categorize them as various barriers represented in economic, technological, informational, and organizational categories (Batwara et al., 2022). About 70% of a product's environmental impacts are determined by decisions made during the design stage (Waage, 2007). Green product design is now seen as the leading strategy for saving resources and reducing adverse eco-effects (Leigh & Li, 2015).

GREEN PRODUCT DESIGN PROCESS

To effectively incorporate environmental protection into industrial design, it's important to investigate new ways and principles. Green design is currently in its early stages of development. It is important to realize that to improve; more designers are needed to comprehend the systematic process and tackle its challenges. Figure 1 shows the green product design process (Hu et al., 2022; Uang & Liu, 2013; Wang, 2016; Zhuo & Shengxue, 2010).

Figure 1. Green product design process

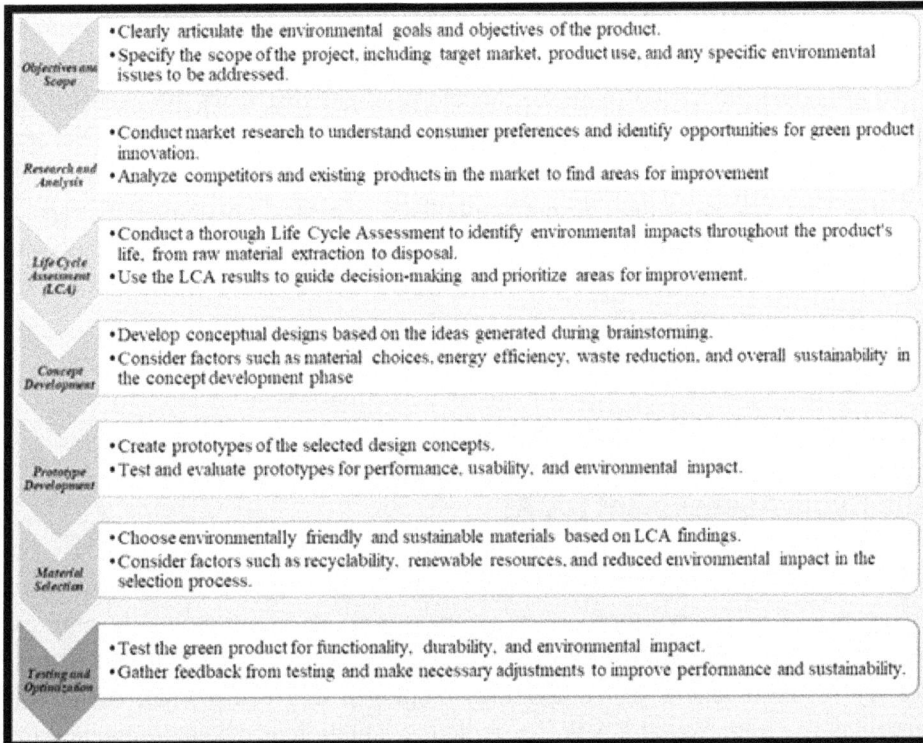

Objectives and Scope	• Clearly articulate the environmental goals and objectives of the product. • Specify the scope of the project, including target market, product use, and any specific environmental issues to be addressed.
Research and Analysis	• Conduct market research to understand consumer preferences and identify opportunities for green product innovation. • Analyze competitors and existing products in the market to find areas for improvement
Life Cycle Assessment (LCA)	• Conduct a thorough Life Cycle Assessment to identify environmental impacts throughout the product's life, from raw material extraction to disposal. • Use the LCA results to guide decision-making and prioritize areas for improvement.
Concept Development	• Develop conceptual designs based on the ideas generated during brainstorming. • Consider factors such as material choices, energy efficiency, waste reduction, and overall sustainability in the concept development phase
Prototype Development	• Create prototypes of the selected design concepts. • Test and evaluate prototypes for performance, usability, and environmental impact.
Material Selection	• Choose environmentally friendly and sustainable materials based on LCA findings. • Consider factors such as recyclability, renewable resources, and reduced environmental impact in the selection process.
Testing and Optimization	• Test the green product for functionality, durability, and environmental impact. • Gather feedback from testing and make necessary adjustments to improve performance and sustainability.

From an innovation point of view in companies, Dangelico stated that top management commitment, collaborative networks, information flows, cross-functional integration, and resource development all play a role in effective green product innovation implementation (Dangelico, 2016). They discussed the development of organizational capabilities, the role of learning and institutional pressures in persuading firms to adopt green product innovations, the barriers that firms face, the benefits of Green product innovations, and the structural changes required to pursue Green product innovations (Khan et al., 2021). According to data, China leads the world in research papers on green product design, with India, US, Italy, Malaysia, Canada, UK, Brazil, and France following (Ghazali et al., 2023).

METHODS AND TOOLS FOR GREEN PRODUCT DESIGN

Conventional methods and tools provide a solid foundation for green product design. By understanding their strengths and limitations, designers can leverage them effectively to create more sustainable products that meet the needs of the environment, society, and the economy. As sustainability continues to evolve, new and innovative approaches will likely emerge, further pushing the boundaries of green design. Green product design is a holistic approach, encompassing various tools and methods throughout the product lifecycle. We present support tools in four categories, namely: Assessment and Planning, Design and Development, Manufacturing and Production, and Marketing and Communication.

ASSESSMENT AND PLANNING

Life Cycle Assessment (LCA)

Life Cycle Assessment (LCA) plays a crucial role in green product design by acting as a powerful tool for identifying, measuring, and minimizing the environmental impact of a product throughout its entire life cycle. Life Cycle Assessment (LCA) is a comprehensive methodology employed in green product design and sustainability initiatives. LCA is a systematic evaluation of the environmental impacts associated with a product or service throughout its entire life cycle, from raw material extraction to end-of-life disposal. This assessment method plays a crucial role in identifying opportunities for improvement, making informed decisions, and advancing sustainable practices (De Luca et al., 2017). Table 2 shows different aspects of LCA integrated with green product design.

Table 2. Examples of green product design with description, key sustainability features, and benefit

Aspect	Description	Benefits
Identification of environmental impacts	Analyzes the environmental impact of a product throughout its lifecycle, from raw material extraction to production, distribution, use, and end-of-life disposal. Identifies hotspots and areas for improvement.	Helps prioritize efforts and resources toward the most impactful areas, leading to more targeted and effective design decisions.
Comparison of design alternatives	Compares the environmental impact of different design options, allowing for objective and data-driven decision-making.	Enables selection of the most sustainable design from an environmental perspective.

continued on following page

Table 2. Continued

Aspect	Description	Benefits
Communication and transparency	Provides a standardized framework for communicating the environmental footprint of a product, building trust and transparency with consumers and stakeholders.	Supports marketing and labeling efforts by providing reliable data to showcase a product's environmental benefits.
Identification of innovation opportunities	Uncovers areas where significant environmental improvements can be made, fostering the development of more sustainable materials, processes, and product features.	Drives innovation and leads to the creation of truly groundbreaking green products.
Supporting compliance and regulations	Helps companies comply with environmental regulations and standards related to product sustainability.	Mitigates risks associated with non-compliance and ensures adherence to environmental laws.
Cost evaluation	Can be used to estimate the environmental cost of a product, alongside traditional financial costs, providing a more complete picture of product sustainability.	Allows for informed decision-making that considers both environmental and economic factors.
Continuous improvement	Provides a baseline for ongoing monitoring and improvement of a product's environmental performance over time.	Enables companies to track progress towards sustainability goals and demonstrate commitment to continuous improvement.

Environmental Footprint Analysis

Environmental footprint analysis (EFA) plays a crucial role in green product design, serving as a quantitative and comprehensive assessment of a product's impact on the environment throughout its entire lifecycle. By understanding this impact, designers can make informed decisions to minimize environmental harm and create more sustainable products. EFA considers various environmental impacts, including:

- Climate change: Carbon footprint (CO_2 equivalent emissions)
- Resource depletion: Water footprint, material footprint
- Air and water pollution: Acidification, eutrophication, ozone depletion
- Land use: Biodiversity loss, soil erosion

Despite its challenges, EFA remains a powerful tool for green product design. By integrating EFA into the design process, designers can make informed decisions, create more sustainable products, and communicate their environmental benefits to consumers. Continuous improvement in data availability, analysis tools, and communication strategies will further enhance the power of EFA in driving sustainable innovation.

Material Selection Tools

In the realm of green product design, choosing the right materials is crucial. A variety of material selection tools empower designers to make informed decisions that prioritize sustainability. These tools offer valuable information on various materials, aiding in the selection of options with lower environmental impact. The material selection tools offer the following support.

- Environmental Data: Access information on a material's embodied energy, water footprint, carbon footprint, recyclability, and other environmental indicators.
- Material Databases: Explore vast databases containing properties and environmental data on diverse materials, from traditional metals and plastics to innovative bio-based options.
- Comparative Analysis: Compare different materials side-by-side based on various criteria, including environmental impact, performance characteristics, cost, and availability.
- Sustainability Filters: Narrow down your search based on specific sustainability goals, such as recycled content, low toxicity, or biodegradability.
- Decision Support Features: Utilize tools that suggest alternative materials based on your requirements and environmental priorities.
- Design for Sustainability Checklists: Frameworks like the Cradle to Cradle and Design for Environment guide designers through key sustainability considerations.

The Cradle to Cradle Philosophy

Cradle to Cradle (C2C) is a design philosophy that aims to create products that are not just less harmful but truly beneficial to the environment and human health. It goes beyond simply minimizing environmental impact, focusing on designing products that are completely safe, recyclable, and even regenerative. "Cradle to Cradle" (C2C) is a design framework that emphasizes the principles of sustainability, environmental health, and the circular economy. Developed by architect William McDonough and chemist Michael Braungart, the Cradle to Cradle approach aims to create products and systems that are not just "less bad" for the environment but actively contribute to ecological and human well-being (Braungart & McDonough, 2009). In the context of green product design, Cradle to Cradle focuses on designing products with a positive impact throughout their entire life cycle. Green product design transcends the traditional "cradle to grave" model, advocating instead for a closed-loop "cradle to cradle" approach. This philosophy, championed by architects

William McDonough and Michael Braungart, envisions products as biological or technical nutrients that circulate within a continuous cycle. Biological nutrients decompose and nourish the earth, returning to the biosphere. Technical nutrients maintain their integrity, being endlessly cycled through disassembly and remanufacturing. This eliminates the concept of waste, fostering a symbiotic relationship between industry and the environment. The key principles of Cradle to Cradle are listed as follows.

- Material Health: All materials used in a product are safe and healthy for humans and the environment, even after their intended use.
- Material Reutilization: Products are designed to be disassembled, and their materials either biodegrade safely or be recycled indefinitely into new, high-quality products.
- Renewable Energy: Manufacturing processes use renewable energy sources to minimize environmental impact.
- Water Stewardship: Water is used responsibly and treated effectively throughout the product lifecycle.
- Social Fairness: Products are produced in a way that respects human rights and promotes social justice.

The benefits of C2C in Green Product Design are as follows.

- Eliminates Waste: Products become resources for future generations, closing the loop on material use.
- Reduces Environmental Impact: Minimizes resource depletion, pollution, and greenhouse gas emissions.
- Promotes Innovation: Drives the development of new, sustainable materials and technologies.
- Improves Human Health: Uses safe and healthy materials, creating safer products and environments.
- Boosts Business Value: Creates products that are durable, recyclable, and attractive to consumers seeking sustainable options.

Design for Environment

Design for the environment (DFE) was created to reduce the environmental effect of goods during their lifetime (Allenby & Fullerton, 1991); (Kuo et al., 2001). DFE is gaining popularity among academics and practitioners due to its ability to address environmental challenges and enhance the environment. Green product design and Design for Environment (DfE) are intertwined concepts. DfE is a specific methodology within green product design that focuses on integrating environmental considerations throughout the entire product lifecycle, from conception to disposal. The examples of DfE include creating packaging that is readily biodegradable or recyclable, using recycled resources in commercial parts, creating modular items that are simple to disassemble for maintenance or upgrading, selecting producing methods and parts that use less energy, creating things that are repairable and made of sturdy materials to give them longer lifespans and planning for simple disassembly and end-of-life handling. DfE emphasizes several key principles mentioned as follows.

- Prevention: Minimize environmental impact by reducing material use, energy consumption, and waste generation during design, manufacturing, and use.
- Lifecycle Assessment: Analyze the environmental impact of a product across its entire lifecycle, identifying hotspots for improvement.
- Material Selection: Choose materials that are environmentally friendly, recyclable, and readily available from sustainable sources.
- Design for Disassembly and Reassembly: Facilitate easy disassembly and repair to extend product lifespan and enable material recovery.
- Energy Efficiency: Design products that minimize energy consumption during use and manufacturing.
- End-of-Life Management: Design for responsible end-of-life options, such as recycling, composting, or safe disposal.

The benefits of DfE in Green Product Design are listed as follows.

- Reduced Environmental Impact: Minimizes resource depletion, pollution, and greenhouse gas emissions.
- Cost Savings: Efficiency improvements can lead to lower production and disposal costs.
- Improved Product Performance: DfE principles often lead to lighter, more durable, and longer-lasting products.
- Enhanced Brand Reputation: Consumers increasingly value sustainable products, and DfE demonstrates a commitment to environmental responsibility.
- Compliance with Regulations: DfE helps companies comply with environmental regulations and avoid potential fines.

DESIGN AND DEVELOPMENT

Biomimicry

Biomimicry, the art and science of learning from and emulating nature's designs, offers a treasure trove of inspiration for green product design. By studying and understanding the ingenious solutions nature has evolved over millions of years, designers can create products that are not only functional and efficient but also environmentally responsible (S. Sharma et al., 2023; S. Sharma & Sarkar, 2019).

Biomimicry involves several key stages:

- Observe: Closely study natural phenomena and organisms, understanding their adaptations and functionality.
- Mimic: Translate those observed principles and strategies into design concepts and solutions.
- Emulate: Apply these concepts to develop products that offer similar benefits while considering human needs and limitations.

Ecodesign

Ecodesign, often used interchangeably with "Design for Environment" (DfE), stands as the cornerstone of green product design. It's more than just a set of tools; it's a strategic approach and philosophy that integrates environmental considerations throughout the entire product lifecycle, from initial conception to end-of-life. The key principles of eco-design are enlisted as follows.

- Prevention: Minimize environmental impact by reducing material use, energy consumption, and waste generation at every stage of design, manufacturing, and use.
- Lifecycle Thinking: Analyze the environmental impact of a product across its entire lifecycle, identifying hotspots for improvement.
- Material Selection: Choose materials with low environmental impact, prioritizing recyclability, biodegradability, and responsibly sourced options.
- Design for Disassembly and Reassembly: Facilitate product disassembly, repair, and refurbishment to extend lifespan and enable material recovery.
- Energy Efficiency: Design products that minimize energy consumption during use and manufacturing.
- End-of-Life Management: Design for responsible end-of-life options, such as recycling, composting, or safe disposal

3D Printing

3D printing provides the following advantages in terms of green product design.

- Reduced material waste: 3D printing uses only the material needed to create the object, unlike traditional manufacturing processes that often generate a lot of scrap.
- Local production: 3D printers can be used to produce products on demand and close to where they are needed, which can reduce transportation emissions.
- Lightweight designs: 3D printing allows for the creation of complex, lightweight designs that can use less material and be more energy-efficient.
- Customization: 3D printing can be used to create products that are customized to individual needs, which can help to reduce waste and resource consumption.

MANUFACTURING AND PRODUCTION

Clean Production Techniques

Clean production, also known as cleaner production, encompasses a set of tools and strategies aimed at reducing the environmental impact of industrial processes. By minimizing resource consumption, waste generation, and pollution, clean production fosters a more sustainable approach to manufacturing, contributing significantly to green product development. The key features of Clean Production Tools are described as follows.

- Process Optimization: Analyze and improve production processes to reduce energy and water consumption, eliminate unnecessary steps, and minimize scrap and waste generation.
- Pollution Prevention: Implement techniques to prevent pollution at its source, avoiding the need for end-of-pipe treatment solutions.
- Material Substitution: Replace hazardous or environmentally harmful materials with safer and more sustainable alternatives.
- Cleaner Technologies: Adopt efficient and environmentally friendly technologies for various production stages, such as low-energy equipment or closed-loop water systems.
- Life Cycle Assessment (LCA): Analyze the environmental impact of a product throughout its lifecycle, identifying hotspots for improvement within manufacturing processes.

Circular Economy Principles

The circular economy offers a revolutionary framework for green product design, shifting away from the traditional "take-make-dispose" model towards a regenerative system where resources are kept in use for as long as possible. By applying circular economy principles, designers can create products that minimize waste, conserve resources, and contribute to a more sustainable future. The key principles of circular economy in green product design are described as follows.

- Design for Longevity: Create products with durable materials, repairable designs, and upgradeable features to extend their lifespan and minimize the need for frequent replacements.
- Modular Design: Design products with modular components that can be easily replaced, repaired, or upgraded, allowing for flexibility and adaptability over time.
- Design for Disassembly: Facilitate product disassembly by using standardized components and tools, enabling efficient material recovery and reuse.
- Material Choice: Utilize recycled, bio-based, and biodegradable materials whenever possible, minimizing reliance on virgin resources and promoting closed-loop material cycles.
- Remanufacturing and Reuse: Design products for remanufacturing and reuse through standardized designs and easy access to spare parts.
- Reverse Logistics: Develop efficient systems for collecting used products and materials, enabling their re-entry into the circular economy.

Life Cycle Cost Analysis

Life cycle cost analysis (LCCA) plays a crucial role in green product design by evaluating the total cost of a product throughout its entire lifecycle, from cradle to grave. This approach goes beyond the initial purchase price and considers all costs associated with the product, including:

- Acquisition: Raw materials, manufacturing, transportation, and distribution.
- Operation and use: Energy consumption, maintenance, repairs, and consumables.
- End-of-life: Disposal, recycling, or remanufacturing.

By understanding the full cost picture, LCCA empowers designers and businesses to create green products that are not only environmentally responsible but also economically viable in the long run.

MARKETING AND COMMUNICATION

Environmental Product Declarations (EPDs)

Environmental Product Declarations (EPDs) serve as vital tools for transparency and credibility in green product design. They act as standardized reports, verified by third-party experts, that disclose the environmental impact of a product across its entire lifecycle, from cradle to grave. This information empowers designers, manufacturers, and consumers to make informed choices for sustainable products. Transparent reports on a product's environmental impact throughout its life cycle. Organizations like UL Environment and Green Seal offer EPD certification.

Eco-Labeling

Eco-labels are powerful tools in green product design, acting as visible certifications that help consumers identify products with reduced environmental impact. By displaying an eco-label, manufacturers demonstrate their commitment to sustainability and provide consumers with an easy way to make informed choices. The eco-labels impact green product design in the following ways.

- Motivating sustainable practices: Manufacturers seek eco-labels as a mark of distinction, driving them to improve product design, source materials responsibly, and optimize production processes to meet specific environmental criteria.
- Guiding consumer choices: Eco-labels simplify complex environmental information, allowing consumers to quickly identify products aligned with their sustainability values.
- Boosting market competitiveness: Products with credible eco-labels often enjoy enhanced consumer preference and market share, providing a competitive advantage for responsible manufacturers.
- Promoting transparency and accountability: Eco-labels encourage transparency throughout the supply chain, holding manufacturers accountable for their environmental claims.

Third-Party Certifications

Third-party certifications like Energy Star and FSC provide consumers with assurance about a product's environmental credentials. Table 3 shows the list of the tools used to implement the methods used in green product design.

Table 3. Methods and tools for supporting green product design

	Aspects	Tools
Life	Analyzes a product's environmental impact throughout its entire life cycle, from cradle to grave.	SimaPro, GaBi, OpenLCA
	Creates products that are easy to disassemble for repair, reuse, and recycling.	CAD software with DfD capabilities: SolidWorks, CATIA, Siemens NX
	Integrates environmental considerations into every stage of the design process.	The Natural Step Framework, Cradle to Cradle, Environmental management systems (EMS), energy monitoring software
	Draws inspiration from nature's efficient and sustainable solutions for product design.	biomimicry databases (AskNature.org, IDEA-INSPIRE), 6W framework
	Understands user needs and behaviors to create products that are durable, repairable, and have long lifespans.	Surveys, interviews, focus groups
	Measures the total environmental impact of a product, considering land, water, and carbon footprints. Tools include PEF and WRI accounting tools	Tools include PEF and WRI accounting tools
	Databases provide environmental data on various materials to aid informed choices.	Granta MI, Material Nexus, EcoInvent, Sphera, BOMcheck
	Clean production tools offer a powerful approach to green product development by optimizing manufacturing processes for sustainability.	Cleaner Production Assessment (CPA), Environmental Footprinting Tools, Material Safety Data Sheets (MSDS), Best Available Techniques (BAT) Reference Documents, UN Environment Programme's Cleaner Production website
	Designing products for disassembly, reuse, and recyclability.	Tools like the Circular Design Guide and the Ellen MacArthur Foundation website offer guidance
	Circular economy promoting sustainability-oriented innovation has a positive impact on financial, environmental, and social performance.	Life Cycle Assessment (LCA): Material Circularity Indicator (MCI), Design for Disassembly (DfD), Product Service Systems (PSS), Remanufacturing and Reverse logistics
3D	Enables rapid prototyping and production with minimal material waste.	Tools like Autodesk Netfab and Stratasys GrabCAD Print help optimize designs for 3D printing.

CHALLENGES TO THE IMPLEMENTATION OF GREEN PRODUCT DESIGN

While the promise of green design is immense, its path is not without challenges. Scaling sustainable materials and technologies remains a hurdle, and shifting consumer mindsets toward circularity requires education and awareness campaigns. However, the opportunities outweigh the challenges. Green design represents a significant growth market, attracting conscious consumers and fostering collaboration between businesses, designers, and scientists. While the potential of green design and innovation for a sustainable future is undeniable, numerous challenges need to be addressed for its widespread adoption and impact.

Economic Challenges

- Higher upfront costs: Sustainable materials and technologies often carry higher initial costs compared to traditional options, making them less competitive in price-sensitive markets.
- Lack of investment: Research and development in green technologies require significant financial resources, and long-term returns might not attract sufficient investment.
- Unstable policy landscapes: Inconsistent government policies and subsidies can create uncertainty and disincentivize long-term investments in green innovation.

Consumer Behavior Challenges

- Limited awareness and knowledge: Consumers might lack sufficient understanding of green products and their benefits, making them hesitant to switch from familiar options.
- Greenwashing and misleading claims: Misleading marketing practices that exaggerate the sustainability of products erode consumer trust in genuine green offerings.
- Short-term priorities and affordability: Consumers often prioritize price and immediate needs over long-term environmental benefits, especially when facing economic constraints.

Systemic Challenges

- Fragmented supply chains: Integrating sustainability practices across complex and geographically dispersed supply chains requires extensive collaboration and transparency.
- Limited infrastructure: Lack of robust recycling, composting, and other end-of-life solutions hinders circular economy practices for green products.
- Standardization and measurement difficulties: Absence of standardized metrics for environmental impact makes it challenging to compare and track progress in green product development.

Organizational Challenges

- Short-termism and shareholder pressure: Businesses might prioritize short-term profits over long-term sustainability goals due to shareholder pressure and quarterly reporting cycles.

- Lack of internal expertise: Companies might lack the knowledge and expertise within their teams to effectively integrate sustainability principles into design and development processes.
- Organizational silos and resistance to change: Internal resistance to change and siloed departmental structures can hinder effective collaboration and implementation of green initiatives.

Societal Challenges

- Social equity considerations: Ensuring that the transition to green products doesn't disproportionately burden disadvantaged communities requires careful social impact assessments and inclusive design approaches.
- Changing consumption patterns: Shifting deeply ingrained consumption habits requires sustained behavioral change initiatives and incentives to promote responsible consumption.
- Lack of public pressure and accountability: Without strong public demand for sustainable products and policies, businesses might lack the motivation to prioritize green design and innovation.

Addressing these challenges necessitates a multi-pronged approach involving collaboration between businesses, governments, consumers, and research institutions. Investing in education, fostering collaboration, implementing supportive policies, and developing clear impact measurement frameworks are crucial steps towards overcoming these hurdles and unlocking the full potential of green product design and innovation for a sustainable future.

CONCLUSION

Green product design and innovation hold the key to unlocking a future where prosperity and environmental well-being coexist. By embracing cradle-to-cradle principles, harnessing ingenuity, and overcoming challenges, we can craft products that nourish the planet and leave a legacy of responsible stewardship for generations to come. This chapter is just a stepping stone on this exciting journey, inviting you to become a co-creator in shaping a greener, more sustainable future through the power of design. The path to a sustainable future seems daunting, but the seeds of positive change are already being sown through green product design and innovation. By addressing economic barriers, shifting consumer behavior, and tackling systemic challenges, we can unlock the immense potential of green products to reshape our world. Imagine a future where products are designed for disassembly and reuse, where materials flow in a closed loop, and where innovation delivers solutions that benefit both people and the planet. This vision is within reach, but it requires a collective effort from businesses, governments, consumers, and researchers. By investing in education and awareness, we empower consumers to make informed choices. By collaborating across industries and disciplines, we foster innovation that transcends traditional limitations. And by holding both ourselves and others accountable, we ensure that progress remains on track. Green product design is not just a trend; it's a crucial shift towards a more sustainable future. By prioritizing the environment throughout the product lifecycle, green design offers numerous benefits for individuals, businesses, and the planet as a whole. As we face growing environmental challenges, green product design has the potential to play a transformative role in creating a more sustainable and responsible future. The journey towards a sustainable future is not without its challenges, but the potential rewards are immeasurable. By embracing green product design and innovation, we can cultivate a world where prosperity and environmental responsibility go hand in hand, leaving a legacy of a healthy planet for generations to come.

REFERENCES

Allenby, B., & Fullerton, A. (1991). Design for Environment: A new strategy for environmental management. *Pollution Prevention Review*, 2(1), 51–61.

Batwara, A., Sharma, V., Makkar, M., & Giallanza, A. (2022). An Empirical Investigation of Green Product Design and Development Strategies for Eco Industries Using Kano Model and Fuzzy AHP. *Sustainability (Basel)*, 14(14), 8735. 10.3390/su14148735

Braungart, M., & McDonough, W. (2009). *Cradle to cradle: Remaking the way we make things*. Vintage Books.

Ciccullo, F., Pero, M., Caridi, M., Gosling, J., & Purvis, L. (2018). Integrating the environmental and social sustainability pillars into the lean and agile supply chain management paradigms: A literature review and future research directions. *Journal of Cleaner Production*, 172, 2336–2350. 10.1016/j.jclepro.2017.11.176

Dangelico, R. M. (2016). Green Product Innovation: Where we are and Where we are Going. *Business Strategy and the Environment*, 25(8), 560–576. 10.1002/bse.1886

De Luca, A. I., Iofrida, N., Leskinen, P., Stillitano, T., Falcone, G., Strano, A., & Gulisano, G. (2017). Life cycle tools combined with multi-criteria and participatory methods for agricultural sustainability: Insights from a systematic and critical review. *The Science of the Total Environment*, 595, 352–370. 10.1016/j.scitotenv.2017.03.28428395257

Dell. (2022). *Concept Luna*. https://www.dell.com/en-us/blog/concept-luna-whats-next/

Durif, F., Boivin, C., & Julien, C. (2010). In search of a green product definition. *Innovative Marketing, 6*(1).

Eyring, G. (1992). *Green Products by Design: Choices for a Cleaner Environment*. Diane Pub Co.

Fairphone. (2023). https://www.fairphone.com/en/

Ghazali, I., Abdul-Rashid, S. H., Dawal, S. Z. M., Irianto, I., Herawan, S. G., Ho, F.-H., Abdullah, R., Abdul Rasib, A. H., & Padzil, N. W. S. (2023). Embedding Green Product Attributes Preferences and Cultural Consideration for Product Design Development: A Conceptual Framework. *Sustainability (Basel)*, 15(5), 4542. 10.3390/su15054542

Graedel, T. E., Comrie, P. R., & Sekutowski, J. C. (1995). Green Product Design. *AT & T Technical Journal*, 74(6), 17–25. 10.1002/j.1538-7305.1995.tb00262.x

Hong, Z., Wang, H., & Gong, Y. (2019). Green product design considering functional-product reference. *International Journal of Production Economics*, 210, 155–168. 10.1016/j.ijpe.2019.01.008

Hu, J., Li, X., Wang, N., & Jiang, B. (2022). Green Product Design. In N. Wang, Q. Jiang, B. Jiang, & Z. He, *Enterprises' Green Growth Model and Value Chain Reconstruction* (pp. 155–183). Springer Nature Singapore. 10.1007/978-981-19-3991-4_7

Iannuzzi, A. (2016). *Greener Products*. CRC Press. 10.1201/b11276

IKEA. (2023). *KUNGSBACKA white kitchen guide*. IKEA. https://www.ikea.com/au/en/rooms/kitchen/

Khan, S. J., Dhir, A., Parida, V., & Papa, A. (2021). Past, present, and future of green product innovation. *Business Strategy and the Environment*, 30(8), 4081–4106. 10.1002/bse.2858

Kuo, T.-C., Huang, S. H., & Zhang, H.-C. (2001). Design for manufacture and design for 'X': Concepts, applications, and perspectives. *Computers & Industrial Engineering*, 41(3), 241–260. 10.1016/S0360-8352(01)00045-6

Leigh, M., & Li, X. (2015). Industrial ecology, industrial symbiosis and supply chain environmental sustainability: A case study of a large UK distributor. *Journal of Cleaner Production*, 106, 632–643. 10.1016/j.jclepro.2014.09.022

Lush Retail. (2024). *Lish shampoo bars*. Lush. https://www.lush.com/us/en_us/c/shampoo-bars

Notpla. (2023). *Notpla Disappearing Packaging*. Notpla. https://www.notpla.com/

Patagonia. (2023). *Gear for a good time and a long time*. Patagonia. https://wornwear.patagonia.com/

Patagonia. (2024). *Black Hole® Bags*. Patagonia. https://www.patagonia.com/shop/gear/bags/black-hole

Philips. (2024). *Philips Hue, smart home lighting made brilliant*. Phillips. https://www.philips-hue.com/en-in

Rau, H., Wu, J.-J., & Procopio, K. M. (2023). Exploring green product design through TRIZ methodology and the use of green features. *Computers & Industrial Engineering*, 180, 109252. 10.1016/j.cie.2023.109252

Sharma, A., & Iyer, G. R. (2012). Resource-constrained product development: Implications for green marketing and green supply chains. *Industrial Marketing Management*, 41(4), 599–608. 10.1016/j.indmarman.2012.04.007

Sharma, S., Gururani, S., & Sarkar, P. (2023). Measuring ideation effectiveness in bioinspired design. *Artificial Intelligence for Engineering Design, Analysis and Manufacturing*, 37, e14. 10.1017/S0890060423000070

Sharma, S., & Sarkar, P. (2019). Biomimicry: Exploring Research, Challenges, Gaps, and Tools. In Chakrabarti, A. (Ed.), *Research into Design for a Connected World* (Vol. 134, pp. 87–97). Springer Singapore. http://link.springer.com/10.1007/978-981-13-5974-3_810.1007/978-981-13-5974-3_8

Shoda, N. (2013). *Barriers to Sustainable Beverage Packaging*. California Polytechnic State University.

Tseng, M.-L., Chiu, A. S. F., Tan, R. R., & Siriban-Manalang, A. B. (2013). Sustainable consumption and production for Asia: Sustainability through green design and practice. *Journal of Cleaner Production*, 40, 1–5. 10.1016/j.jclepro.2012.07.015

Uang, S.-T., & Liu, C.-L. (2013). The Development of an Innovative Design Process for Eco-efficient Green Products. In Kurosu, M. (Ed.), *Human-Computer Interaction. Users and Contexts of Use* (Vol. 8006, pp. 475–483). Springer Berlin Heidelberg. 10.1007/978-3-642-39265-8_53

Waage, S. A. (2007). Re-considering product design: A practical "road-map" for integration of sustainability issues. *Journal of Cleaner Production*, 15(7), 638–649. 10.1016/j.jclepro.2005.11.026

Waldegrave, W., & Davis, S. C. (1987). POLLUTION ABATEMENT TECHNOLOGY AWARD CEREMONY. *Journal of the Royal Society of Arts*, 135(5372), 603–608.

Wang, M.-C. (2016). Development of An Innovative Design Process for Green Products. *Proceedings of International Conference on Artificial Life and Robotics*, (vol. *21*, 108–111). IEEE. 10.5954/ICAROB.2016.OS1-5

Yi-Fei, G. (2017). Green Innovation Design of Products under the Perspective of Sustainable Development. *IOP Conference Series. Earth and Environmental Science*, 51, 012011. 10.1088/1742-6596/51/1/012011

Zhuo, L., & Shengxue, Y. (2010). A Research on Green Product Design Process and Evaluation. *2010 3rd International Conference on Information Management, Innovation Management and Industrial Engineering*, 612–614. 10.1109/ICIII.2010.466

Chapter 6
Green Human Resource Management:
Revealing the Route to Environmental Sustainability

Precious T. Okunhon
University of Northampton, UK

Adejoke Ige-Olaobaju
University of Northampton, UK

ABSTRACT

Green Human Resource Management (GHRM) is an emerging field that focus on integrating HRM functions and practices to support business sustainability. Adopting Green Human Resource Management (GHRM) practices not only contributes to environmental sustainability but also amplifies employee engagement, providing organisations with a distinct competitive advantage. This chapter highlights the importance of GHRM in providing sustainable environmental solutions. This exploratory chapter will provide an introduction to GHRM and its initiatives in promoting environmental sustainability, it will further explore the concepts of GHRM, its advantages, challenges, and limitations. Additionally, the chapter will extensively examine strategies and practices integral to GHRM that contribute to environmental sustainability. Subsequently, recommendations for ensuring the successful implementation of GHRM practices and agenda for future academic research will be discussed. The conclusion of this study will highlight ways in which employees can actively foster and promote sustainable practices.

DOI: 10.4018/979-8-3693-2595-7.ch006

INTRODUCTION

It has become evident among academic scholars in the field of Human Resource Management (HRM), that there is a growing emphasis on incorporating sustainable practices into business strategies and operations. Green Human Resource Management (GHRM) is a growing field of enquiry that represents the notion of applying HRM functions and practices to promote business sustainability (Yong et al., 2020; Zaid, et. al., 2018). The process of utilising green initiatives within people management promotes sustainable and justifiable use of organisational resources. It allows Human Resource (HR) to align its practices and functions while advancing economic and sustainable practices that utilises individual employee's interface to advance environmentally sustainable practices. While previous studies illustrate that GHRM leads to the development of organisational sustainability through the adoption of green practices such as virtual working, job sharing, reduction of paperwork, embracing digitalisation, recycling, online events, providing an energy-efficient work environment, others argue that although these practices sound like little steps, they contribute to sustainable practices such as reduction of employee carbon footprints, lowering costs and environmental protection, which promotes the organisation's goals of environmental sustainability (Ababneh, 2021; Yong et al., 2020; Zaid et. al., 2018; Rani et al., 2014). The above explanation implies that GHRM involves promoting environmentally friendly behaviours among employees, as well as implementing sustainable policies and practices within the organisation. It can therefore be argued that by fostering a culture of environmental awareness and responsibility, organisations would not only reduce their environmental impact but also improve their reputation among customers, investors, and other stakeholders. This study proposes that GHRM offers a pathway to environmental sustainability by integrating green practices into HR processes and fostering a culture of environmental responsibility within organisations. It argues that by aligning HR strategies with sustainability goals, businesses can play a vital role in creating a more sustainable future for our environment (Ababneh, 2021; Yong, et al., 2020; Rani et al., 2014).

To guarantee the alignment of employees' goals within their organisations while facilitating the delivery of sustainable returns to business stakeholders, HR needs to be strategic by conforming to the legal and institutional frameworks. HR needs to direct the focus of businesses to sustainable practices, which empowers stakeholders to contribute to environmental sustainability (Faisal 2023; Lashari et al., 2022). Although previous studies (Daily & Huang 2001; Renwick et al., 2013; Subramanian 2022) argue that the role of HR is vital to the successful development of environmental sustainability. Yet, this can only be achieved if the right people with the right orientation, needed skills and abilities are present within the organisation. Apart from the benefits to the environment, embracing GHRM practices is an additional benefit for organisations because, it improves employee's engagement and gives organisations more competitive advantage over their competitors (Faisal 2023; Mandip 2012). According to the United Nations Environment Program (UNEP,2019), despite the guidelines and agreements among governmental bodies to institute legal and administrative frameworks for ensuring sustainable practices, there is still a huge lack of coordination, corrupt practices, and poor implementation of laid down environmental laws by businesses and their stakeholders. The United Nations, therefore, calls for more examination of business variables such as their culture, values, and missions to echo behaviours and processes that focus on environmental sustainability (UNEP,2019). This explanation suggests that the fundamental assumptions to improving environmental sustainability is GHRM. This study seeks to highlight the importance of GHRM in the provision of environmentally sustainable solutions. It corroborates the arguments by previous studies, which states that the above can be achieved by employing interventions such as green human resource management practices that emphasise the need for organisations to re-evaluate their culture, strategy, and values. It emphasises employees' positive behaviours and attitudes regarding environmental challenges and develops approaches that can be taken towards achieving sustainable practices for the organisation and environment at large (Lashari et al 2022; Masri and Jaaron, 2017; Menezes et al., 2017; Raut et al., 2017; Zaid et al., 2018). The upcoming segments of this exploratory chapter aim to present a comprehensive understanding of Green Human Resource Management (GHRM) concepts, elucidate its advantages for both organizations and the environment, and delve into the challenges and limitations associated with implementing GHRM initiatives. Additionally, the chapter will extensively examine strategies and practices integral to GHRM that contribute to environmental sustainability. Subsequently, recommendations for ensuring the successful implementation of GHRM practices and agenda for future academic research will be discussed. The conclusion of this study will spotlight ways in which employees can actively foster and promote sustainable practices for both organisations and the environment.

LITERATURE REVIEW: THE CONCEPT OF GREEN HUMAN RESOURCE MANAGEMENT

Extensive research among academic studies has delved into the concept of GHRM. While some scholars define it as the fusion of environmental management with Human Resource Management, others view it as utilizing every employee's interaction to advance sustainable practices, thereby enhancing employee awareness and dedication to sustainability related issues (Faisal 2023; Ababneh; Jabbour et. al., 2008; Paillé, 2014). This explanation emphasises the capacity of human resources to leverage employee engagements towards the promotion of environmental sustainability. Recent advancements in GHRM have shed light on various indicators of green practices in the workplace, including virtual work arrangements, job sharing, reduced paper usage, digitalization, recycling initiatives, and the provision of energy-efficient workspaces that safeguards and protects the environment (Faisal 2023; Lashari et al., 2022). Findings from these studies reveal that GHRM initiatives offer substantial benefits to organisations by boosting employee engagement and helping employees gain competitive edge in their respective fields. Based on the above-mentioned insights, this study posits that academic scholars widely agree that the term "green" pertains to the natural environment and the need to keep it 'green'.

A green organisation can therefore be described as a workplace that is environmentally conscious and efficiently managed in terms of its resources and social accountability. When discussing environmental matters, the concept of green management for sustainable development has different interpretations among scholars. However, they all aim to emphasize the need for harmonious balance between organisational growth and wealth creation while protecting the natural environment to ensure future prosperity (Daily and Huang, 2001). Prior to the introduction of GHRM, an organisation's economic performance was the sole measure of corporate success for both the organisation and its shareholders but, this is no longer considered satisfactory. It has become more evident that financial gains must now be accompanied by efforts to minimize environmental impact and enhance awareness of social and environmental considerations (Lee, 2009; Banerjee, 2001; Lee and Ball, 2003). This indicates that green management involves managing an organisation's interactions with, and effects on, the environment (Lee and Ball, 2003). This goes beyond mere compliance with regulations to incorporating proactive measures aimed at preventing pollution, product stewardship, and promoting corporate social responsibility. Embracing sustainability in business involves the conservation of natural resources and the integration of eco-friendly practices into business activities. This shift towards environmentally conscious practices not only benefits the planet, but also positively impacts on employees' work-life. The key point to note is that organisations are increasingly recognizing the importance of green initiatives due to the detrimental effects of traditional industrial practices on the environment and most especially, its effects on climate change which consequently results in the need to implement green practices that mitigates the negative impacts of these practices.

Adopting green business practices offers numerous benefits, such as enhancing brand image and reputation, increasing customer loyalty, and accessing new markets. This implies that by implementing green practices, organisations can effectively promote environmental sustainability and engage their employees in a positive attitude and behaviour (Bangwal and Tiwari, 2015; Paille et al. 2013; Khateeb et. al., 2023). Although the benefits of green business practices are evident, many organisations still face challenges in implementing them. It has become apparent that some organisations struggle to see the feasibility of adopting green practices, while others are concerned about the financial implications of implementing GHRM practices. This suggests that resistance from key stakeholders could pose a threat to the implementation of GHRM practices and as such, to overcome these challenges, organising trainings and workshops to enlighten stakeholders and gaining their support and collaboration while introducing green initiatives is important. This is because, it would improve employee's attitudes and enhance their performance thereby, resulting in cost savings and higher retention rates. Within the context of the organisation and its relevant stakeholder, the role of government regulations is crucial especially in enforcing compliance of organisations with environmental standards (Rani and Mishra, 2014). The key point to note from the above explanation is that GHRM needs to align with HR policies and practices to achieve sustainability goals. It encourages environmentally responsible actions within organisations and most especially, among employees (Bangwal et. al., 2017; Margaretha et. al., 2013).

Previous studies have highlighted the role of HRM policies and practices in promoting sustainable resource utilization within organisations especially, while advocating for green environment (Marhatta et. al. 2013; Zoogah 2011). Green HRM is seen to develop environmentally conscious employees for the benefit of individuals, teams, society, the environment, and the organisation (Opatha et. al, 2014). Several studies suggest that aligning HR practices with sustainability goals helps promote green culture within organisations because it contributes to long-term success (Mandip, 2012; Cherian and Jacob 2012; Bangwal et al., 2018). This study argues that GHRM is essential for organisations because, it helps reduce environmental impact and contributes to a more sustainable future by implementing eco-friendly practices. It enhances organisation's brand image and reputation by attracting customers, investors, and employees who value environmental responsibility. Also, it improves employee morale and engagement by fostering a sense of pride and purpose among them. The key point to note from the above explanations is that GHRM is essential in supporting environmental issues within organisations. It involves implementing HR policies and practices, training employees in environmental protection, and promoting sustainable practices to reduce carbon footprints. It also helps preserve knowledge capital by engaging employees in eco-friendly practices, which creates environmentally friendly and responsive organisation. The next section will examine how organisations incorporate GHRM practices into various HR processes, such as recruitment and selection, training and development, performance management and work-life with the view to promoting environmental sustainability and reduce cost.

GHRM Practices and Strategies Needed to Promote Environmental Sustainability

Academic scholars argue that GHRM practices positively influence employee's ability to engage in green HR behaviours and conducts which, promotes organisation's green goals (Faisal, S., 2023; Dumont et al., 2016). This implies that green organisations through their practices can influence their employee's behaviours such that they reflect upon and have a reconsideration about their job requirements and likely constraints with the view to improving their level of skills and capabilities on their role (Laschinger et al., 2004). Several studies have advocated that employees' feeling of engagement and empowerment increases their commitment to contribute to GHRM practices within their organisation (Kitazawa et al., 2000; Simpson et. al., 2010). Few academic scholars who considered this viewpoint in their studies affirmed that empowered employees feel internally motivated. This translates to positive work-related outcomes such improved work quality, higher level of commitment, self-efficacy, and job satisfaction (Muogbo, 2013; Gutowski et al., 2005). It shows that individual green behaviour encompasses a wide range of behaviours that reflect both in-role (compliance with environmental policies and HRM green initiatives) and extra-role activities (going above and beyond expectations) thereby promoting organisational citizenship behaviours. It proposes that employees are empowered to perform routine tasks and take innovative or pioneering approaches towards environmental stewardship such as engaging in energy-efficient practices and identifying cost-saving opportunities (Dumont et. al.,2017).

Several researchers have proposed theoretical models linking HRM core functions to employee green behaviour in organisations (Jabbour et al. 2015; Pham, Tuckova, and Jabbour 2019; Renwick et al. 2016). These models suggest a direct relationship between HRM practices and individual green behaviour. HRM practices such as recruitment, training, and employee participation have also been examined in relation to organisational outcomes. For example, the Ability-Motivation-Opportunity (AMO) theory suggests that HRM practices enhance individual green behaviour by improving employee skills, motivation, and opportunities (Appelbaum et al. 2000). Studies that have used the AMO theory to examine the significant components of GHRM practices reveal the extent, which it helps in the development of employees' green ability through employee training and development; motivation of employees to green performance management; and provision of green opportunities, aimed at promoting employee engagement and involvement (Appelbaum et al., 2000; Guerci et al., 2016; Masri and Jaaron, 2017; Pinzone et al., 2016; Renwick et al., 2013).

Findings from the above-mentioned studies highlights the extent, which the application of GHRM components enhances employees' ability, motivation, and opportunity to participate in organisation's green activities and effectively contribute to environmental sustainability. In view of all that has been mentioned so far, one may suppose that to effectively promote GHRM practices in organisations, the role of green organisational culture is paramount. This is because, the promotion of green culture in organisations help create a safe and healthy work environment, which supports employee's health and wellbeing (DuBois et. al., 2012; Daily et. al., 2009; Becker, 1998). It can therefore be argued that when the concern for environment is embedded within an organisations' culture, it has the potential to positively influence employee's behaviour and promote organisations' green goals and objectives. The role of HR within this context is to ensure that the organisation's green goals and objectives are effectively communicated, and regular feedback provided to employees about their environmental performances with the view to acknowledge and celebrate innovative things done by employees and also, introduce new measures that would lead to higher levels of employee improvement, participation, inspiration, and involvement (Daily, et al., 2001; Govindarajulu & Daily, 2004; Renwick, et al., 2008; Amrutha & Geetha, 2020; Shafaei et. al., 2020).

It can be argued that green employee training and development needs to be preceded by green job design and analysis, which reveals the extent, which environmentally sustainable organisations are concerned with the creation of new job roles and positions that specifically focus on managing the environmental aspect of their organisation. This foundation forms the basis for job descriptions, which specifies environment-related job duties, detailed information about the role, responsibilities, and requirements. This leads to employer branding because, environment- aware applicants would be encouraged to apply for the advertised roles because it matches their individual values and aspirations to work with environment-friendly organisations hence, allowing suitable applicants to apply for the job(Faisal, S., 2023; Masri & Jaaron, 2017; Renwick, et al., 2013; Guerci, et al., 2016; Shah, 2019; Ansari, et al.,2021; Islam, et al., 2020).

Within the context of an organisation, the aim of training is to educate employees on regulatory and technical standards. It helps them to develop new skills to meet the set standards within their organisation. Incorporating green training in organisations can be a bit challenging as different approaches and various methods are likely to be adopted to achieve organisational green success. A recent study by Daily et al. (2009) found that promoting green training within an organisation is essential for managing environmental effectiveness and performance. According to their study, green training helps educate and raise awareness among employees about environmental issues and equips them with the necessary skills to address these issues in practice. Similar studies that highlight the impact of green training to environmental effectiveness and performance recognise the significant influence of HRM in motivating and inspiring employees (Sarkis et al., 2010; Jabbour et al., 2010).

Fernandez et al. (2003) emphasize that successful implementation of green practices requires environmental awareness and process knowledge among individual employees, which can be achieved through integrating training and development of green HR practices. Their study found a positive correlation between the level of employee green training, environmental effectiveness, and performance. Some examples of green training initiatives considered include Siemens in Germany, which provides daily green training during working hours for employees working with hazardous substances, resulting in 5,000 suggestions for new initiatives. Another example is Imperial Chemical Industries UK, which conducts integrated pollution control training for operators, as well as introductory green training sessions for managers and supervisors. Likewise, Rolls-Royce, Albion Group, and Bristol-Myers Squibb offer environmental training to new and existing employees. Together, these studies provide important empirical insights into the green employee training. In similar vein, the Chartered Institute of Personnel and Development (CIPD) and Klynveld Peat Marwick Goerdeler (KPMG) survey report found that 42% of UK companies provide green training to enhance environmental effectiveness and performance, and reduce environmental degradation (Phillips, 2007; Bangwal et al., 2015).

The key point to note from the above explanation is that green employee training and development is aimed at enhancing employees' knowledge and ability to successfully implement HR practices that takes into consideration, environmental sustainability (Islam, et al., 2020; Jabbour, 2015; Paille et al., 2013). This implies that irrespective of the job task or responsibilities, every employee within the organisation should be encouraged to attend green training with the view to promote individual and collective reflection of their day-to-day activities and its effect on the environment (Renwick, et al., 2013; Jackson, et al., 2011). To measure employees' green activities and output in the long term, green performance management is essential. This is because it would cover environmental related criteria and responsibilities, which enables employees liable to engage positively with environmental issues that advances their environmental performance (Islam, et al., 2020; Renwick, et al., 2013). Taken together, the explanations presented in this chapter indicates the need for HR managers to set specific and realistic green goals and targets, which would be effectively communicated to their team in such a manner that line managers and supervisors are made accountable for deviation of their team from the organisation's set sustainable environmental targets.

The implementation of GHRM as a strategic tool assists organization in achieving their environmental objectives because, it fosters a culture of environmental sustainability and cultivates environmentally conscious employees (Kim et al., 2019; Paillé et al., 2014). While prioritizing environmental protection may increase the workload for employees, it is widely believed that organizations should prioritize environmental stewardship (Chan et. al., 2010). This aligns with the core objectives of GHRM, which aim to promote environmental protection through initiatives that minimize negative impacts and amplify positive contributions to the environment. The study conducted by Chan and Hawkins (2010) shows that when employees are committed to protecting the environment and strive to create a healthier, safer planet, they perceive their actions as making a positive difference. By emphasizing shared environmental goals and values, GHRM fosters a work environment where both employees and employers feel a sense of purpose and contribution because it fosters a sense of meaningfulness in work and ultimately leads to higher levels of job satisfaction (Chan and Hawkins, 2010; Renwick, et al., 2013; Tang et al., 2018; Hanna, et al., 2000; Shafaei et. al., 2020).

Integrating environmental values in recruitment, training, and performance management, organisations can foster a culture of sustainability (Jackson et al. 2011; Yen et al., 2013). Customized training programs can enhance employees' knowledge and awareness of environmental issues (Dumont et. al., 2017). Performance appraisal systems that recognize green behaviours can positively impact on employee's motivation and engagement (Jabbour et. al., 2015; Pinzone et. al., 2016). Employee involvement in green initiatives is expected to improve individual and organisational performance (Jabbour et. al., 2015). While many studies have focused on the impact of individual HRM practices on green behaviour, a holistic approach that considers the combined effect of these practices may provide a better understanding of employees' attitudes and behaviours hence highlighting the need for employees' green work-life (Snape and Redman 2010). The concept of green work-life refers to employees' integration of environmental values, attitudes, and behaviours into both professional and personal life. It involves aligning and balancing eco-friendly practices that promotes environmental sustainability (Muster and Schrader, 2011). The role of GHRM is to encourage environmentally responsible behaviour in all areas of life, thereby, presenting a new approach to employees' environmental consciousness and positive outcomes (Muster and Schrader, 2011). This is because every individual has a unique lifestyle and specific practices that impact on the environment differently (Chuah et al., 2021). The study conducted by Edwards et. al., (2000) illustrates the significant connection that exist between an individual employee's work life and personal life. The findings from their study indicates that it is insufficient to focus solely on employee's green practices in the workplace. It is therefore crucial to recognize and address the interplay that exist between an employee's role at work and in their role in their personal life (Bangwal, et. al., 2017). In a similar study, Muster et. al., (2011) assert that environmental sustainability can only be achieved by incorporating green practices into both the professional and personal lives of individual employees. Collectively, these studies argue that the adoption of environmentally friendly HRM practices is essential for promoting a green work-life balance.

RECOMMENDATIONS

Some essential steps that must be taken to ensure effective implementation of GHRM practices include. Firstly, having a clear environmental GHRM vision. Secondly, providing employees with training to align their goals and visions with their individual and organisational environmental GHRM goals and objectives. Thirdly, assessing the environmental performance of employees is essential because it provides opportunity for tracking progress and promoting good sustainability citizenship behaviour. Finally, recognizing and rewarding employees for their environmental activities is key to sustaining employee motivation and engagement (Daily et. al., 2001; Claire et al., 1996).

Managers need to support the training, empowerment, and rewards process of GHRM, because it aligns with the core characteristics of the job and promotes skill variety, task identity, and task significance. It establishes shared environmental goals and provides training to enhance employees' environmental awareness. Empowering employees to take ownership of their environmental goals and activities gives them a sense of self-sufficiency and independence. Employees ability to evaluate, recognize, and reward their environmental performance enables them to see the concrete influence of their efforts on the environment. Overall, GHRM provides core job characteristics that help employees improve their work behaviour by fostering a sense of purpose and ownership in achieving environmental objectives. This explanation implies that GHRM's ability to shape employees' green work behaviour has significant influence on their personal life thereby, echoing the need to integrate environment conscious practices into all aspects of employees work and life (Bangwal, et. al., 2017; Muster and Schrader, 2011).

The above-mentioned outline of GHRM emphasizes the collaborative effect of combining GHRM practices (Paauwe 2009). For example, training and recruitment practices may influence individual ability, while performance management and remuneration systems may shape employee motivation. The findings from Tang et al., (2018) empirical study reveals how green employee involvement can be enhanced. According to their study, a well-defined green vision would help guide employees and monitor their activities especially when dealing with environmental related aspects of their employment, such as green learning and climate change. Employees involvement in green problem-solving activities during manufacturing and production of goods and services is essential (Yong, et. al., 2020; Chaudhary, 2019). Management's ability to effectively communicate green objectives and encourage employees to lead and be empowered while carrying out their duties at work is important because it enhances employee engagement and commitment (Faisal, S., 2023; Daily, et al., 2001).

FUTURE RESEARCH AGENDA

Potential future research in this field could involve crafting standardized Human Resource Management (HRM) metrics that enhance our understanding and establish a benchmark for organizations to gauge their progress toward sustainable practices. These metrics should encompass the core Green Human Resource Management (GHRM) initiatives discussed in previous sections of this study. To develop such metrics, an in-depth investigation into the commitment and opinions of organizational stakeholders towards sustainable initiatives is crucial. This exploration would effectively identify variables that may moderate the relationship between GHRM and environmental sustainability. Further research endeavours could explore the impact of GHRM initiatives on organizational behaviour, by employing theoretical perspectives such as Organizational Citizenship Behaviour (OCB) and the AMO theory to establish a connection between GHRM practices and employee responses to these initiatives. Consequently, these insights could serve as a foundation for developing sustainable practices that benefit both the organization and the environment. Considering the critical evaluation undertaken in this exploratory chapter, there is a compelling need for an overarching model that scrutinizes the translation of GHRM initiatives into practice and their sustainable outcomes for both the organization and the environment.

Future research endeavours could also delve into the influence of employee engagement on the integration of green culture within organizations. There is potential for specific investigations into strategies aimed at enhancing employee commitment and awareness of environmental sustainability. Leveraging the psychological aspects of employee engagement could be a key focus to facilitating the design of effective HR interventions that fosters a green culture in the workplace. Additionally, exploring the role of organizational leadership in promoting sustainable initiatives is crucial. A comparative study contrasting organisations dedicated to sustainable green practices with those that are not could provide valuable insights into leadership styles and behaviours conducive to cultivating a green workplace culture. This exploration may further illuminate the overarching impact on other organizational variables that contribute to environmental sustainability. Also, taking a global perspective, future studies might delve into variations across different cultural and institutional backgrounds, hereby contributing to a better understanding of regional perspectives and approaches to environmental sustainability. Furthermore, considering the rapid advancements in technology, future research could explore the integration of technology into the Green Human Resource Management (GHRM) process. Employing current digital tools such as data analytics, artificial intelligence, and other technological innovations could empower human resource departments to streamline processes and enhance the effectiveness of sustainable practices within organizations.

CONCLUSION

The field of GHRM has seen a rise in empirical research, and this paper has provided a comprehensive overview of the existing literature on the subject. GHRM pertains to the set of organizational policies, practices, and systems that encourage employees to engage in environment-friendly behaviours that has positive impacts on all stakeholders, including individuals, society, businesses, and the environment. By promoting GHRM, employers and practitioners can establish a crucial connection between employee participation in environmental initiatives and the resulting improvement on environmental performance. Employees can contribute in various ways, such as by promoting efficient use of resources and energy, reducing pollution, managing waste, and promoting recycling. These efforts not only help enhance workplace health and well-being but also minimize environmental harm by reducing waste thereby, enabling employees to play a significant role in enhancing organizational performance and promoting environmental sustainability (Xie, et. al., 2023; Faisal, 2023).

GHRM is a versatile tool that can impact all facets of an organization, yet concerns persist regarding its practicality and fairness. Determining the tangible effects of GHRM implementation remains a challenge. Like any organizational change, introducing GHRM is a complex process that demands unwavering commitment from management and employees at every level. Employee reluctance is common, particularly when there is a lack of understanding or motivation, thus making successful implementation a discouraging task (Al-Romeedy, 2019).

Some of the challenges that may be encountered when implementing GHRM in organisations include ability to recruit environmentally conscious employees, changing employee behaviour, and fostering pro-environmental culture within the organisation. Lack of proper planning and management support of GHRM activities could result in its failure. Other challenges include insufficient information, lack of management support, high costs of implementing GHRM practices and absence of a clear implementation plan and ambiguous green values within the organization, lack of incentives for promoting environmental behaviour, managerial willingness and capability, and financial constraints (Islam et al., 2023; Deshwal, 2015; Fayyazi et al., 2015; Bombiak, 2020; Ren et al., 2018; Likhitkar & Verma, 2017; Jafri, 2012; Kodua et al., 2022). The lack of green education at universities further hinders the establishment of green practices in organizations (Brockett, 2007).

Despite these challenges, successful GHRM implementation is achievable. The role of HRM in fostering green organizational culture is crucial as this requires a supportive work environment and the involvement of top management (Ramasamy, 2017). The importance of transformational leadership in facilitating the greening process has been emphasized (Sakwa, 2018; Roscoe et al., 2019). Academic scholars emphasise the extent to which GHRM strategies such as green recruitment and selection, green training, and green incentives promote positive environmental behaviour among employees in organisations (Khateeb, et. al., 2023; Islam et. al., 2019; Tang et al., 2018). Providing technical and management skills to employees is crucial for implementing green management systems and lastly, creating green awareness and employer branding are essential for attracting environmentally conscious employees who further the organization's green initiatives (Khateeb et al., 2023; Ahmad, 2015).

REFERENCES

Ababneh, O. M. A. (2021). How do green HRM practices affect employees' green behaviours? The role of employee engagement and personality attributes. *Journal of Environmental Planning and Management*, 64(7), 1204–1226. 10.1080/09640568.2020.1814708

Ahmad, S., 2015. Green human resource management: Policies and practices. *Cogent business & management, 2*(1), p.1030817.

Al-Romeedy, B. S. (2019). Green human resource management in Egyptian travel agencies: Constraints of implementation and requirements for success. *Journal of Human Resources in Hospitality & Tourism*, 18(4), 529–548. 10.1080/15332845.2019.1626969

Amrutha, V. N., & Geetha, S. N. (2020). A systematic review on green human resource management: Implications for social sustainability. *Journal of Cleaner Production*, 247, 119131. 10.1016/j.jclepro.2019.119131

Ansari, N. Y., Farrukh, M., & Raza, A. (2021). Green human resource management and employees' pro-environmental behaviours: Examining the underlying mechanism. *Corporate Social Responsibility and Environmental Management*, 28(1), 229–238. 10.1002/csr.2044

Appelbaum, E., Bailey, T., Berg, P., & Kalleberg, A. (2000). *Manufacturing Advantage: Why High-Performance Work Systems Pay off*. Cornell University Press.

Banerjee, S. (2001). Managerial perceptions of corporate environmentalism: Interpretation from industry and strategic implications for organizations. *Journal of Management Studies*, 38(4), 489–513. 10.1111/1467-6486.00246

Bangwal, D., & Tiwari, P. (2015). Green HRM–A way to greening the environment. *IOSR Journal of Business and Management*, 17(12), 45–53.

Bangwal, D., Tiwari, P., & Chamola, P. (2017). Green HRM, work-life and environment performance. *International Journal of Environment. Workplace and Employment*, 4(3), 244–268. 10.1504/IJEWE.2017.087808

Becker, B. (1998). High performance work systems and firm performance: A synthesis of research and managerial implications. *Research in Personnel and Human Resources Management*, 16, 53.

Bombiak, E. (2020). Barierrs to implementing the concept of green human resource management the case of Poland. *European Research Studies Journal, 23*(4).

Brockett, J. (2007). Prepare now for big rise in 'green' jobs'. *People Management*, 13(10), 9.

Chan, E. S., & Hawkins, R. (2010). Attitude towards EMSs in an international hotel: An exploratory case study. *International Journal of Hospitality Management*, 29(4), 641–651. 10.1016/j.ijhm.2009.12.002

Chaudhary, R. (2019). Green human resource management in Indian automobile industry. *Journal of Global Responsibility*, 10(2), 161–175. 10.1108/JGR-12-2018-0084

Cherian, J., & Jacob, J. (2012). A study of Green HR practices and its effective implementation in the organization: A review. *International Journal of Business and Management*, 7(21), 25–33. 10.5539/ijbm.v7n21p25

Chuah, S. C., Mohd, I. H., Kamaruddin, J. N. B., & Md Noh, N. (2021). Impact of Green Human Resource Management Practices Towards Green Lifestyle and Job Performance. *Global Business and Management Research*, 13(4).

Clair, J. A., Milliman, J., & Whelan, K. S. (1996). Toward an environmentally sensitive eco-philosophy for business management. *Industrial & Environmental Crisis Quarterly*, 9(3), 289–326. 10.1177/108602669600900302

Daily, B. F., Bishop, J. W., & Govindarajulu, N. (2009). A conceptual model for organizational citizenship behavior directed toward the environment. *Business & Society*, 48(2), 243–256. 10.1177/0007650308315439

Daily, B. F., & Huang, S. C. (2001). Achieving sustainability through attention to human resource factors in environmental management. *International Journal of Operations & Production Management*, 21(12), 1539–1552. 10.1108/01443570110410892

Deshwal, P. (2015). Green HRM: An organizational strategy of greening people. *International Journal of Applied Research*, 1(13), 176–181.

DuBois, C. L., & Dubois, D. A. (2012). Strategic HRM as social design for environmental sustainability in organization. *Human Resource Management*, 51(6), 799–826. 10.1002/hrm.21504

Dumont, J., Shen, J., & Deng, X. (2017). Effects of green HRM practices on employee workplace green behavior: The role of psychological green climate and employee green values. *Human Resource Management*, 56(4), 613–627. 10.1002/hrm.21792

Edwards, R. J., & Rothbard, P. N. (2000). Mechanisms linking work and family: Clarifying the relationship between work and family constructs. Academy of Management. *Academy of Management Review*, 25(1), 178–199. 10.2307/259269

Faisal, S. (2023). Green human resource management—A synthesis. *Sustainability (Basel)*, 15(3), 2259. 10.3390/su15032259

Fayyazi, M., Shahbazmoradi, S., Afshar, Z. and Shahbazmoradi, M., 2015. Investigating the barriers of the green human resource management implementation in oil industry. *Management science letters, 5*(1), pp.101-108.

Govindarajulu, N., & Daily, B. F. (2004). Motivating employees for environmental improvement. *Industrial Management & Data Systems*, 104(4), 364–372. 10.1108/02635570410530775

Guerci, M., Montanari, F., Scapolan, A., & Epifanio, A. (2016). Green and non-green recruitment practices for attracting job applicants: Exploring independent and interactive effects. *International Journal of Human Resource Management*, 27(2), 129–150. 10.1080/09585192.2015.1062040

Gutowski, T., Murphy, C., Allen, D., Bauer, D., Bras, B., Piwonka, T., Sheng, P., Sutherland, J., Thurston, D., & Wolff, E. (2005). Environmentally benign manufacturing: Observations from Japan, Europe and the United States. *Journal of Cleaner Production*, 13(1), 1–17. 10.1016/j.jclepro.2003.10.004

Hameed, Z., Khan, I. U., Islam, T., Sheikh, Z., & Naeem, R. M. (2020). Do green HRM practices influence employees' environmental performance? *International Journal of Manpower*, 41(7), 1061–1079. 10.1108/IJM-08-2019-0407

Hanna, M. D., Rocky Newman, W., & Johnson, P. (2000). Linking operational and environmental improvement through employee involvement. *International Journal of Operations & Production Management*, 20(2), 148–165. 10.1108/01443570010304233

Islam, M. A., Hunt, A., Jantan, A. H., Hashim, H., & Chong, C. W. (2020). Exploring challenges and solutions in applying green human resource management practices for the sustainable workplace in the ready-made garment industry in Bangladesh. *Business Strategy & Development*, 3(3), 332–343. 10.1002/bsd2.99

Islam, M. A., Jantan, A. H., Yusoff, Y. M., Chong, C. W., & Hossain, M. S. (2023). Green Human Resource Management (GHRM) practices and millennial employees' turnover intentions in tourism industry in malaysia: Moderating role of work environment. *Global Business Review*, 24(4), 642–662. 10.1177/0972150920907000

Jabbour, C. J. C. (2015). Environmental training and environmental management maturity of Brazilian companies with ISO14001: Empirical evidence. *Journal of Cleaner Production*, 96, 331–338. 10.1016/j.jclepro.2013.10.039

Jabbour, C. J. C., & Santos, F. C. A. (2008). Relationships between human resource dimensions and environmental management in companies: Proposal of a model. *Journal of Cleaner Production*, 16(1), 51–58. 10.1016/j.jclepro.2006.07.025

Jackson, S. E., Renwick, D. W., Jabbour, C. J., & Muller-Camen, M. (2011). State-of-the-art and future directions for green human resource management: Introduction to the special issue. *German Journal of Human Resource Management*, 25(2), 99–116. 10.1177/239700221102500203

Jafri, S. (2012). Green HR practices: An empirical study of certain automobile organizations of India. *Human Resource Management*, 42(4), 6193–6198.

Khateeb, F.R. and Nabi, T., 2023. Green Human Resource Management: A Review of Two Decades of Research. *Management Research & Practice, 15*(2).

Kitazawa, S., & Sarkis, J. (2000). The relationship between ISO 14001 and continuous source reduction programmes. *International Journal of Operations & Production Management*, 20(2), 225–248. 10.1108/01443570010304279

Kodua, L. T., Xiao, Y., Adjei, N. O., Asante, D., Ofosu, B. O., & Amankona, D. (2022). Barriers to green human resources management (GHRM) implementation in developing countries. Evidence from Ghana. *Journal of Cleaner Production*, 340, 130671. 10.1016/j.jclepro.2022.130671

Laschinger, H. K. S., Finegan, J. E., Shamian, J., & Wilk, P. (2004). A longitudinal analysis of the impact of workplace empowerment on work satisfaction. *Journal of Organizational Behavior: The International Journal of Industrial. Journal of Organizational Behavior*, 25(4), 527–545. 10.1002/job.256

Lashari, I. A., Li, Q., Maitlo, Q., Bughio, F. A., Jhatial, A. A., & Rashidi Syed, O. (2022). Environmental sustainability through green HRM: Measuring the perception of university managers. *Frontiers in Psychology*, 13, 1007710. 10.3389/fpsyg.2022.100771036467149

Lee, K. H. (2009). Why and how to adopt green management into business organizations: The case study of Korean SMEs in manufacturing industry. *Management Decision*, 47(7), 1101–1121. 10.1108/00251740910978322

Lee, K. H., & Ball, R. (2003). Achieving Sustainable Corporate Competitiveness: Strategic Link between Top Management''s (Green) Commitment and Corporate Environmental Strategy. *Greener Management International*, 2003(44), 89–104. 10.9774/GLEAF.3062.2003.wi.00009

Likhitkar, P. & Verma, P. (2017). Impact of green HRM practices on organization sustainability and employee retention. *International journal for innovative research in multidisciplinary field, 3*(5), 152-157.

Mandip, G. (2012). Green HRM: People management commitment to environmental sustainability. *Research Journal of Recent Sciences, ISSN,* 2277, 2502.

Margaretha, M., & Saragih, S. (2013). *Developing New Corporate Culture through Green Human Resource Practice,* International Conference on Business, Economics, and Accounting 20 – 23 March 2013, Bangkok – Thailands.

Marhatta, S., & Adhikari, S. (2013). Green HRM and sustainability. *International eJournal of Ongoing Research in Management & IT*. www.asmgroup.edu.in/incon/publication/incon13-hr-006pdf

Masri, H. A., & Jaaron, A. A. (2017). Assessing green human resources management practices in Palestinian manufacturing context: An empirical study. *Journal of Cleaner Production,* 143, 474–489. 10.1016/j.jclepro.2016.12.087

Menezes, E., Filho, S., & Drigo, E. (2017). Analysis of organisational and human factors in the local production arrangement of the hotel chain to avoid social and environmental impacts, case study of Maragogi, Alagoas, Brazil. In *Advances in Social & Occupational Ergonomics:Proceedings of the AHFE 2016 International Conference on Social and Occupational Ergonomics,* (pp. 421-433). Springer International Publishing.

Muogbo, U. (2013). The impact of employee motivation on organisational performance (A study of some selected firms in Anambra state Nigeria). *The International Journal of Engineering and Science,* 2, 70–80.

Muster, V., & Schrader, U. (2011). Green work-life balance: A new perspective for Green HRM. *Zeitschrift Fur Personalforschung,* 25(2), 140–156. 10.1177/239700221102500205

Opatha, H. H., & Arulrajah, A. A. (2014). Green Human Resource Management: Simplified general reflections. *International Business Research,* 7(8), 101–112. 10.5539/ibr.v7n8p101

Paauwe, J. (2009). HRM and performance: Achievements, methodological issues and prospects. *Journal of Management Studies,* 46(1), 129–142. 10.1111/j.1467-6486.2008.00809.x

Paillé, P., Chen, Y., Boiral, O., & Jin, J. (2014). The impact of human resource management on environmental performance: An employee-level study. *Journal of Business Ethics,* 121(3), 451–466. 10.1007/s10551-013-1732-0

Pham, N. T., Tučková, Z., & Jabbour, C. J. C. (2019). Greening the hospitality industry: How do green human resource management practices influence organizational citizenship behavior in hotels? A mixed-methods study. *Tourism Management*, 72, 386–399. 10.1016/j.tourman.2018.12.008

Phillips, L. (2007). Go green to gain the edge over rivals. *People Management*, 23(August), 9.

Pinzone, M., Guerci, M., Lettieri, E., & Redman, T. (2016). Progressing in the change journey towards sustainability in healthcare: The role of 'Green'HRM. *Journal of Cleaner Production*, 122, 201–211. 10.1016/j.jclepro.2016.02.031

Ramasamy, A. (2017). *A study on implications of implementing green HRM in the corporate bodies with special reference to developing nations.*

Rani, S., & Mishra, K. (2014). Green HRM: Practices and strategic implementation in the organisations. *International Journal on Recent and Innovation Trends in Computing and Communication*, 2(11), 3633–3639.

Raut, R. D., Narkhede, B., & Gardas, B. B. (2017). To identify the critical success factors of sustainable supply chain management practices in the context of oil and gas industries: ISM approach. *Renewable & Sustainable Energy Reviews*, 68, 33–47. 10.1016/j.rser.2016.09.067

Ren, S., Tang, G., & Jackson, E, S. (. (2018). Green human resource management research in emergence: A review and future directions. *Asia Pacific Journal of Management*, 35, 769–803. 10.1007/s10490-017-9532-1

Ren, S., Tang, G., & Jackson, E, S. (. (2018). Green human resource management research in emergence: A review and future directions. *Asia Pacific Journal of Management*, 35, 769–803. 10.1007/s10490-017-9532-1

Renwick, D., Redman, T., & Maguire, S. (2008). Green HRM: A review, process model, and research agenda. *University of Sheffield Management School Discussion Paper, 1*(1), 1-46.

Renwick, D. W., Redman, T., & Maguire, S. (2013). Green human resource management: A review and research agenda. *International Journal of Management Reviews*, 15(1), 1–14. 10.1111/j.1468-2370.2011.00328.x

Sakwa, S. M. (2018). *Factors affecting implementation of green human resource practices in the civil service in Kenya* [Doctoral dissertation, University of Nairobi].

Sarkis, J., Gonzalez-Torre, P., & Adenso-Diaz, B. (2010). Stakeholder pressure and the adoption of environmental practices: The mediating effect of training. *Journal of Operations Management*, 28(2), 163–176. 10.1016/j.jom.2009.10.001

Shafaei, A., Nejati, M., & Yusoff, Y. M. (2020). Green human resource management: A two-study investigation of antecedents and outcomes. *International Journal of Manpower*, 41(7), 1041–1060. 10.1108/IJM-08-2019-0406

Shah, M. (2019). Green human resource management: Development of a valid measurement scale. *Business Strategy and the Environment*, 28(5), 771–785. 10.1002/bse.2279

Subramanian, P. & Jaganathan, A. (2022). Promoting Environment Sustainability Through Green HRM: The Socially Responsible Organizations. *Ushus Journal of Business Management, 21*(2), 01-13.

Tang, G., Chen, Y., Jiang, Y., Paillé, P., & Jia, J. (2018). Green human resource management practices: Scale development and validity. *Asia Pacific Journal of Human Resources*, 56(1), 31–55. 10.1111/1744-7941.12147

UNEP. (2019). *Global Environment Outlook – GEO-6: Healthy Planet*. Healthy People. 10.1017/9781108627146

Xie, H., & Lau, T. C. (2023). Evidence-Based Green Human Resource Management: A Systematic Literature Review. *Sustainability (Basel)*, 15(14), 10941. 10.3390/su151410941

Yong, J. Y., Yusliza, M. Y., Ramayah, T., Chiappetta Jabbour, C. J., Sehnem, S., & Mani, V. (2020). Pathways towards sustainability in manufacturing organisations: Empirical evidence on the role of green human resource management. *Business Strategy and the Environment*, 29(1), 212–228. 10.1002/bse.2359

Zaid, A. A., Jaaron, A. A., & Bon, A. T. (2018). The impact of green human resource management and green supply chain management practices on sustainable performance: An empirical study. *Journal of Cleaner Production*, 204, 965–979. 10.1016/j.jclepro.2018.09.062

Zoogah, D. (2011). The dynamics of Green HRM behaviors: A cognitive social information processing approach. *Zeitschrift für Personalforschung*, 25(2), 117–139. 10.1177/239700221102500204

Chapter 7
Green Human Resource Management and Sustainable Performance Management

Malvern Chiboiwa
University of Hertfordshire, UK

Elizabeth Babafemi
http://orcid.org/0000-0003-4111-0225
University of Hertfordshire, UK

Felicia Momoh Oseghale
Alvan Ikoku University of Education, Nigeria

Raphael Oseghale
http://orcid.org/0000-0001-6557-9488
University of Hertfordshire, UK

ABSTRACT

This study investigates how GHRM facilitates sustainable performance management. Specifically, the study investigates the role of specific GHRM activities on responsible production/consumption, climate action, clear water/sanitation, and sustainable cities. Drawing on secondary literature, the ability, motivation, and opportunity (AMO) model of HRM and the United Nations sustainability development goals (SDG) framework, the study suggests that GHRM can bolster sustainable performance management and thus lead to the attainment of responsible production/consumption, climate action, clear water/sanitation, and sustainable cities. The study extends the

DOI: 10.4018/979-8-3693-2595-7.ch007

GHRM literature by uncovering how the SDG framework and the AMO model can interact to facilitate the development and deployment of green skills and bolster the attainment of the environmental dimension of SDGs through green leadership. The practical implications of the findings were discussed.

INTRODUCTION

Organisations play a vital role in society, such as creating employment opportunities and contributing towards societal development. However, the negative effects of their activities, such as resource depletion as well as water, air, and land pollution, should not be ignored (Ogbeibu et al., 2023). Thus, over the last few decades, organisations are increasingly facing pressure from stakeholders such as the government, non-governmental organisations, society at large, and employees to adopt environmentally friendly policies (Mousa & Othman, 2020). For example, the United Nations' (UN's) Sustainable Development Goals (SDGs) evidence the universal call to protect the planet. According to Pucik et al. (2023), climate change is arguably the most significant environmental challenge facing humanity, and organisations are facing increasing pressure to be active players in creating solutions to rather than continuing to be part of the problem. If left unchecked, climate change can affect different ecosystems through storms and other types of natural disasters. Therefore, more studies on how organisations can design and execute green practices to achieve the SDGs are required (Pucik et al., 2023).

Many studies suggest that green human resources management (GHRM) is one of the most important approaches to the achievement of the SDGs (Ogbeibu et al., 2023). Green human resources management is defined as 'a set of guidelines and initiatives that inspire environmentally focused behaviours among employees so that they can use their creativity to achieve green innovation outcomes, thus aiding the global cause to engender environmental sustainability' (Ogbeibu et al., 2020, p. 3). Traditional human resources management (HRM) tends to focus on practices that enable organisations to improve their overall business performance and generally do not emphasise the achievement of the SDGs (Ogbeibu et al., 2023). While studies on the relationship between GHRM and environmental performance ('[an] organisation's capability to decrease the waste it produces, as well as its air emission, and limiting the use of hazardous materials, while minimising the frequency of environmental accidents', Mousa & Othman, 2020, p. 4) are becoming more prevalent, some studies (Khaskhely et al., 2022; Martins et al. 2021; Mousa & Othman, 2020; Ogbeibu et al., 2023; Zaid et al., 2018) examine the relationship between GHRM and sustainable performance management. The current study defines 'sustainable performance management' as including activities that managers undertake (or process) to accomplish set sustainable development goals (Park & Krause, 2021). However, previous studies in this area examine sustainable performance (social, environmental, and economic performance) as an outcome or goal (Khaskhely et al., 2022) rather than as a process (Hernu, 2022) that leads to a sustainable outcome.

Drawing on the UN's SDGs framework (Pucik et al., 2023); the ability, motivation, and opportunity (AMO) theory of HRM; and a developmental literature review, the current study explores the influence of GHRM on sustainable performance management. In particular, it investigates the role of specific GHRM activities in relation to specific SDGs (responsible production or consumption; climate action; clear water and sanitation; and sustainable cities). It focuses on the environmental dimensions of the SDGs, as there is extant research on how HRM can facilitate the achievement of the social dimensions of SDG, such as by reducing social inequalities (Liu et al., 2022). Using the SDGs framework, the study extends the literature on AMO by highlighting how specific GHRM practices influence the attainment of specific SDGs.

The following section sets out the theoretical framework of the chapter. It outlines how the SDGs framework builds on the AMO model. In the section thereafter, GHRM practices and sustainable performance management are discussed in detail. The final section takes the form of a conclusion.

METHODOLOGY

An integrated literature review methodology is adopted in the current study to critically examine and synthesise the available literature in order to obtain new insights (Lubbe et al., 2020). The method is appropriate for this study as it has allowed the researchers to go beyond an analysis of extant research findings and provide new insights on the subject: GRHM and sustainable performance management. Below is a visual representation of the stages of the integrated literature review employed in this study.

Figure 1. An outline of the approach taken to the literature search

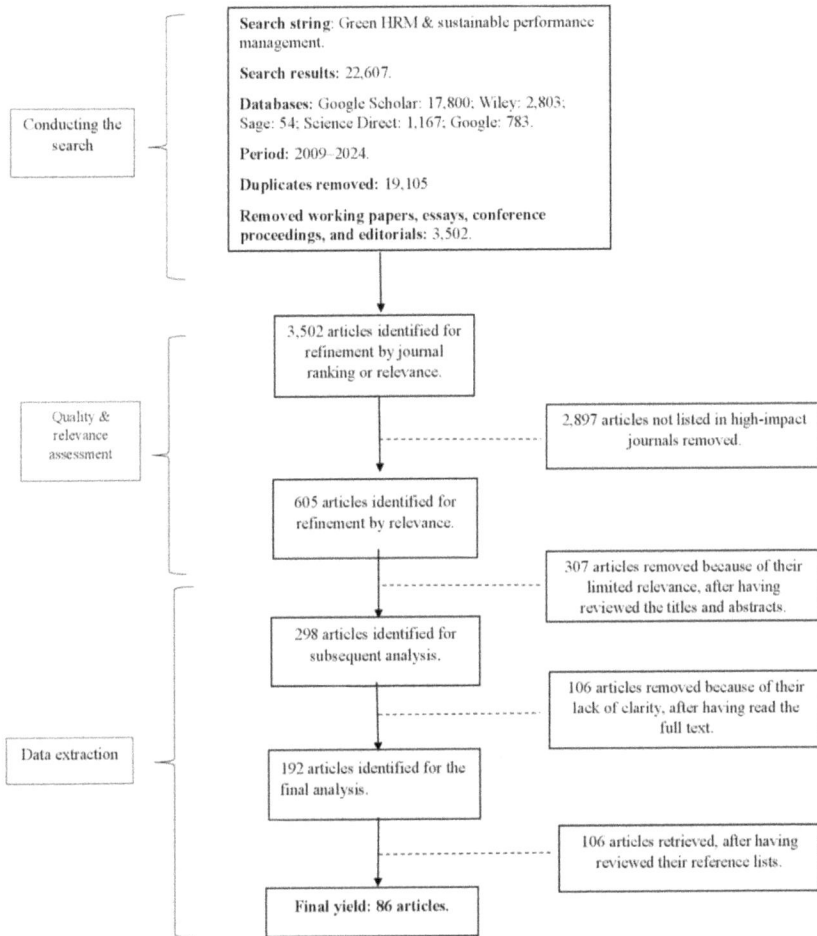

	Search string: Green HRM & sustainable performance management.
	Search results: 22,607.
Conducting the search	Databases: Google Scholar: 17,800; Wiley: 2,803; Sage: 54; Science Direct: 1,167; Google: 783.
	Period: 2009–2024.
	Duplicates removed: 19,105
	Removed working papers, essays, conference proceedings, and editorials: 3,502.

3,502 articles identified for refinement by journal ranking or relevance.

Quality & relevance assessment

2,897 articles not listed in high-impact journals removed.

605 articles identified for refinement by relevance.

307 articles removed because of their limited relevance, after having reviewed the titles and abstracts.

298 articles identified for subsequent analysis.

106 articles removed because of their lack of clarity, after having read the full text.

Data extraction

192 articles identified for the final analysis.

106 articles retrieved, after having reviewed their reference lists.

Final yield: 86 articles.

The first stage of the integrated literature review involves preparing the review plan, in which inclusion and exclusion criteria as well as keywords are determined. The keywords for the literature search were 'green human resources management' and 'sustainable performance management', consistent with the goal of the review.

Using the identified keywords, the researchers conducted a thorough search of relevant databases – online library platforms, like Google, Google Scholar, Wiley, Sage, and Science Direct. The search string focused on GHRM, the SDGs, and sustainable performance management. The search excluded material like conference proceedings, editorials, essays, and working papers. The period of our search was

2009–2024, and the search yielded 3,502 articles, which the researchers imported into Mendeley.

In the second stage, the researchers evaluated the significance and quality of these 3,502 articles. The researchers refined the results further by considering the impact level of the journal as well as the title, abstract, and relevance of each article. A comprehensive review of full articles on studies solely contributing to GHRM and sustainable performance management followed. Furthermore, the researchers assessed the relevance of each article by considering the link between GHRM and sustainable performance management. In addition, the researchers further compartmentalised GHRM and sustainable performance management in order to determine the elements on which the study should focus, such as green hiring, green training, green rewards, green performance evaluations, the performance management cycle, and sustainable performance management. The combined quality and relevance criteria led to the researchers' selection of articles for the present study. Ultimately, the researchers also checked and included articles from other studies that they deemed relevant. The integrated literature review included 86 articles, as shown in Fig. 1 above.

Theoretical Underpinnings

In 2015, the UN developed the SDGs framework as part of its 2030 agenda for sustainable development. The SDGs framework is a global blueprint on how different stakeholders, including businesses, can protect the planet and its people while creating value for customers in order to make a profit (Pucik et al., 2023). At the heart of the SDGs are the 17 goals that provide significant insights into specific areas that businesses can emphasise in addressing issues of extreme poverty, inequality, and protecting the planet. Figure 2 below shows the UN's SDGs.

Figure 2. The 17 United Nations sustainable development goals

Source: United Nations, 2024

As shown in Figure 2, the 17 SDGs cover social, environmental, and economic issues (Pucik et al., 2023). However, the present study considers only the environmental SDGs: goals 6, 7, 11, 12, and 13, as there is limited research on these goals as they relate to HRM (Ansari et al., 2021; Jabbour et al., 2010; Ogbeibu et al., 2023). Most of the research on SDG and HRM (sustainable) focuses on how HRM can facilitate the achievement of the social (and economic) dimension of SDG, such as reducing inequalities (Adisa et al., 2017; Adisa et al., 2019; Adisa et al., 2021; Aragon-Correa et al., 2013). Businesses must focus on climate change, resource depletion, and environmental degradation to address goals 6, 7, 11, 12, and 13 (environmental dimension). For example, businesses need to achieve carbon neutrality; manufacture products with zero waste; and offer significant maintainability, reparability, and reprocessing opportunities at the end of a product's life – all in order to promote clean sanitation, clean energy, sustainable cities, responsible production and consumption, and climate action (Pucik et al., 2023). While the SDG framework provides a clear roadmap in relation to the areas on which businesses should focus – addressing climate change, resource depletion, and environmental degradation – it does not explain how businesses should develop and engage employees with green skills and competencies in order to meet these goals. The current study does not depend on the SDG framework alone but rather combines it with the AMO model in order to close this research gap.

The ability, motivation, and opportunity model of HRM is mainly applied in research on GHRM and environmental performance (Chams & Garcia-Blandon, 2019; Renwick et al., 2013). According to the AMO model, the sustainability outcomes of a firm can be enhanced when three factors are present: 1. the ability of the business to develop the green knowledge and skills required to engage in the sustainable production process; 2. the necessary motivation and willingness to engage in sustainable practices driven by organisational remuneration and rewards for green behaviour and initiatives; and 3. the business's provision of opportunities for employees to engage and be involved in green and sustainable work processes and contribute to green and sustainable activities in order to foster an eco-friendly atmosphere inside and outside the organisation (Chams & Garcia-Blandon, 2019; Renwick et al., 2013).

Scholars of the AMO model note that GHRM helps create the AMO that employees require in relation to climate change, resource depletion, and environmental degradation (Chams & Garcia-Blandon, 2019; Pucik et al., 2023; Renwick et al., 2013). Ogbeibu et al. (2023) contend that proactive environmental strategies such as GHRM help businesses drive the human capital, motivation, and engagement that employees need in order to enhance the organisation's environmental performance. However, the current literature on GHRM does not extensively explore the environmental dimension of the SDG framework. This study therefore integrates the SDG framework with the AMO model of HRM to explain the relationship between GHRM and sustainable performance management.

Green Human Resource Management

Green practices help organisations achieve sustainable competitiveness and – if appropriately pursued – ensure a productive work environment that benefits the natural ecosystem (Atiku & Fapohunda; 2020; Martins et al., 2021). Organisations have become aware of the need to have a sustainable environment and are continuously transforming themselves in line with environmentally sustainable goals by adopting and implementing environmentally sustainable practices (Atiku, 2019; 2020), including GHRM, which involves embedding environmental sustainability initiatives into HRM practices. Green hiring, green training, green rewards, and green performance management are examples of such practices. Below is a discussion of how these abovementioned GHRM practices contribute towards a sustainable environment.

Green Hiring

Green hiring, sometimes also referred to as 'green recruitment', involves hiring employees who have the knowledge, skills, behaviour, and approach to engage in environmental management activities (Yong et al., 2020). For example, green hiring focuses on the recruitment of applicants who are responsive to environmental issues and keen to be proactive in enhancing a firm's environmental performance (Tang et al., 2018).

The practice of green hiring contributes to the achievement of SDGs, such as the attainment of the goals in relation to the enhancement of the sustainability of cities and communities; sustainable production; climate action; clean water and sanitation; and the preservation of life below sea level and on land (Ojo et al., 2022). Green hiring is one of the efforts that businesses make to reduce environmental degradation and promote smart cities. Research shows that forms of automated recruitment processes, such as green interviews (through platforms like Zoom, Microsoft Teams, and Skype) and online job advertisements (through platforms like LinkedIn, company websites, and other e-career portals) reduce environmental pollution as they negate the need for paper-based job applications and result in the production of fewer gas emissions (which would otherwise be produced from cars when the applicants are travelling to and from interviews) (Dahlmann et al., 2019; Adewumi & Ntshangase, 2022).

Furthermore, the evolving nature of jobs, increasingly incorporating green responsibilities, means that the recruitment and selection criteria for employees also changes. In this context, the selection of employees therefore relates to whether or not they possess green traits, green competencies, and the ability to fit into a green organisational culture. Such green competencies are the capabilities possessed by an individual in relation to how they are able to ensure their organisation's environmental sustainability, and these competencies include the prospective employees' attitudes, awareness, skills, abilities, and knowledge in relation to environmental issues (Priansa, 2016).

Having employees who have green attitudes should be a prerequisite for organisations aiming to attain the SDGs. Green attitudes focus on an individual's positive actions or feelings towards environmental performance. Green attitudes are therefore key factors that cause the pro-environmental behaviour necessary for ensuring overall environmental sustainability (Atiku, 2019; Kim et al., 2019). When hiring employees, organisations should look for candidates who display green attitudes. Selection questions can be used to measure the level at which a prospective employee displays green attitudes. This is particularly important, because it reduces the costs associated with making efforts to change employees' attitudes and behaviours in order to foster a green organisational culture.

Hiring employees with what is known as green awareness enhances environmental sustainability. It increases organisational efficiency, innovation, and environmental performance (Masri & Jaaron, 2017). According to Gadenne et al. (2009), green awareness is a competency that enables employees to think about the impact of their actions on the environment and possibly mitigate any negative impacts.

'Green skills' are the abilities, attitudes, values, and knowledge required for developing and supporting a sustainable and resource-efficient society (United Nations Industrial Development Organisation [UNIDO], 2022). Similarly, the European Centre for the Development of Vocational Training (2012) defines 'green skills' as the values, attitudes, abilities, and knowledge required to support a society intending to reduce the effects of human activity on the environment.

Furthermore, 'green knowledge' is an important attribute of environmental sustainability (Atiku, 2020). Fryxell and Lo (2003) define 'green knowledge' as the facts that one has about the natural environment and the entire ecosystem. It is concerned with one's capacity to understand environmental issues and be able to act upon them. Green hiring, if done correctly, provides the foundation upon which other green HRM practices (such as green training and development; green performance management; and green rewards) can be based. When practising green hiring, organisations should focus on identifying the candidates who possess green competencies. Their selection questions should be able to elicit the level of green competencies and understanding of the candidates. Green job descriptions and advertisements should be sufficiently written in order to attract candidates who have green competencies. Thus, organisations are increasingly aligning their job roles and job descriptions with environmental issues, and when they advertise positions, they are becoming more focused on recruiting the candidates who are the most focused on environmental issues (Chaudhary, 2018).

Green Training

Green training plays a key role in the attainment of the global SDGs, and it also gives a competitive advantage to organisations. Green training involves equipping employees with the necessary skills, abilities, and knowledge to aid in the reduction of environmental degradation, conservation of energy, sufficient resource utilisation, and the reduction of waste during the production process (Lawal & Olawoyin, 2021). Green training therefore enhances employees' knowledge and skills in relation to environmental sustainability by creating positive attitudes among employees and encouraging their participation in green initiatives. In other words, green training equips employees with the competencies necessary for attaining the sustainability objectives of an organisation. Organisations that are conscious of their green objectives undertake green training needs analyses in order to identify and establish appropriate training initiatives that will help them achieve their green objectives (Arulrajah et al., 2015). Equipping employees with green knowledge and skills is important, because employees with these competencies perform their jobs in an environmentally friendly way. Green training programmes improve employees' knowledge and understanding of the significance of addressing green issues, making them more concerned about increasing their awareness of environmental protection (Fields & Atiku, 2017; 2018; Joshi & Dhar, 2020). Green training therefore lays the foundation for a green organisational culture.

Ramlee (2015) explains that training and education in green skills are important for transforming an economy into a green economy and for ensuring the clean environment required for overall economic growth. A workforce endowed with green skills supports organisations' achievement of their sustainable goals and the green economy at large. The proliferation of green technology is contributing to the fourth industrial revolution, characterised by new green jobs. Equipping employees with green skills is thus a necessity for a sustainable green economy. The establishment of educational institutions and government initiatives are therefore crucial for cultivating the necessary green skills for a green economy.

According to Heong et al. (2016), economies are increasingly shifting to the use of green-conscious economic models in the job market. As such, new green jobs have emerged. For example, in the automobile manufacturing industry, new jobs focusing on fuel-efficient technologies have emerged. Such technology is in line with global SDGs like reducing pollution; ensuring cities and communities are sustainable; ensuring energy is affordable and clean; and promoting life on land and below sea level. These changes in the job market have resulted in changes in workforce skills. Thus, organisations need to train employees in order to improve their green skills. Training institutions also need to revise or change their curricula to include green training. In this way, the goal of making economies green can be achieved, as the labour market will have employees with green skills (Teixeira et al., 2016).

The development of employees' skills is not enough for achieving green goal. Managers of organisations should possess green leadership skills. Leaders who have green skills are able to consider and deal with sustainability issues from a strategic point of view. Consequently, their organisations will benefit from their improved competitive advantage gained through a reduction in costs, a good corporate reputation, a culture of continuous improvement, and increased employee motivation (Yang et al., 2010). Promoting environmental sustainability therefore requires the commitment of all employees, regardless of their rank. As such, it is crucial for leaders to have the necessary green skills and green leadership skills to enhance employees' commitment as well as strengthen the culture necessary for implementing environmental management initiatives (Teixeira et al., 2016; Perron et al., 2006).

Green Rewards

GHRM practices support pro-environmental behaviour, which can be defined as the routines and actions that individuals implement to promote sustainable practices and reduce their negative impact on the environment (Woo, 2021). Offering employees green rewards is one way of embedding pro-environmental behaviour in organisational culture. Green rewards are the financial and non-financial rewards given to employees in order to encourage their pro-environmental behaviour. When employees are rewarded for their pro-environmental efforts, they are likely to continue engaging in behaviours that support their organisation in attaining its environmental performance objectives (Saeed et al., 2019). According to the AMO model, employees are motivated to perform well when they are rewarded for exhibiting appropriate or valued behaviours (Rayner & Morgan, 2018). When organisations recognise and appraise employees' green behaviour as well as connect it to their pay or promotional opportunities, employees feel encouraged and become more involved in the green initiatives at their organisations (Dumont et al., 2017).

Rewarding green competencies is important for attracting, engaging, and retaining employees and for achieving sustainable objectives. Employees feel valued when they know that organisations recognise the green skills and other competencies they have acquired over the years. Renwick et al. (2013) state that organisations should consider competence-based reward schemes for employees who acquire the essential green competencies for mitigating environmental threats, workplace accidents, and illegal gas emissions. In order to attain the global SDGs, such as the promotion of smart cities and communities; the provision of affordable and clean energy; and industry, innovation and infrastructure, organisations and governments need to implement policies that reward skills that increase sustainable development. In the words of Olanipekun et al. (2017), green rewards serve as a motivational factor for stakeholders in the construction industry, such as contractors, consultants, designers, and private developers, to spearhead the construction of green buildings and infrastructures. This initiative can be seen in Australia, where the government provides incentives for the promotion of the construction of green buildings (Steinfeld et al., 2011). Likhitkar and Verma (2017) argue that employees should be given green rewards for acquiring new environmental management skills. When employees use their green skills and knowledge to innovate, they need to be rewarded with awards and bonuses for using their skills and displaying environmentally friendly behaviour (Suharti & Sugiarto, 2020). Furthermore, green reward schemes require managerial support. When managers take the lead in supporting green initiatives, employees also value the initiatives. A study undertaken by Cantor et al. (2012) reveals that not only do employees value the presence of pro-environmental rewards, but they also value managerial and organisational support. Dumont (2015) adds that managers should specifically reward employees who are engaging in green behaviours in order to demonstrate their intention to foster green culture and encourage employees to engage in green behaviours, which will contribute to the achievement of those goals.

Green Employee Involvement

Working towards the attainment of organisational sustainable development goals requires higher levels of employee involvement. Green employee involvement is when employees involve themselves in activities and actions promoting environmental sustainability. Danirmala and Prajogo (2022) define 'green employee involvement' as the provision of employees with the opportunities to learn green skills and participate in environmentally friendly strategies developed to combat environmental problems. Thus, it is a strategy used by organisations to include employees in environmental sustainability issues. Renwick et al. (2013) state that green employee involvement is a management and work practice designed to provide employees with opportunities to contribute to environmental sustainability issues.

Some researchers argue that employees who possess important knowledge and expertise regarding environmental sustainability initiatives should be highly involved in spearheading such initiatives (Tang et al., 2018; Renwick et al., 2013; Sanyal & Haddock-Millar, 2018). The achievement of green objectives relies heavily on the performance of employees, as they are responsible for implementing most environmental sustainability initiatives. Notably, employees are responsible for the daily implementation of environmental sustainability initiatives such as turning off the lights, recycling waste material, powering down electronics, and efficient resource utilisation (Boiral et al., 2015). The implementation of these activities mostly depends on employees; therefore, managers must focus on involving employees in environmental sustainability tasks (Renwick et al., 2013).

Renwick et al. (2013) argue that there are three important processes through which green employee involvement influences environmental sustainability: 1. harnessing employees' tacit knowledge, as they are responsible for implementing the production processes; 2. making contributions regarding environmental improvements; and 3. developing and maintaining a culture that supports environmental sustainability goals. Other researchers suggest several ways through which employee involvement can be encouraged, such as forming problem-solving groups, celebrating low-carbon champions, implementing work-based recycling schemes, producing and distributing green bulletins proposing eco-initiatives, establishing green teams, and instituting other formal and informal communication channels to promote green employee involvement (Tang et al. 2018; Renwick et al., 2013).

Green Performance Appraisal

Green performance appraisal is another significant GRHM practice for promoting sustainable development. According to Ardiza et al. (2021), green performance appraisal involves evaluating the progress of employees in achieving a green environment. In other words, green performance appraisals focus on assessing employees' performance based on set green standards. Anton (2016) posits that it considers the results produced by employees when they engage in actions related to greening the environment. The idea behind green performance appraisals is therefore to motivate employees to display the necessary behaviours for the attainment of green organisational goals.

Deshwal (2015) comments that organisational performance management systems should incorporate green targets in their key performance areas, which can then be decoded into green behaviour and performance standards, which act as measures of performance at all levels in the organisation. Saeed et al. (2019) state that organisations can develop green performance indicators or criteria upon which all members of the organisation can be evaluated. These indicators can include tasks related to ensuring that environmental responsibilities, including workplace incidents, are adequately addressed.

Organisations aiming to achieve green excellence conduct employee performance evaluations based on their adherence to green practices and standards (Yong et al., 2019). Essentially, these evaluations focus on measuring the extent to which green knowledge, skills, attitudes, and awareness are being used and displayed in organisations. Organisations then reward or use other interventions to reinforce the use of these competencies. In other words, green performance appraisals impact how green competencies are utilised and how green rewards are distributed in organisations. Through green performance appraisals, when employees lack green competencies, relevant green training and development can be prescribed. This approach not only motivates employees but also fosters a green organisational culture. The motivation component of the AMO model suggests that continuous evaluation and feedback regarding the attainment of green goals motivates employees to involve themselves in tasks related to green goals (Reinholt et al., 2011).

Most researchers argue that GHRM is a means of achieving sustainable performance management (Park & Krause, 2021; Noordiatmoko & Riyadi, 2023), which can be defined as performance management that focuses on the accomplishment of developmental outcomes and leads to sustainable development (Park & Krause, 2021). As much as performance management deals with operational aspects of HRM, like performance planning, employee appraisals, and giving rewards, it is different to sustainable performance management in that sustainable performance management focuses on the attainment of sustainable developmental goals (Noor-diatmoko & Riyadi, 2023).

Sustainable Performance Management

Sustainable performance management involves the work-related activities that managers undertake to achieve their established sustainable development goals (Park & Krause, 2021), consistent with the global SDGs. Sustainable performance management goes beyond the usual performance management practices. Thus, traditional HRM may not be able to effectively bolster sustainable performance management. Notably, sustainable performance management is committed to the principles of the global SDGs in all aspects of operations. This means considering not just profit but also the business's impact on the environment and society. In this case, it is crucial that environmentally friendly practices are aligned with the global SDGs; specifically, responsible production or consumption; climate action; clean water and sanitation; the enhancement of the sustainability of cities and communities; and the promotion of people's good health and wellbeing.

This management approach focuses on the use of metrics, including key performance indicators (KPIs) related to responsible production or consumption; climate action; the provision of clean water and sanitation; and the enhancement of the sustainability of cities and communities in setting employees' work objectives and assessing their performance against these objectives to drive organisation environmental performance. This comprehensive approach includes setting clear goals and supporting employees through training, rewards, and engagement in order to achieve such goals through strategies such as management by objectives.

The Performance Management Cycle

The performance management cycle is the process of planning, measuring, evaluating, rewarding, and adjusting employees' tasks and activities to achieve business goals. From a sustainability perspective, the performance management cycle integrates the global SDGs into the process, and this process needs to be planned and executed with the support of senior managers. The process usually begins with setting objectives that are realistic, measurable, and achievable (Pauwe et al., 2023). Examples of such business objectives might be to limit the organisation's carbon footprint by promoting employees' use of public transportation, cycling, and carpooling to promote sustainable cities and communities. Then, these objectives can be developed into KPIs that can be used to measure progress and identify any issues.

The next step in the performance management cycle involves monitoring and supporting employees in order to achieve set objectives – monitoring if the employees are working sufficiently smart and hard to meet the specific KPIs. If they are not, greater support and interventions can be introduced in the form of training and development; rewards; and engagement. Rewards, both financial and non-financial, are necessary for motivating employees and maintaining their high performance. They can reinforce proactive behaviours that lead to the achievement of goals such as responsible consumption or production and can lead to changes in unacceptable behaviour.

In the final stage of the performance management cycle, employees' performance is measured to assess their contribution towards the sustainable objectives set in the first instance through the collection of qualitative and quantitative data on the performance of the individual employees. This can be further deployed to make informed HR decisions. For employees who may not have fully adhered to sustainability goals, the performance management cycle provides a structured approach to address areas of improvement, and it may be necessary to discipline such employees, giving them constructive feedback (Ciemleja & Lace, 2011). Learning what went right and needs improvement as well as facilitating continuous improvement during the next performance cycle are strategic issues. In the case of sustainable performance management, it is argued herein that the focus of the three main stages discussed above is how to facilitate environmental performance.

Why Use Sustainable Performance Management?

A critical goal of sustainable performance management is that it supports the design and implementation of GHRM. Sustainable performance management helps managers set clear sustainability objectives (in relation to combating climate change, resource depletion, and environmental degradation) and support the achievement thereof correctly through training, rewards, and engagement (Pucik et al., 2023). Moreover, sustainable performance management facilitates the collection of data that can help an organisation understand and improve its HRM strategy; for example, sustainable performance management can help managers identify training, rewards, and engagement gaps (Ciemleja & Lace, 2011) and suggest possible ways on how to address the gaps to enhance sustainable performance. Organisations that engage in sustainable performance management can identify the training needs of managers so that they can become leaders who are committed to sustainability and thus achieve organisational goals in relation to responsible production or consumption; climate action; clean water and sanitation; the enhancement of the sustainability of cities and communities; and promoting good health and wellbeing (Montiel et al., 2021). When it is implemented correctly, sustainable performance management provides managers with insights into how to develop the appropriate green competencies among employees and motivate them to act in environmentally responsible ways in order to drive environmental performance (Dumont et al., 2017; Morioka & de Carvalho, 2016).

Notably, organisations that engage in sustainable performance management can ensure that they achieve the SDGs in relation to climate action; clean water and sanitation; and the enhancement of the sustainability of cities and communities – and even exceed them (Ciemleja & Lace, 2011). This is crucial for organisations to safeguard and enhance their brand image as well as protect them against fines and legal proceedings. On the one hand, when an organisation complies with environmental regulations, it can achieve higher profits and a healthier return on investment, demonstrating the business case for achieving the UN's SDGs than it would otherwise be able to do (Montiel et al., 2021). For example, when a business is known to engage in environmentally sustainable practices, it can generate more customers, leading to greater profit maximisation than it would otherwise achieve (Pucik et al., 2023). On the other hand, sustainable performance management can help organisations attract and retain talent in the competitive job market (Renwick et al., 2013). Workplaces that embrace environmental sustainability are characterised by low employee turnover rates, which results in low staff turnover costs, which can significantly contribute to overall HR cost savings. Retaining staff is necessary in order for an organisation to avoid challenges such as skills shortages. Moreover, high levels of job satisfaction at an organisation mean that there is also a low level of absenteeism there. There is evidence that employees have high levels of job satisfaction in a sustainable workplace, and this can lead to enhanced performance, which can benefit the wider organisation (Renwick et al., 2023). This in turn can generate other positive developments, such as improved collaboration and innovation, which will benefit the organisation in terms of product development and the quality of outputs.

Finally, a manager who engages in sustainable performance management is able to gain a thorough understanding of whether their organisation is attaining its goals through HR, particularly goals related to climate action, as exemplified by Morioka and de Carvalho (2016). Moreover, the indicators provide managers with insights into the effectiveness of their organisation's environmental sustainability strategies, including its HRM strategy. The current study argues that while sustainable performance management is useful for achieving the abovementioned goals – it can only do so through well-designed and well-implemented GHRM practices. While traditional HRM plays an important role in performance management processes, the current study argues that it may not be able to support sustainable performance management. The reason is that traditional HRM is limited in terms of its support for environmental performance according to Ogbeibu et al. (2023). Thus, the study now explores the impact of GHRM on sustainable performance management.

The Relationship between Green Human Resource Management and Sustainable Performance Management

The concept of GHRM plays an important role in shaping an organisation's commitment to sustainability, and it focuses on environmentally responsible practices in HRM. When seeking to drive sustainability goals, sustainable performance management can facilitate the planning, monitoring, and improvement of GHRM efforts.

Hiring high-quality employees who have green values and skills is a key challenge in the current war for talent (Renwick et al., 2013). Thus, efforts to drive sustainable performance management through GHRM should start with the green hiring of employees who have green human capital and values after setting clear environmental objectives, such as combating climate change. Through green hiring efforts, organisations seeking to combat climate change, resource depletion, and environmental degradation through their production and work processes can attract employees with green human capital (Atiku, 2020). According to Renwick et al. (2013), organisations can use green hiring practices in order to attract qualified employees who have the necessary green skills to drive the work and production processes relevant to combating climate change, resource depletion, and environmental degradation. Job descriptions and person specifications should clearly describe the green aspects of the relevant job in terms of how to work and produce in ways that combat climate change, resource depletion, and environmental degradation in order to promote the establishment of smart cities and to obtain the human capital required to achieve the same.

According to Debrah et al. (2018), employers should help new and long-serving employees sharpen their skills, regardless of their experience level, to help them develop firm-specific human capital. Thus, the current study suggests that organisations seeking to achieve the environmental dimension of the SDGs should design and implement green training programmes to help all employees develop the green skills they require to work in ways that will combat resource depletion, climate change, and environmental degradation (Pucik et al., 2023), because having a positive impact on the environment is now regarded as a responsibility for all employees. Notably, training in the development of skills in energy-efficient technologies is important (Ogbeibu et al., 2023). Research undertaken by Renwick et al. (2013) suggests that environmental knowledge and values are important predictors of personal behaviours. Consistent with the AMO model, developing environmental leaders is also important for driving sustainable performance management through GHRM. In the words of Renwick et al. (2013), it is important for organisations to develop green leadership skills among their managers, who should possess personal environmentally friendly values. With this, managers can provide relevant support for setting clear work objectives with employees, providing relevant training and engagement at work. It is anticipated that with the right managerial and leadership support, employees can take sustainability initiatives (Renwick et al., 2013).

In order to effectively practice sustainable performance management, skilled employees should be provided with opportunities to engage in green activities and work processes that will address resource depletion, climate change, and environmental degradation. This can be achieved through green employee involvement practices, such as a green suggestion scheme that enables employees to suggest ideas that can drive the development of smart cities. During green engagement, leaders with green skills can monitor employees to provide relevant support in the form of real-time feedback. In this way, mistakes are not made before they are identified during appraisals in the last stage in the sustainable performance management cycle (Oseghale et al., 2018).

Additionally, green reward schemes should be implemented in order to encourage the green behaviours and values that are appreciated by organisations. In this vein, Ogbeibu et al. (2023) posit that employees are motivated to apply their green knowledge and competencies at work in order to tackle issues of climate change, resource depletion, and environmental degradation. In line with the AMO model, having the relevant green skills and competencies is not enough. Rather, the transfer of such skills and learning to the workplace is critical (Debrah et al., 2018) for driving relevant SDGs. In the last stage of sustainable performance management, green appraisals can be deployed to measure employees' contributions to the attainment of the relevant sustainable development goals as outlined in the employees' work objectives. Following appraisal, employees who are contributing to the attainment of sustainable development goals should be rewarded accordingly. However, those who are not contributing much to the attainment of these goals should be supported by giving them timely feedback and by providing them with effective green training and development. The use of constant reprimands and cautions should be avoided. According to Ogbeibu et al. (2023) and Renwick et al. (2013), the use of negative reinforcement does not educate staff.

DISCUSSION

This study sought to explore the influence of GHRM on sustainable performance management and to identify the role of specific GHRM activities on the achievement of SDGs, such as the enhancement of the sustainability of cities and communities; sustainable production; climate action; clean water and sanitation; and the preservation of life below sea level and on land. The study found that GHRM (not traditional HRM) can drive sustainable performance management through green hiring, training, involvement, and performance appraisals to develop and deploy the green competencies and behaviour required to address climate change, resource depletion, and environmental degradation. However, the study also uncovered that sustainable performance management is a key driver of GHRM. The need for sustainable performance management will drive organisations to implement effective GHRM practices to bolster the attainment of key sustainable development goals.

Driven by the need to pursue sustainability goals, the study findings show that organisations hire green talent/human capital. However, the study revealed that having employees with relevant green competencies is not enough in the sustainable performance management process. Rather, organisations must ensure that the right green reward systems and support systems, including green appraisal systems and leadership, are in place in order to drive an effective sustainable performance management process through GHRM. Regarding green reward schemes, the extant literature suggests that many organisations only provide green rewards for their leaders and senior workers at the expense of other employees (Renwick et al., 2013). While previous work suggests that this approach is effective (Renwick et al., 2013), the current study found that such practices may be counterproductive. All employees are responsible for driving sustainability in the workplace and should therefore be encouraged, through green reward schemes, to deploy their green skills. According to the AMO model, having the relevant skills (green, in this case) is not enough (Debrah et al., 2018), as deploying such skills and behaviours is equally important.

CONCLUSION

From a theoretical standpoint, the current study advances the extant literature on GHRM in that it highlights how GHRM influences sustainable performance management and vice versa. While existing literature provides insights into how traditional HRM drives performance management, this is the first study to consider how GHRM drives sustainable performance management. We extend the UN's SDG framework by articulating how the AMO model can facilitate the development and deployment of green skills and bolster the attainment of the environmental dimension of the SDGs through green leadership. Practically speaking, organisations seeking to address climate change, resource depletion, and environmental degradation to drive smart cities should draw on these GHRM practices to develop an effective sustainable performance management system. Based on this discussion, the researchers have developed a conceptual framework highlighting the sustainable performance management and GHRM nexus (Figure 3).

Figure 3. The green human resources management and sustainable performance management nexus

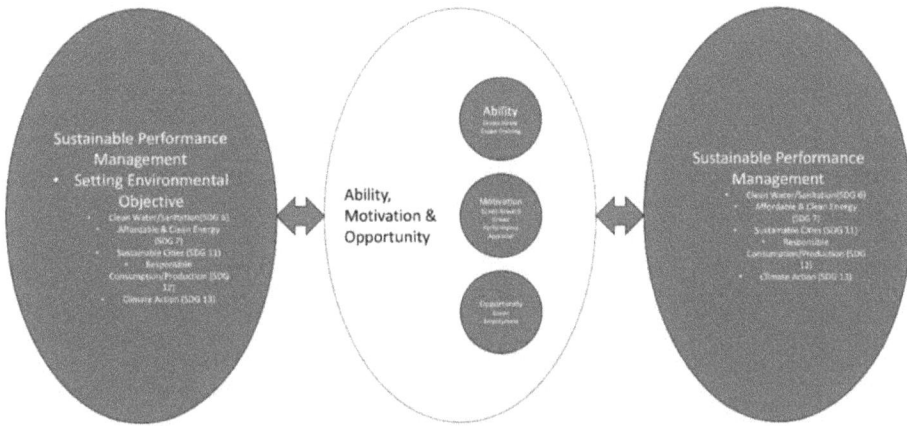

Limitations and Directions for Future Research

While the current study makes several key contributions to the current literature on GHRM, it has one major limitation that future research can address: The study has focused on theory expansion without providing any empirical evidence to support the claims made. Thus, it is suggested that future work on GHRM and sustainable performance management should collect empirical data to test the arguments presented in this study.

REFERENCES

Adewumi, S. A., Ajadi, T., & Ntshangase, B. (2022). Green human resource management and green environmental workplace behaviour in the Thekwini municipality of South Africa. *International Journal of Research in Business and Social Science, 2147-4478, 11*(4), 159–170.

Adisa, T., & Abdulkareem, I. (2017). Long working hours and the challenges of work-life balance: The case of Nigerian medical doctors. *British Academy of Management*, 2017(October), 1–12.

Adisa, T., Abdulkareem, I., & Isiaka, S. (2019). Patriarchal hegemony: Investigating the impact of patriarchy on women's work-life balance. *Gender in Management*, 34(1), 19–33. 10.1108/GM-07-2018-0095

Adisa, T. A., Aiyenitaju, O., & Adekoya, O. D. (2021). The work–family balance of British working women during the COVID-19 pandemic. *Journal of Work-Applied Management*, 13(2), 241–260. 10.1108/JWAM-07-2020-0036

Ansari, N. Y., Farrukh, M., & Raza, A. (2021). Green human resource management and employees' pro-environmental behaviours: Examining the underlying mechanism. *Corporate Social Responsibility and Environmental Management*, 28(1), 229–238. 10.1002/csr.2044

Anton, A. (2016). Green human resource management practices: A review. *Sri Lankan Journal of Human Resource Management*, 5(1), 1–16.

Aragon-Correa, J. A., Martin-Tapia, I., & Hurtado-Torres, N. E. (2013). Proactive environmental strategies and employee inclusion: The positive effects of information sharing and promoting collaboration and the influence of uncertainty. *Organization & Environment*, 26(2), 139–161. 10.1177/1086026613489034

Ardiza, F., Nawangsari, L. C., & Sutawidjaya, A. H. (2021). The influence of green performance appraisal and green compensation to improve employee performance through OCBE. *International Review of Management and Marketing*, 11(4), 13–22. 10.32479/irmm.11632

Arulrajah, A. A., Opatha, H. H. D. N. P., & Nawaratne, N. N. J. (2015). Green human resource management practices: A review. *Sri Lankan Journal of Human Resource Management*, 5(1), 1–16. 10.4038/sljhrm.v5i1.5624

Atiku, S. O. (2019). Institutionalizing Social Responsibility Through Workplace Green Behavior. In Atiku, S. (Ed.), *Contemporary Multicultural Orientations and Practices for Global Leadership* (pp. 183–199). IGI Global. 10.4018/978-1-5225-6286-3.ch010

Atiku, S. O. (2020). Knowledge Management for the Circular Economy. In Baporikar, N. (Ed.), *Handbook of Research on Entrepreneurship Development and Opportunities in Circular Economy* (pp. 520–537). IGI Global. 10.4018/978-1-7998-5116-5.ch027

Atiku, S. O. (Ed.). (2020). *Human Capital Formation for the Fourth Industrial Revolution*. IGI Global. 10.4018/978-1-5225-9810-7

Atiku, S. O., & Fapohunda, T. (Eds.). (2021). *Human Resource Management Practices for Promoting Sustainability*. IGI Global. 10.4018/978-1-7998-4522-5

Boiral, O., Paillé, P., & Raineri, N. (2015). The nature of employees' pro-environmental behaviors. *The Psychology of Green Organizations*, 1(March), 12–32. 10.1093/acprof:oso/9780199997480.003.0002

Cantor, D. E., Morrow, P. C., & Montabon, F. (2012). Engagement in environmental behaviors among supply chain management employees: An organizational support theoretical perspective. *The Journal of Supply Chain Management*, 48(3), 33–51. 10.1111/j.1745-493X.2011.03257.x

Chams, N., & García-Blandón, J. (2019). On the importance of sustainable human resource management for the adoption of sustainable development goals. *Resources, Conservation and Recycling*, 141(February), 109–122. 10.1016/j.resconrec.2018.10.006

Chaudhary, R. (2018). Can green human resource management attract young talent? An empirical analysis. *Evidenced-Based HRM*, 6(3), 305–319. 10.1108/EBHRM-11-2017-0058

Ciemleja, G., & Lace, N. (2011). The model of sustainable performance of small and medium-sized enterprises. *The Engineering Economist*, 22(5), 501–509.

Dahlmann, F., Branicki, L., & Brammer, S. (2019). Managing carbon aspirations: The influence of corporate climate change targets on environmental performance. *Journal of Business Ethics*, 158(1), 1–24. 10.1007/s10551-017-3731-z

Danirmala, L., & Prajogo, W. (2022). The mediating role of green training to the influence of green organizational culture to green organizational citizenship behavior and green employee involvement. *International Journal of Human Capital Management*, 6(1), 66–75. 10.21009/IJHCM.06.01.6

Debrah, Y. A., Oseghale, R. O., & Adams, K. (2018). Human capital development, innovation and international competitiveness in Sub-Saharan Africa. In Adeleye, I., & Esposito, M. (Eds.), *Africa's Competitiveness in the Global Economy* (pp. 219–248). Palgrave Macmillan. 10.1007/978-3-319-67014-0_9

Deshwal, P. (2015). Green HRM: An organizational strategy of greening people. *International Journal of Applied Research*, 1(13), 176–181.

Dumont, J. (2015). *Green human resource management and employee workplace outcomes* [Thesis, University of South Australia].

Dumont, J., Shen, J., & Deng, X. (2017). Effects of green HRM practices on employee workplace green behavior: The role of psychological green climate and employee green values. *Human Resource Management*, 56(4), 613–627. 10.1002/hrm.21792

European Centre for the Development of Vocational Training. (2012). *A Strategy for Green Skills? A Study on Skill Needs and Training Has Wider Lessons for Successful Transition to a Green Economy: Briefing Report*. European Centre for the Development of Vocational Training.

Fields, Z., & Atiku, S. O. (2017). Collective Green Creativity and Eco-Innovation as Key Drivers of Sustainable Business Solutions in Organizations. In Fields, Z. (Ed.), *Collective Creativity for Responsible and Sustainable Business Practice* (pp. 1–25). IGI Global. 10.4018/978-1-5225-1823-5.ch001

Fields, Z., & Atiku, S. O. (2018). Collaborative Approaches for Communities of Practice Activities Enrichment. In Baporikar, N. (Ed.), *Knowledge Integration Strategies for Entrepreneurship and Sustainability* (pp. 304–333). IGI Global. 10.4018/978-1-5225-5115-7.ch015

Fryxell, G. E., & Lo, C. W. (2003). The influence of environmental knowledge and values on managerial behaviours on behalf of the environment: An empirical examination of managers in China. *Journal of Business Ethics*, 46(1), 45–69. 10.1023/A:1024773012398

Gadenne, D. L., Kennedy, J., & McKeiver, C. (2009). An empirical study of environmental awareness and practices in SMEs. *Journal of Business Ethics*, 84(1), 45–63. 10.1007/s10551-008-9672-9

Heong, Y. M., Sern, L. C., Kiong, T. T., & Mohamad, M. M. B. (2016). The role of higher order thinking skills in green skill development. *MATEC Web of Conferences, 70*, 05001. EDP Sciences.

Hernu, R. (2022). *Sustainable performance: How to make success last?* [Post]. LinkedIn. https://www.linkedin.com/pulse/sustainable-performance-how-make -success-last-raphaelle-hernu

Jabbour, C. J. C., Santos, F. C. A., & Nagano, M. S. (2010). Contributions of HRM throughout the stages of environmental management: Methodological triangulation applied to companies in Brazil. *International Journal of Human Resource Management*, 21(7), 1049–1089. 10.1080/09585191003783512

Joshi, G., & Dhar, R. L. (2020). Green training in enhancing green creativity via green dynamic capabilities in the Indian handicraft sector: The moderating effect of resource commitment. *Journal of Cleaner Production*, 267, 121948. 10.1016/j.jclepro.2020.121948

Khaskhely, M. K., Qazi, S. W., Khan, N. R., Hashmi, T., & Chang, A. A. R. (2022). Understanding the impact of green human resource management practices and dynamic sustainable capabilities on corporate sustainable performance: Evidence from the manufacturing sector. *Frontiers in Psychology*, 13, 1–17. 10.3389/fpsyg.2022.84448835846624

Kim, Y. J., Kim, W. G., Choi, H. M., & Phetvaroon, K. (2019). The effect of green human resource management on hotel employees' eco-friendly behavior and environmental performance. *International Journal of Hospitality Management*, 76, 83–93. 10.1016/j.ijhm.2018.04.007

Lawal, I. O., & Olawoyin, F. S. (2021). Green Human Resources and Sustainable Business Solutions. In Atiku, S., & Fapohunda, T. (Eds.), *Human Resource Management Practices for Promoting Sustainability* (pp. 264–277). IGI Global. 10.4018/978-1-7998-4522-5.ch015

Likhitkar, P., & Verma, P. (2017). Impact of green HRM practices on organization sustainability and employee retention. *International Journal for Innovative Research in Multidisciplinary Field*, 3(5), 152–157.

Liu, J., Gao, X., Cao, Y., Mushtaq, N., Chen, J., & Wan, L. (2022). Catalytic effect of green human resource practices on sustainable development goals: Can individual values moderate an empirical validation in a developing economy? *Sustainability (Basel)*, 14(21), 14502. 10.3390/su142114502

Lubbe, W., ten Ham-Baloyi, W., & Smit, K. (2020). The integrative literature review as a research method: A demonstration review of research on neurodevelopmental supportive care in preterm infants. *Journal of Neonatal Nursing*, 26(6), 308–315. 10.1016/j.jnn.2020.04.006

Martins, J. M., Aftab, H., Mata, M. N., Majeed, M. U., Aslam, S., Correia, A. B., & Mata, P. N. (2021). Assessing the impact of green hiring on sustainable performance: Mediating role of green performance management and compensation. *International Journal of Environmental Research and Public Health*, 18(11), 5654. 10.3390/ijerph1811565434070535

Masri, H. A., & Jaaron, A. A. (2017). Assessing green human resources management practices in Palestinian manufacturing context: An empirical study. *Journal of Cleaner Production*, 143, 474–489. 10.1016/j.jclepro.2016.12.087

Montiel, I., Cuervo-Cazurra, A., Park, J., Antolín-López, R., & Husted, B. W. (2021). Implementing the United Nations' sustainable development goals in international business. *Journal of International Business Studies*, 52(5), 999–1030. 10.1057/s41267-021-00445-y34054154

Morioka, S. N., & de Carvalho, M. M. (2016). A systematic literature review towards a conceptual framework for integrating sustainability performance into business. *Journal of Cleaner Production*, 136, 134–146. 10.1016/j.jclepro.2016.01.104

Mousa, S. K., & Othman, M. (2020). The impact of green human resource management practices on sustainable performance in healthcare organisations: A conceptual framework. *Journal of Cleaner Production*, 243, 118595. 10.1016/j.jclepro.2019.118595

Noordiatmoko, D., & Riyadi, B. S. (2023). The Analysis of Sustainable Performance Management of Government Institution in Indonesia: A Public Policy Perspective. *International Journal of Membrane Science and Technology*, 10(3), 1146–1157. 10.15379/ijmst.v10i3.1684

Ogbeibu, S., Emelifeonwu, J., Pereira, V., Oseghale, R., Gaskin, J., Sivarajah, U., & Gunasekaran, A. (2023). Demystifying the roles of organisational smart technology, artificial intelligence, robotics, and algorithms capability: A strategy for green human resource management and environmental sustainability. *Business Strategy and the Environment*, 33, 369–388. 10.1002/bse.3495

Ogbeibu, S., Emelifeonwu, J., Senadjki, A., Gaskin, J., & Kaivo-oja, J. (2020). Technological turbulence and greening of team creativity, product innovation, and human resource management: Implications for sustainability. *Journal of Cleaner Production*, 244, 118703. 10.1016/j.jclepro.2019.118703

Ojo, A. O., Tan, C. N. L., & Alias, M. (2022). Linking green HRM practices to environmental performance through pro-environment behaviour in the information technology sector. *Social Responsibility Journal*, 18(1), 1–18. 10.1108/SRJ-12-2019-0403

Olanipekun, A. O., Xia, B., Hon, C., & Hu, Y. (2017). Project owners' motivation for delivering green building projects. *Journal of Construction Engineering and Management*, 143(9), 04017068. 10.1061/(ASCE)CO.1943-7862.0001363

Oseghale, O. R., Mulyata, J., & Debrah, Y. A. (2018). Global talent management. In Manchando, C., & Davim, J. P. (Eds.), *Organizational Behaviour and Human Resource Management* (pp. 139–155). Springer. 10.1007/978-3-319-66864-2_6

Paauwe, J., Boon, C., Boselie, P., & Den Hartog, D. (2013). *Reconceptualizing fit in strategic human resource management: 'Lost in translation?'* In Human Resource Management and Performance: Achievements and Challenges. John Wiley & Sons Ltd.

Park, A. Y., & Krause, R. M. (2021). Exploring the landscape of sustainability performance management systems in US local governments. *Journal of Environmental Management*, 279, 111764. 10.1016/j.jenvman.2020.11176433360650

Perron, G. M., Côté, R. P., & Duffy, J. F. (2006). Improving environmental awareness training in business. *Journal of Cleaner Production*, 14(6–7), 551–562. 10.1016/j.jclepro.2005.07.006

Priansa, D. J. (2016). The influence of E-WOM and perceived value on consumer decisions to shop online at Lazada. *Ecodemic Journal of Management and Business Economics*, 4(1), 117–124.

Pucik, V., Bjorkman, I., Evans, P., & Stahl, G. K. (2023). *The global challenge: Managing people across borders.* Edward Elgar Publishing. 10.4337/9781035300723

Ramlee, M. (2015). Green and sustainable development for TVET in Asia. *The International Journal of Technical and Vocational Education*, 11(2), 133–142.

Rayner, J., & Morgan, D. (2018). An empirical study of 'green' workplace behaviours: Ability, motivation, and opportunity. *Asia Pacific Journal of Human Resources*, 56(1), 56–78. 10.1111/1744-7941.12151

Reinholt, M., Pedersen, T., & Foss, N. J. (2011). Why a central network position isn't enough: The role of motivation and ability for knowledge sharing in employee networks. *Academy of Management Journal*, 54(6), 1277–1297. 10.5465/amj.2009.0007

Renwick, D. W. S., Redman, T., & Maguire, S. (2013). Green human resource management: A review and research agenda. *International Journal of Management Reviews*, 15(1), 1–14. 10.1111/j.1468-2370.2011.00328.x

Saeed, B. B., Afsar, B., Hafeez, S., Khan, I., Tahir, M., & Afridi, M. A. (2019). Promoting employee's proenvironmental behavior through green human resource management practices. *Corporate Social Responsibility and Environmental Management*, 26(2), 424–438. 10.1002/csr.1694

Sanyal, C., & Haddock-Millar, J. (2018). Employee engagement in managing environmental performance: A case study of the planet champion initiative, McDonalds UK and Sweden. In Renwick, D. W. S. (Ed.), *Contemporary Developments in Green Human Resource Management Research: Towards Sustainability in Action* (pp. 39–56). Taylor and Francis. 10.4324/9781315768953-3

Steinfeld, H., & Gerber, P. (2010). Livestock production and the global environment: Consume less or produce better? *Proceedings of the National Academy of Sciences of the United States of America*, 107(43), 18237–18238. 10.1073/pnas.101254110720935253

Suharti, L., & Sugiarto, A. (2020). A qualitative study of green HRM practices and their benefits in the organization: An Indonesian company experience. *Business: Theory and Practice*, 21(1), 200–211. 10.3846/btp.2020.11386

Tang, G., Chen, Y., Jiang, Y., Paillé, P., & Jia, J. (2018). Green human resource management practices: Scale development and validity. *Asia Pacific Journal of Human Resources*, 56(1), 31–55. 10.1111/1744-7941.12147

Teixeira, A. A., Jabbour, C. J. C., de Sousa Jabbour, A. B. L., Latan, H., & De Oliveira, J. H. C. (2016). Green training and green supply chain management: Evidence from Brazilian firms. *Journal of Cleaner Production*, 116, 170–176. 10.1016/j.jclepro.2015.12.061

United Nations Industrial Development Organisation (UNIDO). (2022) *Sustainable development goals*. UN. https://www.unido.org/unido-sdgs

Woo, E. J. (2021). The necessity of environmental education for employee green behaviour. *East Asian Journal of Business Economics*, 9(4), 29–41.

Yang, C. L., Lin, S. P., Chan, Y. H., & Sheu, C. (2010). Mediated effect of environmental management on manufacturing competitiveness: An empirical study. *International Journal of Production Economics*, 123(1), 210–220. 10.1016/j.ijpe.2009.08.017

Yong, J. Y., Yusliza, M. Y., Ramayah, T., Chiappetta Jabbour, C. J., Sehnem, S., & Mani, V. (2020). Pathways towards sustainability in manufacturing organizations: Empirical evidence on the role of green human resource management. *Business Strategy and the Environment*, 29(1), 212–228. 10.1002/bse.2359

Zaid, A. A., Jaaron, A. A. M., & Bon, A. T. (2018). The impact of green human resource management and green supply chain management practices on sustainable performance: An empirical study. *Journal of Cleaner Production*, 204, 965–979. 10.1016/j.jclepro.2018.09.062

ADDITIONAL READING

Paille, P. (Ed.). (2022). *Green human resource management research: Issues, trends, and challenges*. Sustainable Development Goals Series. Palgrave Macmillan. 10.1007/978-3-031-06558-3

Yusliza, M. Y., & Renwick, D. W. S. (Eds.). (2024). *Green human resource management: A view from global South countries*. Springer. 10.1007/978-981-99-7104-6

KEY TERMS AND DEFINITIONS

Green Hiring: Green hiring involves hiring employees who have the necessary knowledge, skills, and behaviour for attaining environmentally sustainable goals.

Green Human Resource Management: Green human resource management involves embedding environmental sustainability initiatives into human resource management practices.

Green Performance Appraisal: This involves evaluating an employee's performance based on their adherence to and achievement of set sustainable development performance indicators.

Green Rewards: These are rewards given to employees for displaying and achieving sustainable development performance indicators.

Green Training: Green training involves equipping employees with the necessary knowledge and skills to perform sustainable tasks in an organisation.

Sustainable Development Goals: They provide the blueprint for the achievement and attainment of a sustainable future.

Sustainable Performance Management: These are the activities that management performs to achieve set sustainable development goals.

Chapter 8
Green Marketing Strategies for Indonesia's Micro, Small, and Medium Enterprises

Vanessa Gaffar

Universitas Pendidikan Indonesia, Indonesia

Tika Koeswandi

Universitas Pendidikan Indonesia, Indonesia

ABSTRACT

Indonesia as a developing country has a long history of implementing the development transition to become a developed country. During this transition process, social and environmental phenomena also emerged as a result of this agenda. Environmental awareness campaigns for micro, small, and medium enterprises (MSMEs) in Indonesia require a different approach compared to other types of businesses. With the unique characteristics of MSMEs, a special strategy and practice is needed. This chapter explains the implementation of green marketing of MSMEs from the internal and external environment. A case study is also offered to describe how a green B2b Indonesia MSME has succeed in running sustainable business practice. This chapter offers a matrix to visualize the best positioning mapping for MSMEs to effectively and efficiently implement green marketing in Indonesia and any other developing countries. The chapters set the tone for the rest of the chapters examining the implications of the issues discussed for waste management and life cycle assessment for sustainable business practice.

DOI: 10.4018/979-8-3693-2595-7.ch008

INTRODUCTION

Indonesia as a developing country has a long history of implementing the development transition to become a developed country. During this transition process, Indonesia slowly experienced economic growth, structural changes in the economic sector, and an increase in people's living standards (Sutikno, 2020). Unfortunately, at the same time, social and environmental phenomena are also emerging as a result of this agenda. One important change is environmental degradation, namely the exploitation of natural resources resulting in environmental damage and loss of biodiversity. The Indonesian government encourages the business sector to implement strategies and practices that support environmental sustainability and aim to reduce negative impacts on nature (Handajani, Husnan & Rifai, 2019). In the last 10 years, both Ministries, state-owned enterprises, financial institutions and public companies have attempted to document environmental sustainability movements through Sustainability Reports with GRI (Global Reporting Initiative) standards. What remains a challenge is its implementation in the Micro, Small and Medium Enterprises business sector.

Environmental awareness campaigns for Micro, Small and Medium Enterprises (MSMEs) in Indonesia require a different approach compared to other types of businesses. MSMEs in Indonesia have different characteristics and uniqueness compared to other types of business. First, the number of MSMEs fluctuates every year. According to a private survey institute called Katadata, it is stated that in 2024, there are 63.9 million micro business units, 193.9 thousand small business units, 44.7 thousand medium business units and 5.5 thousand large business units. Interestingly, Muhammad (2023) stated that the proportion has not changed much in the last 10 years, so that in the future, the increase in the number of MSMEs is predicted to increase by 10 – 25%. Second, technological developments in Indonesia which support the escalation of MSMEs. The phenomenon of industry 4.0 and Covid-19 is accelerating technological development for MSMEs in Indonesia. Technology is used by MSMEs not only as a promotional medium, but also used by MSMEs to facilitate business operations. The most widely used technology is the use of social media, AI and e-commerce. However, in its use it is known that MSMEs also encounter challenges, as shown in figure 1 below:

Figure 1. Challenges of MSMEs Indonesia in adopting digital technology

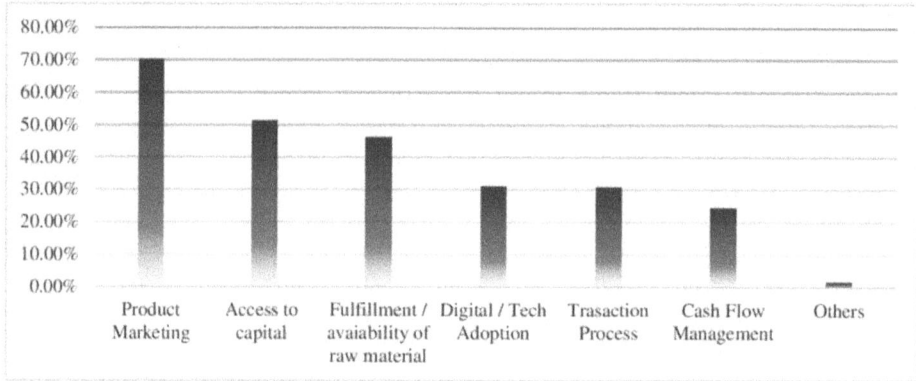

Source: MSME Empowerment Report 2022

Figure 1 shows these challenges according to the MSME Empowerment Report 2022 consisting of product marketing (70.2%), access to capital (51.2%), fulfillment / availability of raw materials (46.3%), digital / tech adoption (30 .9%), transaction process (30.8%), cash flow management (24.3%) and others (1.7%). This data shows that marketing products through technology is a problem that must be resolved. This is in line with the results of research by Nurfitriya (2022) which states that the digital literacy level of MSMEs is currently still at a moderate level (not yet advanced), so literacy and coaching programs must continue to be ongoing to solve this problem. Even in this type of green-based business and green marketing, it means that not all MSMEs can market their products via digital.

Lastly, there are still many MSME products that are still tied to cultural heritage values. The relationship between cultural heritage products and green marketing can be quite complex depending on the context. On the one hand, cultural heritage products often have historical, cultural and traditional values that are important to preserve. However, the production, marketing and distribution processes of these products may not be environmentally friendly. In the context of green marketing, cultural heritage products can be considered as products that have added value in terms of sustainability. For example, if the production process uses natural ingredients or traditional, environmentally friendly techniques, the product can be sold with a strong sustainability narrative. However, there is also a risk of greenwashing, namely when manufacturers or marketers use sustainability narratives to attract customers without any real support for environmentally friendly practices. Therefore, it is important for manufacturers and marketers of cultural heritage products to ensure that the sustainability claims they make are supported by appropriate practices. The three phenomena above are in line with what was conveyed by Yacob et al (2021), so a special strategy and practice is needed that is able to mediate the uniqueness of MSME and the environmental sustainability agenda. The agenda is "green marketing".

This chapter explains how green marketing is implemented in Indonesia which is analyzed from the internal environment (business commitment, marketing programs and strategies and implementation) and external (government role and policies), a case study of how local communities support this implementation so that acceptable to Indonesian consumers. This chapter also visualises the stages of consumer awareness towards green marketing and how best positioning mapping is for businesses to be able to effectively and efficiently implement green marketing in Indonesia and developing countries. The chapters set the tone for the rest of the chapters examining the implications of the issues discussed for waste management and life cycle assessment for sustainable business practice.

METHODOLOGY

This chapter used a descriptive qualitative method. It is a method used to explore and understand the characteristics and qualities of a phenomenon (Hignett & McDermott, 2015). It involves the collection and analysis of data in the form of words, images, or other non-numerical forms of information (Myers, 2019). The goal is to provide a rich and detailed account of the phenomenon under study, allowing for further research questions to be developed and informing policy or practice (Vohra & Arora, 2021). This chapter collected secondary data through literature studies. The process begins with defining the research objectives and identifying relevant sources of secondary data, which can include government publications related Small Medium Entreprises, academic journals related green marketing in Small Medium Enterprises, books, Small Medium Enterprises reports, and online databases. Once the data sources are identified, the researcher evaluates their reliability, validity, and relevance to the research questions. Data is then collected through methods such as literature reviews, content analysis, and data mining techniques. After collection, the data is organized and analyzed using qualitative method. Finally, the findings are interpreted and used to draw conclusions and make recommendations.

GREEN MARKETING FOR MSMES IN INDONESIA

This section will discuss 1) the definition of green marketing and how it is connected with MSMEs especially in Indonesia; 2) the green marketing commitment of MSMEs in Indonesia; 3) the green marketing strategy and program of MSMEs in Indonesia; 4) the green marketing and con; and 5) the case study of the Siklus Mutiara Nusantara.

Green Marketing in MSMES

According to Dangelico & Vocalelli (2017) marketing is a fundamental part in the process of implementing pro-environmental sustainability strategies and practices. The role of marketing in this context is to communicate, educate and motivate customers, as well as build a company image that focuses on sustainability. In response to practical demands, the concept of green marketing was known as a concerned with all marketing activities that have served to help cause environmental problems and that may serve to provide a remedy for environmental problems (Dangelico & Vocalelli, 2017). In the long term, green marketing practices in the business environment aim to extend the life of the business and sustainable production and consumption (Gaffar & Koeswandi, 2021). Ismail (2023) explains that implementing green marketing cannot be separated from bringing green products, green service and green performance to MSME business practices. It is known that there are seven green marketing mix indicators consisting of green product, green price, green place, green promotion, green consumer, green process and green physical evidence.

Green Products

Durif, Boivin, & Julien (2010) explained that green products can be defined into three perspectives, namely academic, industrial and consumer perspectives. From an academic perspective, it is known that Durif, Boivin, & Julien (2010) conducted a bibliometric study and found 35 definitions of green products from an academic perspective. It is known that the keywords that often appear in this definition are: environment (30 occurrences); product(19); maximize(13); reduce(11); life-cycle (11); design (10), and resources (10). Thus, Durif, Boivin, & Julien (2010) draw the conclusion that a green product according to academics is a product whose design and/or attributes (and/or production and/or strategy) use recycling (renewable/ toxic-free/biodegradables) resources and which improves environmental impact or reduces environmental toxic damage throughout its entire life cycle. Meanwhile from an industrial perspective, Durif, Boivin, & Julien (2010) found differences with the academic view, namely in the academic literature, the topic of certification only appears in one of the 35 definitions identified in academic literature whereas in the industrial literature certification is granted by an official entity is a sine qua non condition in defining a green product. Second, the notion of animal protection appears many times in the industrial perspective, in the sense that a green product should not have been tested on animals. Third, in the industrial perspective, a green product is generally a product that must respect the "3 R" ("reduce", "reuse", and "recycle").

Meanwhile, from a consumer perspective, Durif, Boivin, & Julien (2010) stated that consumers associate green products with green household cleaning products that are mainly a "biodegradable product", "non-toxic for nature", "with minor impact on the environment", "safe for the planet". In a 3R move, only 2 of the "3 R's" ("reduce" and "reuse") are associated. Meanwhile, 3 AI generators define a green product as an environmentally friendly product or eco-friendly product, is a product that has been designed, manufactured, packaged, and distributed in a way that minimizes its environmental impact. Green products aim to reduce pollution, conserve resources, and promote sustainability throughout their entire life cycle. These products are typically made from sustainable materials, use energy-efficient manufacturing processes, generate minimal waste, and can be recycled or disposed of in an environmentally responsible manner. From the four definitions, there is one common thread that can be concluded that green products are a designated product that can encompass a wide range of consumer goods, including household items, personal care products, clothing, electronics, and building materials. In addition to benefiting the environment, green products can also offer health benefits to consumers by reducing exposure to harmful chemicals and toxins. By choosing green products, consumers can contribute to a more sustainable future and help mitigate the negative impacts of climate change. Overall, green products play a crucial role in promoting environmental sustainability and encouraging responsible consumption patterns among individuals and businesses alike.

Green Price

Sinambela et al (2022) explain that green prices are related to how consumers get more value so that consumers can get product benefits in the long term. In determining prices, businesses need to be careful because there are consumers who understand green value and there are also those who don't. However, businesses don't need to worry, because there are consumer segmentations that already understand this (Chekima et al, 2016) (Hong, Wang & Yu, 2018) (Sana, 2020). So the green price refers to the price of a financial instrument or commodity that has been adjusted to reflect its environmental impact or sustainability characteristics. There are 3 indicators according to Hashem & Al-Rifai (2011), namely 1) affordable price; 2) price according to benefits; 3) the price includes environmental conservation efforts. This concept is often used in the context of green finance, which aims to promote environmentally friendly investments and projects. Green pricing mechanisms can include factors such as carbon emissions, water usage, waste generation, and other environmental considerations. In the financial markets, green pricing can be applied to various assets, including green bonds, renewable energy certificates, carbon credits, and sustainable investment funds. By incorporating environmental factors into pricing models, investors can support sustainable initiatives and companies while also potentially achieving financial returns. Overall, green pricing plays a crucial role in promoting sustainability and addressing environmental challenges through the allocation of capital towards environmentally responsible projects and businesses.

Green Place

A green place typically refers to a location that is environmentally friendly, sustainable, and promotes the conservation of natural resources (Chaturvedi et al, 2024); (Kosatica, 2024); (Tamim & Akter, 2024). These places are often characterized by their efforts to reduce carbon emissions, protect biodiversity, and promote eco-friendly practices. Green places can include parks, gardens, eco-friendly buildings, sustainable communities, and other environmentally conscious spaces (Chaturvedi et al, 2024). In relation to business, green places often become tourist destinations for both green consumers and non-green consumers. This destination will provide an experience in consuming products while preserving culture, for example the experience of planting, picking fruit and consuming fruit in the same place (Chaturvedi et al, 2024); (Kosatica, 2024); (Tamim & Akter, 2024). For MSMEs, a green place means that MSMEs can provide outlets with interiors made from environmentally friendly materials or even recycled products.

Green Promotion

Boztepe (2012) states that green promotion is a strategic step in preparing and implementing green-based marketing of products and services through advertising campaigns, promotions, public relations and other marketing tools are adopted to be 'green'. Green in this promotion can be in the form of ways that uphold environmental values or messages that are built that show empathy and positive messages for consumers. In addition, this type of promotion aims to raise awareness about sustainability issues, encourage consumers to make eco-conscious choices, and highlight the environmental benefits of certain products or services.

Green Consumers

Naini et al (2024) green consumers are known as an eco-conscious-consumers who seek to make purchasing decisions that are environmentally friendly and socially responsible. Green consumers are concerned about the impact of their consumption habits on the environment and society, and they strive to support products and companies that prioritize sustainability, ethical practices, and social responsibility (Suhartanto et al, 2024). Gaffar & Koeswandi (2021) stated that the green behaviour showed by the customers started with the awareness. It refers to the level of consciousness or knowledge that individuals, organizations, or society as a whole have about environmental issues, sustainability, and the impact of human activities on the planet (Atiku, 2020; Shehawy & Khan, 2024). It involves an understanding of the importance of protecting the environment, conserving natural resources, reducing pollution, and promoting sustainable practices in everyday life and business operations. Green awareness is often associated with efforts to promote environmental education, raise public awareness about environmental issues, and encourage environmentally friendly behavior and decision-making (Atiku, 2019). It plays a crucial role in driving change towards more sustainable lifestyles, industries, and policies (Suhartanto et al, 2024). Shehawy & Khan (2024) add that after customer aware they tend to have a green attitude. This is when customer started to implement the green behaviour to their daily basis. Naini et al (2024) found that eight major factors infuencing green consumer behavior, out of which green habit, green culture awareness and attitude, interpersonal infuence, and green purchase intention/ behavior emerged as the most signifcant factors. These consumers often choose products that are organic, locally sourced, fair trade, cruelty-free, energy-efficient, recyclable, or made from renewable resources. By supporting eco-friendly products and companies, green consumers aim to reduce their carbon footprint, minimize waste generation, conserve natural resources, and promote a more sustainable way of living (Fields & Atiku, 2017). Gunawan et al (2024) found that in Indonesia, the young consumers tend to be more critical towards buying green product. They concern to the packaging first rather than the product quality. It was believed that social media marketing, influencer use, and trend tracking, as reasons for being involved in the decision-making process.

Green Process

A green process refers to a method or technique that aims to minimize negative environmental impacts and promote sustainability throughout its lifecycle. Green processes are designed to reduce resource consumption, waste generation, and pollution emissions while maximizing efficiency and productivity. These processes often incorporate renewable energy sources, eco-friendly materials, and innovative technologies to achieve environmental goals. Green processes can be applied across various industries, including manufacturing, agriculture, construction, transportation, and energy production. By adopting green processes, businesses and organizations can contribute to environmental conservation efforts, comply with regulations, enhance their public image, and potentially reduce costs in the long run. Overall, green processes play a crucial role in advancing sustainable development practices and addressing global environmental challenges such as climate change, biodiversity loss, and resource depletion (Gaffar & Koeswandi, 2021). One approach that can explain how a green process can be implemented in business is by visualizing the butterfly diagram in the circular economy as follows:

Figure 2. Butterfly diagram of circular economy

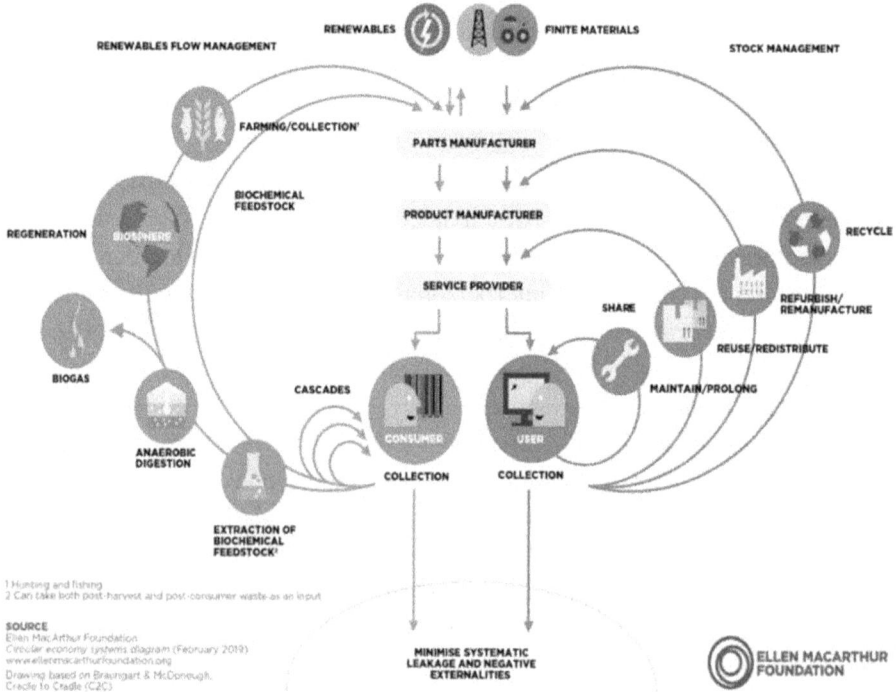

Source: Ellen MacArthur Foundation (2015)

Figure 2 shows a butterfly diagram which in the context of a circular economy refers to a visual representation that depicts the flow of materials and resources in a closed loop system. This diagram displays a 'green process' of production, consumption and waste management cycles that are interconnected like butterfly wings. In a circular economy, green processes highlight the concept of circularity, where resources are used efficiently, recycled and reused to minimize waste and impact on the environment. This visualization serves as a tool for policymakers, businesses, and researchers to understand and analyze the complexity of resource flows in a circular economy. By visualizing the interconnected nature of material cycles, stakeholders can identify opportunities to increase resource efficiency, reduce waste generation and encourage sustainable practices. This process emphasizes closing the loop on material flows and transitioning towards a more sustainable and regenerative economic model.

Green Physical Evidence

Green physical evidence refers to tangible materials or objects that are used in environmental investigations to gather information about pollution, contamination, or other environmental issues. This type of evidence can include soil samples, water samples, water samples, vegetation samples, and physical artifacts found at a site. Green physical evidence plays a crucial role in environmental science and forensic investigations by providing valuable data that can help identify sources of pollution, assess environmental impacts, and support legal cases related to environmental violations. In MSMEs, green physical can be translated as how MSMEs can create a marketing environment that associates a green business and products. For example, choosing natural colors, namely green, yellow and orange or using names that are closely related to nature.

The Green Marketing Commitment of MSMEs in Indonesia

At the 2022 Green Economy Indonesia Summit, the Minister of Cooperatives and Small and Medium Enterprises of the Republic of Indonesia - Mr Teten Masduki, as reported by Antara newspaper, revealed that many young people are already running environmentally friendly businesses, for example using materials from wood waste to make watches and eyeglass frames. Based on the survey results, around 95 percent of MSMEs expressed interest in environmentally friendly business practices, with women-owned businesses showing stronger interest. Another 90 percent said they were interested in implementing inclusive business practices, which is an important component of the Sustainable Development Goals (SDG) agenda. This is supported by the statement of the Minister of Trade of the Republic of Indonesia - Mr Zulkifli Hasan who is currently attending the 2022 MSME Summit on "Encouraging MSMEs to Go Global through Innovation, Digitalization and Green Business". As reported by Pressrelease (2023), the digital economy and the green economy are two megatrends that are changing the direction of world economic development with new market mechanisms and new business models. Consumers are increasingly aware of environmental issues, so product development for micro, small and medium enterprises (MSMEs) must be adapted to world trends so that Indonesian products can compete in the global market.

In 2023, it is known that there will be an increase in interest in implementing environmentally friendly businesses. As reported by Kompas, the Ministry of Co-operatives and Small and Medium Enterprises of the Republic of Indonesia said that the current trend of young entrepreneurs is starting to shift to environmentally friendly businesses, including the electric motor vehicle industry. Deputy Secretary for SMEs at the Ministry of Cooperatives and SMEs, Koko Haryono, said that 84 percent of young entrepreneurs are interested in environmentally friendly business-es. Then, 58 percent started businesses to improve the environment, and around 56 percent produced eco-friendly clothing, low-carbon products and waste reduction systems. So, it can be concluded that the commitment referred to in implementing green marketing for MSMEs in Indonesia is still in its initial stages. MSMEs are still in the 'awareness' stage and entering implementation (Kartawinata et al, 2020); (Rahmawati et al, 2022); (Suasana & Ekawati, 2018). However, Indonesian MSMEs have the potential to have a strong and long-term commitment to green marketing (Kartawinata et al, 2020); (Rahmawati et al, 2022); (Suasana & Ekawati, 2018).

The Green Marketing Strategy and Program of MSMEs in Indonesia

From research conducted by Thoibah, Arif, & Harahap (2022), Rahmawati et al (2022), Romli, Safitri, & Yustitia (2023), Yaputra et al (2023) it is known that green MSMEs are divided into 2, namely 'born green ' and 'adopted green'. 'Born green' are MSMEs that since their founding have implemented green principles from operational to marketing. Meanwhile, 'adopted green' are MSMEs that started as commercial MSMEs and then switched to green MSMEs and carried out green operations and marketing in stages. Thoibah, Arif, & Harahap (2022) examine how the green marketing strategy implemented by Indonesian MSMEs is of the 'born green' type, meaning that from the start MSMEs have focused on green business and marketing. The strategies prepared for most of MSMEs in Indonesia are green product strategy and green marketing startegy, meanwhile the

Green Products Strategy

The products offered have the characteristics of Tangibility, Warranty and Re-liability. Tangibility in the context of green marketing refers to the extent to which the product or service being promoted can physically or concretely provide environ-mental benefits. This can include various aspects, such as the use of environmentally friendly materials, energy efficiency, waste reduction, or other positive impacts on the environment. For example, the products sold are handicraft products from recycled materials, clothes from recycled materials, or foods made from organic ingredients.

Meanwhile, warranties in green marketing can be a strong instrument to increase consumer confidence in products sold with sustainability claims. This guarantee can cover various aspects, such as guarantees regarding the sustainability of raw materials, product safety or environmental performance. For example, recycling information or the product's beneficial impact on the environment is explained in detail on the packaging and in marketing. Meanwhile, in the context of green marketing, reliability refers to the consistency and reliability of sustainability claims made by a product or brand. This reliability includes the extent to which consumers can trust that the product or service being promoted is truly environmentally friendly and has the promised impact on the environment (Aprianti et al, 2021). This means that MSMEs voice the extent to which the product has received recognition and tests regarding its reliability in green business.

Green Promotion Strategy

The forms of green promotion carried out include Green Advertisement and Eco-Labelling. Green advertising is carried out by designing forms of advertising and formulating messages about the benefits of the environment and how the product benefits in the long term. The communication used in this promotion is also descriptive and persuasive. This aims to provide education, awareness and green consciousness to consumers. So, even though they are interested in using MSME products, they are given information about a type of consumption that is different from the type of commercial advertising. Consumption of green products is of course not a type of fast consumption, but rather sustainable consumption. Meanwhile eco-labelling or ecological labels, is a label system used to identify products or services that have a lower environmental impact than similar products or services. These labels help consumers to identify environmentally friendly products and enable them to make more sustainable choices. This type of strategy is carried out by Indonesian MSMEs to ensure that every logo, design, jingle and branding of the product is associated with the product's eco-image. So, the products being sold are at the top of consumers' minds. However, as quoted from Antara news (2023), it was stated that eco-labels are a challenge for MSMEs due to the high cost of obtaining them.

Furthermore, the implementation of green marketing that has been carried out by MSMEs in Indonesia has had quite a significant impact, one of which is providing opportunities for MSMEs to carry out international marketing (Thoibah, Arif, & Harahap, 2022). This is because the green consumer segmentation also has a larger niche market than in Indonesia. Another impact is also felt by the community around where the MSMEs are established, where there is an increase in awareness among the community regarding good use of the environment through green marketing.

Government Supports in Implementing Green Program

The implementation of green marketing for MSMEs in Indonesia has been supported by government regulations and policies that support its implementation. Based on the mandate of article 28H of the 1945 Constitution of the Republic of Indonesia, Law Number 32 of 2009 concerning Environmental Protection and Management, and Law Number 23 of 2014 concerning Regional Government, Bali Province has become one of the provinces in Indonesia which seriously realizes the vision and mission of the Sustainable Development Goals by mainstreaming the environment into various aspects of policy and activities in the business and other sectors. Under the Bali Green Province program, the Province of Bali creates a clean, healthy, comfortable, green, sustainable and beautiful environment based on Tri Hita Karana to support the achievement of a progressive, safe, peaceful and prosperous Bali. The Regional Development of Bali Province is based on Balinese culture which is imbued with Hinduism and the Tri Hita Karana concept to improve community welfare by balancing economic goals, cultural preservation and the environment (Suarna, 2018). In utilizing natural resources, Bali has various local wisdoms that reflect the cultural strengths that underlie attitudes and behavior as well as people's lifestyles (Dalem, et al. 2007). So, the success of Bali Province in realizing green behavior is the result of integrity between religion, custom, culture and government support which influences the mindset, awareness, emotions and commitment of the community to focus on realizing the 'green and sustainability' movement.

Meanwhile in West Java Province, there are areas called Jakarta, Bogor, Depok, Tangerang, Bekasi and Karawang which are industrial area cities. Many Integrated Companies have built factories there, resulting in a lot of waste, both B3 waste (hazardous and non-recyclable waste, such as oil, gas, etc.) and non-B3 (hazardous but recyclable waste, such as plastic, food, etc.). Non-B3 waste has the opportunity to be processed by MSMEs into environmentally friendly value-added products. Management practices are regulated by Regulation of the Minister of Environment and Forestry of the Republic of Indonesia Number 19 of 2021. The central and local governments also take part in carrying out monitoring and evaluation, as stated in article 46 which states that: (1) Ministers, governors, regents/guardians The city, in accordance with its authority, carries out monitoring at least 1 (one) time in 1 (one) year on the activities and mass balance of non-B3 waste management. (2) The implementation of monitoring as intended in paragraph (1) is carried out in the context of improving the performance of non-B3 Waste management implementation. And also, Article 47 which states that (1) The implementation of non-B3 Waste management activities as intended in Articles 4 to Article 40 must be reported to the Minister, governor, regent/mayor in accordance with their authority at least 1 (one) time in 1 (one year. (2) The report as intended in paragraph (1) is submitted electronically via the page https://plb3.menlhk.go.id. (2) The report as intended in paragraph (1) is submitted electronically via the page https://plb3.menlhk.go.id. (3) The report as intended in paragraph (1) contains at least: a. name of non-B3 waste; b. non-B3 Waste code; c. amount of non-B3 waste generated every month; d. mass balance of non-B3 waste management; and e. type of non-B3 Waste management activity. (4) The mass balance for non-B3 waste management as intended in paragraph (2) letter d consists of: a. Non-B3 waste resulting from the production process; b. Stored non-B3 waste; and c. Non-B3 waste that has been managed. (5) The report as intended in paragraph (2) is prepared using the format as stated in Appendix X which is an inseparable part of this Ministerial Regulation. So,it can be concluded that the implementation of green marketing in Indonesia has received attention from the government starting from how MSMEs manage their business to external ecosystem support from related parties. So, green marketing starts with building concern, awareness and clarifying green campaigns for consumers in Indonesia.

Consumer Education on Green Marketing

Reynolds et al (2024) proposed that consumer education on green marketing is crucial for ensuring that consumers make informed and environmentally responsible purchasing decisions. Wagner (2002) believe that a good consumer education reflects on the consumer's purchasing process from pre- to post-purchase where consumers will first think about the impact of the goods consumed on the environment before buying them. According to Gaffar et al (2021), currently Indonesian consumers are at the stage of caring about environmental issues, green marketing and circular economy movement campaigns. In MSME products, consumers are increasingly seeking out green products as they become more aware of the environmental consequences of their purchasing decisions. Companies are also responding to this demand by developing and marketing products that meet stringent environmental standards and certifications.

In Indonesia, the value of 'being green' is given to the consumers since they were in elementary school. There is a program called 'Adiwiyata School'. Adiwiyata School is a government program implemented by the Indonesian Ministry of Environment and Forestry (KLHK) in 2019 concerning the Environmental Care and Culture Movement in madrasas. The aim of the Adiwiyata Program is to create good conditions for schools to become places of learning and awareness for the school community, so that in the future the school community can take responsibility for efforts to save the environment and sustainable development (Atiku & Anane-Simon, 2022; Fields & Atiku, 2018, Wulandari, 2018). In relation to consumerism, this program is expected to give an impact to the young consumers in creating a green eco-consiousness since their early age.

According to Natakoesoemah & Adiarsi (2020), in the case of Indonesian millennials, it was found that there was an influence of environmental knowledge and perceived consumer effectiveness on purchasing behavior of eco-friendly products, especially green MSME products. This means that to create effective marketing and purchase, it is necessary to increase consumer knowledge about the importance of environmental conservation. This is reinforced by the findings from research by Raharja & Chan (2021) which states that young people's perceptions about the importance of green consumer behavior practices and their willingness to buy green products are related to attitudes towards the environment, the seriousness of environmental problems, and knowledge of environmental problems. Environmental Knowledge Problems are the strongest predictor of a desire to buy green products, followed by Consumer Attitude. The lowest predictor is Perceived Seriousness of Environmental Problems.

Thus, green marketing and consumer education are crucial components of sustainability, serving to raise awareness, change behavior, and drive demand for environmentally friendly products and practices. These efforts not only inform individuals about pressing environmental issues like climate change and resource depletion but also empower them to make sustainable choices in their daily lives (Natakoesoemah & Adiarsi,2020; Gaffar et al, 2021; Reynolds et al,2024). By highlighting the benefits of eco-friendly options and providing information on their impact, green marketing and consumer education can effectively influence consumer behavior and encourage the adoption of more sustainable practices (Atiku & Abatan, 2021; Gaffar et al, 2021; Reynolds et al,2024). Moreover, these initiatives can incentivize MSMEs to improve their sustainability practices, innovate in sustainable technologies, and contribute to a more sustainable economy (Natakoesoemah & Adiarsi,2020; Gaffar et al, 2021; Randa & Atiku, 2021; Reynolds et al,2024). Overall, green marketing and consumer education are integral to fostering a culture of sustainability and ensuring a greener, more sustainable future for all.

The Case of Indonesia B2B Green MSMEs: Siklus Mutiara Nusantara

Siklus Mutiara Nusantara is an Indonesian MSME that operates in green business and also implements green marketing. This MSMEs is chosen as how this MSMEs can strongly build and operate in the center of Indonesia's the most industrial center. This Founded in 2017, this MSME has a vision, namely "to become a consulting service MSME that is in accordance with safety, environmental health, and appropriate in providing solutions to individuals and companies". This MSME also has a vision, namely providing services of excellent quality that are tailored to customer needs and ensuring customer satisfaction, providing planning and development of safety and security management systems as well as providing safety and security equipment, and is committed to implementing occupational health and safety, environmental sustainability and corporate social responsibility. In terms of core business, this company focuses on providing services in managing non-B3 waste. This waste is a type of waste that can be recycled into renewable energy. The motivation that supports MSMEs in starting this business is based on concern about the areas around the cities of Jakarta, Depok, Tangerang and Bekasi which have many factories that produce a lot of industrial waste. So far, this waste has been thrown into a landfill. The waste is removed, then piled up and then buried in the ground. This waste disposal technique is not completely effective. The result creates piles of rubbish and pollutes the environment. On this basis, Siklus Mutiara Nusantaraoffers a green business-based business concept with the main focus being waste treatment management. This MSME commitment to running a green business is reflected in the company logo which is visualized in the image below:

Figure 3. The logo of Siklus Mutiara Nusantara

Source: Siklus Mutiara Nusantara Profile Document (2024)

Figure 3 shows a logo in the form of a cycle which means sustainable, a circle shape which reflects the nature of MSMEs which have dynamic rotation and movement. An odd number also indicates completeness, cycle, continuation, or unity. Meanwhile, the green color represents MSMEs' concern for the environment, harmonization and revitalization. The color yellow shows optimistic values and creativity and brings happiness to the world. Meanwhile, the blue color represents the value of trust and loyalty in MSMEs towards partners and also their commitment to waste management and the environment. The values reflected in the logo are well represented and through the results of the SWOT analysis you can find out what the strengths, weaknesses, threats and opportunities are in this business.

Table 1. SWOT analysis of Siklus Mutiara Nusantara

Strengths:	Weaknesses:
Providing collection, recycling, treatment and disposal services for non - hazardous waste (Non B3). Doing collaborative research internal and external with the expert on waste management to achieve green sustainable waste handling in responsible way, compliance, more effective and efficient Providing consultancy services to the client on waste management system: - Environmental compliance - Legal & permit licenses -Corporate Social Responsibility on environmental aspect - Green and sustainable waste management	Dependence on niche markets: Sustainable businesses may have a more limited market compared to conventional businesses. This can lead to dependence on niche markets which can be risky if those markets suddenly change. Strict regulations: Regulations imposed by the Ministry of the Environment as well as local governments can increase operational costs and reduce business flexibility. High capital costs: require a larger initial investment than conventional businesses. This can be an obstacle for people or companies that do not have large capital.
Opportunities Growing Market: More and more environmentally conscious consumers are looking for environmentally friendly products and services. This creates opportunities for sustainability businesses to expand market share and attract new customers. Product and Service Innovation: Sustainability businesses can leverage innovation to develop more environmentally friendly products and services, which can increase the competitiveness and relevance of the business in an ever-changing marketplace.	Threats Market Challenges: Intense competition or shifting consumer preferences can threaten business growth and impact demand for environmentally friendly products or services. Environmentally Unfriendly Technology: Technological advances that are not environmentally friendly may make sustainable business products or services less competitive or irrelevant.

Source: Data Processing Results (2024)

Table 1 shows that this business has strong strengths in providing green business. The highly upheld value of creating a more sustainable and environmentally friendly economy is reflected in the service programs. The strengths possessed are internal to the company, but weaknesses are found in external factors of the company. This business is very dependent on the niche market. This business is B2B based, so the target market is more limited compared to conventional businesses. This business also has the characteristic that its consumers have a high level of loyalty. Because of this, if the target consumers partner with other businesses in the future, then consumers in the niche market will be disadvantaged. Another possibility is, if target consumers can manage it themselves, then there will be fewer consumers in the niche market. Because of this consideration, what causes the shortcomings of this business is its dependence on a niche market which can be risky if the market suddenly changes. Meanwhile, a good opportunity for business is that there is a development in the mindset and awareness of consumers regarding waste management. However, the threat is found that there are many competitors who are also involved in this field. This competitor problem is the main threat, this is because some competitors use camouflage where in their profile, they claim that they are a waste treatment business, but in practice they only re-design it and sell it again at a cheaper price to the market. For this reason, an overview of the competitive environment and the position of MSMEs as distribution and waste management treatment partners is depicted in the following picture:

Figure 4. The logo of Siklus Mutiara Nusantara

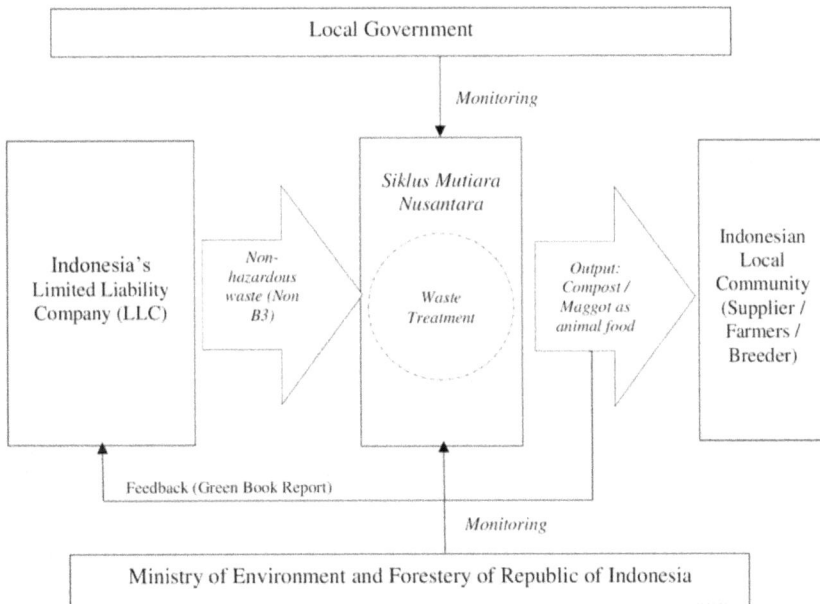

Source: Data Processing Results (2024)

Figure 4 shows the position of the MSMEs Siklus Mutiara Nusantara as a party that provides services in carrying out waste management treatment. Figure 2 above shows that the position of Siklus Mutiara Nusantara is an intermediary between the recipient, production and promotion of waste products produced by LLC-based companies. The waste given to MSMEs is then processed to produce output in the form of products that can be remarketed, namely compost fertilizer and maggots as animal food. The resulting output is then resold, with MSME ownership rights, to local communities such as farmers or breeders. The feedback given to partners/clients is in the form of a report document. In this process it is also known that there are government parties such as the local government and the Ministry of Environment and Forestry of the Republic of Indonesia monitoring the processes carried out by MSMEs in waste treatment.

The commitment of Siklus Mutiara Nusantara MSMEs in conducting green business can be said to be high. This is proven by the fact that since 2017, this MSME has successfully served more than 15 LLC-type companies. LLC is the abbreviation of Limited Liability Company. This is a form of legal entity commonly used in Indonesia for companies with limited shareholder liability according to the capital paid in. In this context, a Limited Liability Company is similar to a Limited Liability Company (LLC) in other countries. Business service activities in MSMEs are described as follows:

Figure 5. One of Siklus Mutiara Nusantara's business services

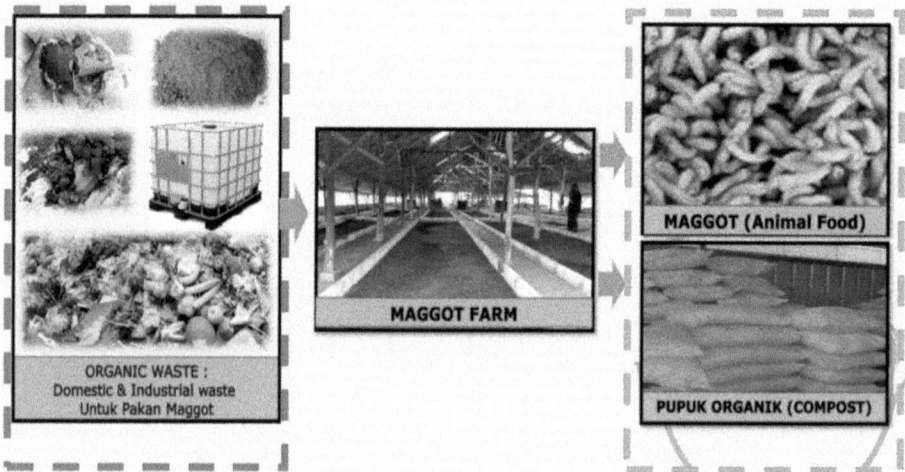

Source: Siklus Mutiara Nusantara Profile Document (2024)

Figure 5 shows the process of how business services regarding waste treatment management are carried out. The process is carried out from the left to the right image, where in the first stage, the Siklus Mutiara Nusantarareceives organic waste in the form of domestic and industrial waste (for example: ice cream waste, food waste with an expiration date of less than 12 months). This waste is then given to maggots. Maggots are larvae or caterpillars of flies. Adult flies will lay eggs in wet or rotting organic material, and from these eggs will hatch larvae called maggots. Maggots are often found in places with high humidity and decomposing organic material, such as trash cans, animal waste, or other rotting organic material. In this business, maggots act as organic waste decomposing agents, because they are able to decompose organic material into pupae and then into adult flies. Adult flies will reproduce and lay eggs. The carcasses of adult flies and egg shells have high protein and can be used for livestock feed such as cattle feed and organic fertilizer. From this decomposition process, it can be seen that the waste produced can be decomposed without hurting the environment. Waste decomposes circularly and this business has succeeded in running a circular economy. In carrying out green-based marketing, Siklus Mutiara Nusantarahas the following marketing strategy design:

Figure 6. Siklus Mutiara Nusantara marketing strategy canvas

WHY HOW WHAT	WHO	CHANNELS	ERRAND
What problem are we solving? Processing, collecting, recycling, giving treatment and disposaling non-hazardous waste (Non B3).	*What does the value chain look like from us to the customer?* 1. Environmental Awareness: Consumers will see the green business value chain as one that pays close attention to the environment, from raw materials to environmentally friendly production and distribution processes.	*Use the customer journey/channel canvas* Consumer experience in ensuring marketing services arrive and are available. MSMEs will carry out brand awareness through eco labels and environmental marketing.	*Use the customer journey/message canvas* The green message conveyed by MSMEs is in the form of making it easier for companies to manage waste by MSMEs as third parties.
How do we do that differently than others? Integreted & commited pro-environment process. Customized and documented.	2. Transparency: Consumers want transparency in the value chain, including information about raw materials, production processes and the environmental impact of the products produced.	**PRIORITIES** *Set priorities in ongoing business or new business.* The business priority is to ensure that waste obtained from clients can be decomposed properly according to the desired procedures and applicable government regulations. Apart from that, MSMEs also ensure that the evaluation and monitoring carried out can take the form of monitoring with clients or also with government parties such as the Ministry of Environment and Forestry.	
What product or service do we provide? Waste treatment management for Industrial Businessn to create a value added and keeping the waste in circular.	3. Innovation: Consumers tend to value businesses that innovate to reduce their environmental footprint, for example by using new technology or more efficient production processes.	**ON GOING** *You set priorities for current products or companies based on the company goals.* MSMEs carry out green monitoring and evaluation by involving client partners and also the government. MSMEs also promote green reports to the public so that clients gain green awareness and empathy from their consumers.	**NEW** *For startups and new business (e.g. product launches), the priority is always first at the conversion moment (decision phase).* MSMEs will carry out brand awareness through ecolabels and campaigns.
	Which target groups or persona do we appeal to? LLC with B2B category. LLC is the abbreviation of Limited Liability Company. This is a form of legal entity commonly used in Indonesia for companies with limited shareholder liability according to the capital paid in.	**REPEAT FOR EACH AUDIENCE** The marketing strategy carried out is a personal approach by approaching it through eco brands and environmental advertisements.	

Source: Results of Interview Data Processing (2024)

Figure 6 shows that the marketing strategy implemented by MSMEs is a green marketing strategy that promotes pro-environmental eco labels by promoting products with added value. This is in line with what was conveyed by MSME owners who stated that in the case of green B2B-based MSMEs, the marketing approach taken cannot be through social media. This is because social media will be easily imitated by competitors. So, the marketing program is:

"We provide education to end users, namely suppliers or farmers and breeders regarding the use of products (produced from green processes) and the destruction of environmentally friendly product packaging. Green business needs to be delivered with green marketing." – Owner of Siklus Mutiara Nusantara (March, 2024).

However, the interviewee said that, in the case of his study, the success of B2B-based green marketing could not be separated from the closeness/chemistry that was built between the MSME owners and the LLC company partners. Even though competition is tight, this business is expected to grow and attract many interested people in the coming year.

SOLUTIONS AND RECOMMENDATIONS

Indonesia MSMEs Positioning Mapping for Green Marketing

A positioning map is a visual tool used to map the relative position of various brands or products in consumers' minds based on certain attributes (Wuni, 2022). In the context of implementing green marketing, a positioning map can help MSMEs understand their position in an environmentally friendly market and determine appropriate marketing strategies. This is important because 1) the green consumer segmentation which is still developing in Indonesia, especially dominated by millennials; 2) the ability of MSMEs to focus on green business is still in the initial stage and does not yet have a strong commitment; 3) Types of MSMEs based on 'born green' and 'adopted green' were found. So, before actually entering the 'green' area, MSMEs need to ensure where their business is positioned in the picture below:

Figure 7. Indonesia MSMEs positioning mapping for green marketing

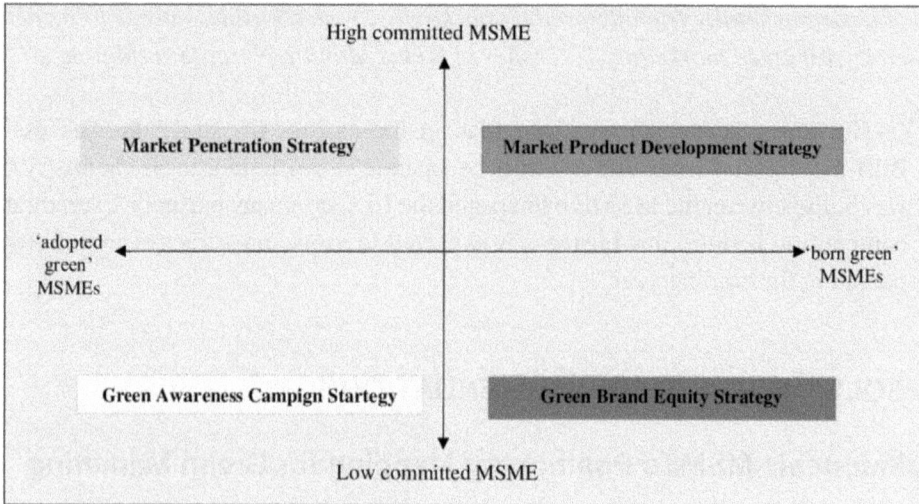

Source: Author's Data Processing Results (2024)

Figure 7 shows an overview of the positioning mapping of Indonesian MSMEs. If MSMEs are known to be 'adopted green' with low commitment to green marketing, then these MSMEs can carry out a 'green awareness campaign strategy' whose function is to slowly educate target consumers. If MSMEs are 'born green' with low commitment, then MSMEs can carry out a green brand equity strategy, meaning that MSMEs focus on managing the image of the products being marketed. If MSMEs are 'adopted green' with high commitment, then MSMEs can carry out a market penetration strategy. This means that MSMEs can start entering the green market niche both at the national and international levels. Lastly, if MSMEs are 'green born' with high commitment, then MSMEs can carry out a market product development strategy. This means that products can be marketed to expand their market share, increase sales, and reduce dependence on existing markets. Positioning mapping in green marketing can be beneficial of MSMEs in a way that the process of identifying and analyzing the competitive landscape within the sustainable or environmentally friendly product market (Adel, 2021). This tool is essential for companies looking to understand the market dynamics, differentiate their offerings from competitors, and effectively communicate their green credentials to consumers (Wuni, 2022). By conducting positioning mapping, companies can gain insights into consumer preferences, competitor strategies, and market trends, which can help them develop targeted marketing campaigns and product positioning strategies.

FUTURE RESEARCH DIRECTIONS

As the future research direction, this chapter suggests exploring how green MS-MEs are able to utilize the green technologies and their role in enabling marketers to create more environmentally friendly products and services. In addition, research regarding green ecosystem in green MSMEs is also valuable consideration as a future research direction.

CONCLUSION

For MSMEs in Indonesia, implementing green marketing is an archipelago of business and consumer education processes that require time and integrated support from various parties. The commitment referred to in implementing green marketing for MSMEs in Indonesia is still in its initial stages. MSMEs are still in the 'awareness' stage and entering implementation. However, Indonesian MSMEs have the potential to have a strong and long-term commitment to green marketing. In Indonesia, the value of 'being green' is given to the consumers since they were in elementary school. There is a program called 'Adiwiyata School'. Adiwiyata School is a government program implemented by the Indonesian Ministry of Environment and Forestry (KLHK) in 2019 concerning the Environmental Care and Culture Movement in madrasas. The aim of the Adiwiyata Program is to create good conditions for schools to become places of learning and awareness for the school community, so that in the future the school community can take responsibility for efforts to save the environment and sustainable development. In relation to consumerism, this program is expected to give an impact to the young consumers in creating a green eco-consiousness since their early age. Realizing the long-term goals of this strategy requires a business ecosystem that cares and understands the value and price of 'sustainability'. Future investment commitments that are paid 'highly' at this time are still a challenge for MSMEs in production and consumers in consumption. In fact, complying with MSMEs in using green marketing is a real step in MSMEs' contribution to the development transition which can reduce operational costs and increase profitability. So, a more conceptual and concrete explanation is needed regarding the extent to which green marketing stages can be accepted and implemented in the MSME business sector which will later be able to help MSMEs visualize the position of MSMEs in competing in emerging markets.

REFERENCES

Adel, H. M. (2021). Mapping and assessing green entrepreneurial performance: Evidence from a vertically integrated organic beverages supply chain. *Journal of Entrepreneurship and Innovation in Emerging Economies*, 7(1), 78–98. 10.1177/2393957520983722

Aprianti, V., Hurriyati, R., Gaffar, V., & Wibowo, L. A. (2021). The effect of green trust and attitude toward purchasing intention of green products: A case study of the green apparel industry in Indonesia. *The Journal of Asian Finance. Economics and Business*, 8(7), 235–244.

Atiku, S. O. (2019). Institutionalizing Social Responsibility Through Workplace Green Behavior. In Atiku, S. (Ed.), *Contemporary Multicultural Orientations and Practices for Global Leadership* (pp. 183–199). IGI Global. 10.4018/978-1-5225-6286-3.ch010

Atiku, S. O. (2020). Knowledge Management for the Circular Economy. In Baporikar, N. (Ed.), *Handbook of Research on Entrepreneurship Development and Opportunities in Circular Economy* (pp. 520–537). IGI Global. 10.4018/978-1-7998-5116-5.ch027

Atiku, S. O., & Abatan, A. A. (2021). Strategic Capabilities for the Sustainability of Small, Medium, and Micro Enterprises. In Ayandibu, A. (Ed.), *Reshaping Entrepreneurship Education With Strategy and Innovation* (pp. 17–44). IGI Global. 10.4018/978-1-7998-3171-6.ch002

Atiku, S. O., & Anane-Simon, R. (2022). Stimulating Creativity and Innovation Through Apt Educational Policy. In Fields, Z. (Ed.), *Achieving Sustainability Using Creativity, Innovation, and Education: A Multidisciplinary Approach* (pp. 113–133). IGI Global. 10.4018/978-1-7998-7963-3.ch006

Boztepe, A. (2012). Green marketing and its impact on consumer buying behavior. *European Journal of Economic & Political Studies*, 5(1).

Chaturvedi, P., Kulshreshtha, K., Tripathi, V., & Agnihotri, D. (2024). Investigating the impact of restaurants' sustainable practices on consumers' satisfaction and revisit intentions: A study on leading green restaurants. *Asia-Pacific Journal of Business Administration*, 16(1), 41–62. 10.1108/APJBA-09-2021-0456

Chekima, B., Wafa, S. A. W. S. K., Igau, O. A., Chekima, S., & Sondoh, S. L.Jr. (2016). Examining green consumerism motivational drivers: Does premium price and demographics matter to green purchasing? *Journal of Cleaner Production*, 112, 3436–3450. 10.1016/j.jclepro.2015.09.102

Damarayudha, T. R., Sadat, A. M., & Febrillia, I. (2023). Pengaruh Green Marketing Mix Terhadap Purchase Intention Dengan Environmental Knowledge Sebagai Variabel Moderator: Survei Pada Toko Furniture Modern. *Indonesian Journal of Economy, Business. Entrepreneurship and Finance*, 3(2), 306–322.

Dangelico, R & Vocalelli, D. (2017). "Green Marketing": an analysis of definitions, strategy steps, and tools through a systematic review of the literature. *Journal of Cleaner Production*. S0959652617316372–. 10.1016/j.jclepro.2017.07.184

Durif, F., Boivin, C., & Julien, C. (2010). In search of a green product definition. *Innovative Marketing, 6*(1).

Ellen MacArthur Foundation. (2015). *Schools of thought – industrial ecology.* Ellen MacArthur Foundation. https://www.ellenmacarthurfoundation.org/circular-economy/schools-of-thought/cradle2cradle

Fields, Z., & Atiku, S. O. (2017). Collective Green Creativity and Eco-Innovation as Key Drivers of Sustainable Business Solutions in Organizations. In Fields, Z. (Ed.), *Collective Creativity for Responsible and Sustainable Business Practice* (pp. 1–25). IGI Global. 10.4018/978-1-5225-1823-5.ch001

Fields, Z., & Atiku, S. O. (2018). Collaborative Approaches for Communities of Practice Activities Enrichment. In Baporikar, N. (Ed.), *Knowledge Integration Strategies for Entrepreneurship and Sustainability* (pp. 304–333). IGI Global. 10.4018/978-1-5225-5115-7.ch015

Fitrianingrum, A. (2020). *Greenwashing, does it Work Well for Indonesian Millennials Buyers.*

Gaffar, V., & Koeswandi, T. (2021). Climate Change And The Sustainable Small And Medium-Sized Enterprises. In *Handbook Of Research On Climate Change And The Sustainable Financial Sector* (pp. 171–189). IGI Global. 10.4018/978-1-7998-7967-1.ch011

Gaffar, V., Rahayu, A., Adi Wibowo, L., & Tjahjono, B. (2021). The adoption of circular economy principles in the hotel industry. *Journals and Gaffar, Vanessa and Rahayu, Agus and Adi Wibowo, Lili and Tjahjono, Benny, The Adoption of Circular Economy Principles in the Hotel Industry (June 30, 2021). Reference to this paper should be made as follows. Gaffar*, V, 92–97.

Gunawan, A. I., Amalia, F. A., Ramadhan, M., & Bansah, P. F. (2024). Exploring The Reasons Of Indonesian Young Adult Consumers Toward Sustainably Packaged Food & Beverages Product. [JMI]. *Journal of Marketing Innovation*, 4(1). 10.35313/jmi.v4i1.106

Handajani, L., Husnan, L. H., & Rifai, A. (2019). Kajian Tentang Inisiasi Praktik Green Banking Pada Bank BUMN di Indonesia. *Jurnal Economia Review of Business and Economics*, 15(1), 1–16.

Hariyanto, O. I. B. (2019). *Customer Green Awareness and Eco-Label for Organic Products.* In: 2019 International Conference of Organizational Innovation (2019 ICOI).

Hashem, T. N., & Al-Rifai, N. A. (2011). The influence of applying green marketing mix by chemical industries companies in three Arab States in West Asia on consumer's mental image. *International Journal of Business and Social Science*, 2(3).

Hignett, S., & McDermott, H. (2015). Qualitative methodology. *Evaluation of human work*, 119-138.

Hong, Z., Wang, H., & Yu, Y. (2018). Green product pricing with non-green product reference. *Transportation Research Part E, Logistics and Transportation Review*, 115, 1–15. 10.1016/j.tre.2018.03.013

Ismail, I. J. (2023). The role of technological absorption capacity, enviropreneurial orientation, and green marketing in enhancing business' sustainability: Evidence from fast-moving consumer goods in Tanzania. *Technological Sustainability*, 2(2), 121–141. 10.1108/TECHS-04-2022-0018

Jennah, H., & Ismail, A. (2023). Pengaruh Green Marketing Mix Terhadap Purchase Decision Dalam Menggunakan Eco Friendly Product. *Journal of Trends Economics and Accounting Research*, 3(4), 390–398. 10.47065/jtear.v3i4.636

Kartawinata, B. R., Maharani, D., Pradana, M., & Amani, H. M. (2020, August). The role of customer attitude in mediating the effect of green marketing mix on green product purchase intention in love beauty and planet products in indonesia. In *Proceedings of the International Conference on Industrial Engineering and Operations Management* (Vol. 1, pp. 3023-3033).

Katadata Insight Center. (2024). *Digitalisasi UMKM.* KIC.

Kosatica, M. (2024). Semiotic landscape in a green capital: The political economy of sustainability and environment. *Linguistic Landscape*, 10(2), 136–165. 10.1075/ll.23016.kos

Muhammad, N. (2023, October 13). *Usaha Mikro Tetap Merajai UMKM, Berapa Jumlahnya?* [webpage]. Diakses pada https://databoks.katadata.co.id/datapublish/2023/10/13/usaha-mikro-tetap-merajai-umkm-berapa-jumlahnya

Myers, M. D. (2019). *Qualitative research in business and management.*

Naini, S. R., Mekapothula, R. R., Jain, R., & Manohar, S. (2024). Redefining green consumerism: A diminutive approach to market segmentation for sustainability. *Environmental Science and Pollution Research International*, 1–17. 10.1007/s11356-023-31717-938180668

Natakoesoemah, S., & Adiarsi, G. R. (2020). The Indonesian Millenials Consumer Behaviour on Buying Eco-Friendly Products: The Relationship Between Environmental Knowledge and Perceived Consumer Effectiveness. *International Journal of Multicultural and Multireligious Understanding*, 7(9), 292–302.

Novela, S., & Hansopaheluwakan, S. (2018). Analysis of Green Marketing Mix Effect on Customer Satisfaction using 7p Approach. *Pertanika Journal of Social Sciences & Humanities*.

Nurfitriya, M., Fauziyah, A., Koeswandi, T. A. L., Yusuf, I., & Rachmani, N. N. (2022). Peningkatan Literasi Digital Marketing UMKM Kota Tasikmalaya. *Acitya Bhakti*, 2(1), 57. 10.32493/acb.v2i1.14618

Raharja, S. U. J., & Chan, A. (2021). Youth's Green Consumer Behavior: A Study In Citarum Watersehd West Java Indonesia. *AdBispreneur: Jurnal Pemikiran dan Penelitian Administrasi Bisnis dan Kewirausahaan*, 6(3).

Rahmawati, M., Pratiwi, S. R., Devi, C., & Nainggolan, Y. T. (2022). Penerapan Strategi Green Marketing Di Tengah Pandemi Covid-19. *Jurnal Ekonomika*, 13(01), 1–18. 10.35334/jek.v13i0.2410

Randa, I. O., & Atiku, S. O. (2021). SME Financial Inclusivity for Sustainable Entrepreneurship in Namibia During COVID-19. In Baporikar, N. (Ed.), *Handbook of Research on Sustaining SMEs and Entrepreneurial Innovation in the Post-COVID-19 Era* (pp. 373–396). IGI Global. 10.4018/978-1-7998-6632-9.ch018

Reynolds, M., Salter, N., Muranko, Ż., Nolan, R., & Charnley, F. (2024). Product life extension behaviours for electrical appliances in UK households: Can consumer education help extend product life amid the cost-of-living crisis? *Resources, Conservation and Recycling*, 205, 107527. 10.1016/j.resconrec.2024.107527

Romli, N. A., Safitri, D., & Yustitia, P. (2023). Strategi Komunikasi Pemasaran Hijau Dalam Pemberdayaan Kewirausahaan Masyarakat Mat Peci. *IKRA-ITH HUMANIORA: Jurnal Sosial dan Humaniora*, 7(3), 59-71.

Sana, S. S. (2020). Price competition between green and non green products under corporate social responsible firm. *Journal of retailing and consumer services, 55*, 102118.

Shehawy, Y. M., & Khan, S. M. F. A. (2024). Consumer readiness for green consumption: The role of green awareness as a moderator of the relationship between green attitudes and purchase intentions. *Journal of Retailing and Consumer Services*, 78, 103739. 10.1016/j.jretconser.2024.103739

Sinambela, E. A., Azizah, E. I., & Putra, A. R. (2022). The Effect of Green Product, Green Price, and Distribution Channel on The Intention to Repurchasing Simple Face Wash. *Journal of Business and Economics Research (JBE), 3*(2), 156-162.

Suarna, I. W. (2018). Bali dalam Tarikan Pembangunan Berkelanjutan. *Jurnal Bali Membangun Bali*, 1(3), 199–206. 10.51172/jbmb.v1i3.31

Suasana, I. G. A. K. G., & Ekawati, N. W. (2018). Environmental commitment and green innovation reaching success new products of creative industry in Bali. *The Journal of Business and Retail Management Research*, 12(4). 10.24052/JBRMR/V12IS04/ART-25

Suhartanto, D., Dean, D., Amalia, F. A., & Triyuni, N. N. (2024). Attitude formation towards green products evidence in Indonesia: Integrating environment, culture, and religion. *Asia Pacific Business Review*, 30(1), 94–114. 10.1080/13602381.2022.2082715

Sutikno, A. N. (2020). Bonus demografi di indonesia. *VISIONER: Jurnal Pemerintahan Daerah Di Indonesia*, 12(2), 421–439.

Tamim, M. S., & Akter, L. (2024). Green marketing impact on youth purchasing: Bangladesh district-wise study on consumer intentions. *Annals of Management and Organization Research*, 5(3), 205–217. 10.35912/amor.v5i3.1818

Thoibah, W., Arif, M., & Harahap, R. D. (2022). Implementasi Green Marketing Pada UMKM Upaya Memasuki Pasar Internasional (Studi Kasus pada Creabrush Indonesia). *Jurnal Ekonomika Dan Bisnis*, 2(3), 798–805.

Vohra, N. D., & Arora, H. (2021). *Quantitative techniques in management.* McGraw Hill.

Wagner, S. A. (2002). *Understanding green consumer behaviour: A qualitative cognitive approach.* Routledge. 10.4324/9780203444030

Wulandari, N. P. D. (2018). Between Eco-Education And Critical Thinking: The Application of Emancipatory Learning on Gaining The Awareness of Environmental Problems. In *Bali. In Proceding-International Seminar Culture Change and Sustainable Development in Multidisciplinary Approach: Education, Environment, Art, Politic, Economic, Law, and Tourism* (pp. 126–132). Udayana University.

Wuni, I. Y. (2022). Mapping the barriers to circular economy adoption in the construction industry: A systematic review, Pareto analysis, and mitigation strategy map. *Building and Environment*, 223, 109453. 10.1016/j.buildenv.2022.109453

Yacob, S., Erida, E., Machpuddin, A., & Alamsyah, D. J. M. S. L. (2021). A model for the business performance of micro, small and medium enterprises: Perspective of social commerce and the uniqueness of resource capability in Indonesia. *Management Science Letters*, 11(1), 101–110. 10.5267/j.msl.2020.8.025

Yaputra, H., Risqiani, R., Lukito, N., & Sukarno, K. P. (2023). Pengaruh Green Marketing, Sustainable Advertising, Eco Packaging/Labeling Terhadap Green Purchasing Behavior (Studi Pada Kendaraan Listrik). [IMA]. *Journal of Indonesia Marketing Association*, 2(1), 71–90.

Yusiana, R., Widodo, A., & Sumarsih, U. (2021). Integration Consumer Response during the Pandemic Covid-19 on Advertising: Perception Study on Eco Labeling and Eco Brand Products Eco Care. *Inclusive Society and Sustainability Studies*, 1(2), 45–56. 10.31098/issues.v1i2.708

Chapter 9
Green Organizational Culture and Sustainable Development:
Nurturing Environmental Responsibility in Businesses

Omolola Ayobamidele Arise
MANCOSA, South Africa

Meshel Muzuva
http://orcid.org/0009-0005-4284-5727
MANCOSA, South Africa

ABSTRACT

As society grapples with environmental challenges, businesses are increasingly compelled to align their operations with principles of sustainability. One pivotal mechanism for fostering sustainable development is the cultivation of a green organizational culture; an organizational tenet that embeds environmental responsibility within its values, norms, and behaviours. This chapter investigates the transformative role of Green Organizational Culture (GOC) within businesses in advancing Sustainable Development (SD), highlighting the imperative shift from profit-centric motives towards fostering ecological balance and social equity. Employing a qualitative synthesis of existing literature, this chapter navigates through the theoretical frameworks that underpin GOC, highlighting its significance in driving environmentally responsible behaviours within businesses. By examining case studies and scholarly works, the chapter identifies and analyzes the core components and characteristics of GOC, the strategic implementation strategies for fostering such a culture, and its profound impact on SD. The chapter systematically outlines how a commitment to

DOI: 10.4018/979-8-3693-2595-7.ch009

environmental responsibility at all organizational levels can be achieved through aligning with external institutional norms and effectively managing internal processes for change. It provides strategic recommendations for embedding a green culture within organizational practices, thereby contributing to sustainable development goals. It concludes with strategic insights for organizations seeking to navigate the complexities of environmental sustainability, emphasizing the long-term benefits of such endeavours for ensuring corporate success and legitimacy in a globally conscious marketplace.

INTRODUCTION

In recent decades, there has been a growing recognition among business leaders worldwide of the increasing risks and responsibilities associated with escalating environmental degradation. Human economic activities, such as industrialization and consumption, have played a significant role in issues like climate change, biodiversity loss, pollution, deforestation, and resource depletion (Whiteman et al., 2013; Upadhyay, 2020). For example, climate change has led to extreme weather events, rising sea levels, disruption of natural cycles, and mass species extinction (IPCC, 2022). Pollution from sources such as manufacturing and agriculture poses threats to public and ecosystem health. Deforestation for development and agriculture has negative impacts on natural capital and habitats. These interconnected challenges are putting immense strain on global systems. As a result, the business landscape is undergoing a paradigm shift where success is no longer solely measured in financial terms but increasingly tied to an organization's commitment to environmental stewardship.

As contemporary society grapples with the complex challenges posed by climate change, depletion of natural resources, and other environmental crises, it is crucial for corporations to critically evaluate their operational practices to foster a sustainable future. These entities are facing greater scrutiny and are expected to integrate sustainability principles into their operational frameworks. This involves not only reducing their ecological footprint but also making meaningful contributions to societal well-being. A valuable strategy for corporations to achieve these objectives is to cultivate a green organizational culture. Such a culture is characterized by the integration of environmental stewardship into the organization's core values, norms, and behaviours, as discussed by Liu and Lin (2020). In this paradigm, a commitment to environmental responsibility permeates all levels of the workforce, driving concerted efforts to minimize the organization's environmental impact and exert a positive influence on a global scale. This chapter explores the importance of fostering a green organizational culture within business entities as a crucial strategy for promoting sustainable development. It is supported by a comprehensive review of relevant literature, including the works of scholars such as Denison (1990), Cameron and Quinn (2006), Jerónimo et al. (2020), and Umair, Mrugalska, and Al Shamsi (2023), which provide insights into how organizational culture influences the attitudes and behaviours of employees.

Ultimately, by focusing on how organizations can integrate environmental stewardship into their core values and norms, the chapter explored the transformative impact of a green organizational culture on sustainability practices. This exploration is critical, as fostering such a culture is increasingly recognised as a fundamental component for achieving positive environmental outcomes. In doing so, the study contributes to a deeper understanding of the strategic alignment between organizational culture and environmental sustainability. The ensuing sections will further clarify this relationship, presenting empirical findings and discussing their implications for theory and practice.

The chapter starts with an introduction into the concept of organizational culture, then the overview of green organizational culture and its critical role in furthering sustainability and sustainable development within the business world. The discussion highlights the significance of promoting a green organizational culture as a basis for encouraging environmentally responsible practices in business operations. Moreover, the importance of fostering a green organizational culture is emphasized to motivate environmentally friendly practices in business operations. In addition, theoretical background supporting the interactions between organizational culture and sustainable development; and how businesses can harness their benefits were highlighted. The narrative also explores how organizational culture interacts with sustainable development, tackling the challenges and barriers that might arise and proposing potential solutions. By concluding with strategic recommendations, the chapter aims to provide insights into fostering a green organizational culture, contributing to the broader goals of sustainable development.

THE CONCEPT OF ORGANIZATIONAL CULTURE (OC)

The critical importance of organizational culture not just shapes internal operations but also defines a company's strategic direction and its capacity to adapt to and lead within its market Nimfa et al. (2021). Organizational culture encompasses the shared values, beliefs, and norms that define the internal environment of an organization and influence the behaviours and attitudes of its members (Weber and Martensen, 2021). The authors considered OC as a guiding framework that dictates how employees interact, make decisions, and perceive their roles and the expectations placed upon them. A well-aligned organizational culture not only fosters a positive work environment but also boosts productivity, employee satisfaction, and overall organizational efficacy. Alvesson and Sveningsson (2015) in support, note that culture is a concept that is deeply embedded within the organization's fabric, making it both a powerful ally in strategic alignment and a formidable barrier to change. The complexity of managing and sustaining organisational culture according to the authors, is highlighted by its dual character as both a benefit and a challenge. Thus, to achieve long-term success with strategic changes, it is essential to have a comprehensive grasp of the current cultural framework and to take an innovative approach to cultural transformation, striking a balance between alignment and adaptability.

Fietz and Günther (2021) provide a comprehensive examination of organizational culture, describing it as a complex blend of values, practices, symbols, and behaviours that permeate every corner of an organization. They argue that these cultural elements are not confined to the structural limits of the organization; rather, they significantly shape how employees perceive their roles, interact with one another, and approach decision-making tasks. This perspective highlights the expansive impact of organizational culture, suggesting that it is a crucial determinant of the workplace environment and a foundational influence on employee engagement and operational effectiveness. The role of organizational culture as emphasized by Fietz and Günther (2021), is also pivotal in driving environmental management and sustainability. Such integration emphasizes organizational culture as a critical determinant in achieving enhanced employee engagement, operational effectiveness, and the successful implementation of sustainability practices.

OVERVIEW OF GREEN ORGANIZATIONAL CULTURE (GOC)

A culture that values innovation and openness to change is crucial for promoting sustainable practices and environmental stewardship. Hence, nurturing a supportive cultural environment is vital for achieving long-term environmental goals. According to Olafsen et al. (2021), a flexible organizational culture plays a crucial role in creating an environment that supports sustainable change by enhancing change self-efficacy and minimizing negative personal impacts on employees. This highlights the interdependence between organizational culture, managerial actions, and the successful implementation of sustainability objectives.

Baumgartner et al. (2007) emphasized the integration of zero-emission initiatives to foster a Green Organizational Culture (GOC), where environmental sustainability is a core value. This shift aligns corporate procedures and regulations with environmental objectives, promoting sustainability. By incorporating Baumgartner et al.'s (2007) insights, organizational culture can be viewed as a complex blend of shared beliefs, values, practices, symbols, and behaviours that deeply influence every aspect of an organization. The integration of zero-emission strategies fosters a GOC, embedding environmental consciousness into the organization's DNA and increasing commitment to sustainable development.

To further foster the discussion on the impact of organizational cultures on business outcomes, one can draw on studies by Deirmentzoglou et al. (2020) and, Ergün and Tasgıt, (2013), using the Competing Values Framework (CVF) which assesses the influence of different cultures (clan, adhocracy, hierarchy, market) on sustainable development in Greece and innovation performance in Turkish hotels. These studies reveal that while hierarchical cultures promote economic sustainability through structured processes, and adhocracy cultures facilitate environmental sustainability and innovation through their flexibility and creativity, the roles of clan and market cultures are more complex. Clan culture, with its emphasis on internal community and teamwork, potentially enhances social sustainability and internal innovation, though evidence is mixed. Market culture, focused on competitiveness and external achievements, drives innovation aimed at improving market share and competitiveness, yet its impact on social sustainability is less clear and may require further exploration. This discussion underlines the importance of aligning organizational culture with strategic objectives, whether aiming for sustainability or enhanced innovation, and prompts a deeper investigation into how organizations can balance these cultural attributes to meet dynamic market demands and global challenges. Such a discussion is crucial for understanding the delicate role of organizational culture in achieving diverse and context-specific business outcomes.

This chapter supports the adhocracy organizational culture view as it is advantageous for adopting green organizational initiatives due to its inherent flexibility, innovation, and risk-taking capabilities. Adhocracy fosters a creative environment that encourages rapid adaptation to environmental changes and regulatory demands, making it ideal for developing and implementing sustainable technologies and practices (Fietz and Günther, 2021). This culture type enhances the organization's ability to quickly integrate new environmental information and technologies, empowering employees across levels to contribute to eco-friendly solutions (Olafsen et al., 2021). Additionally, the dynamic and collaborative nature of adhocracy facilitates cross-functional integration of sustainability into all business operations, not just isolated departments (Martela, 2019).. Ultimately, adopting an adhocracy organizational culture not only aligns with but actively promotes the goals of environmental stewardship, positioning an organization to lead in sustainability and innovate within the green market space.

Adopting a green organizational culture is an effective strategy for businesses to promote sustainability, with a focus on environmental stewardship as a core priority (Hakim, 2023). This approach aligns the organization around ecologically sustainable values and encourages employees to consider environmental factors in their daily operations and decision-making processes (Norton et al., 2015). Wang (2019) explored the influence of Green Organizational Culture (GOC) on manufacturing entities and its role in enhancing green performance and gaining a competitive advantage. The research argued that an organization's commitment to environmental sustainability, as reflected through its culture, is a driving force behind the adoption of green innovations. These innovations not only mitigate environmental impacts but also serve as strategic tools for outperforming competitors and capturing a larger market share. Wang's study examined Taiwanese manufacturing firms and provided empirical evidence that a well-developed GOC fosters a preference for green innovative practices. These practices enhance the firm's environmental effectiveness and solidify its competitive position, emphasizing the strategic importance of integrating sustainability into the fabric of the organization for driving innovation and distinguishing the firm in a competitive market. The concept of green organizational culture emerged due to the pressing need for environmental sustainability (Harris and Crane, 2002; Afum et al., 2020; Shah et al., 2021). It involves shared assumptions, norms, and behaviours that demonstrate a commitment to ecological responsibility. Green organizational culture places emphasis on integrating sustainability into a company's core practices and values as a foundational aspect.

SUSTAINABLE DEVELOPMENT IN THE CURRENT BUSINESS LANDSCAPE

In the evolving business landscape, sustainable development has evolved from being a peripheral concern to becoming a central strategy for companies around the globe. This shift reflects a growing recognition of the limited availability of natural resources, the consequences of climate change, and the societal demand for ethical business practices. According to the Brundtland Commission, sustainable development is "development that meets the needs of the present without compromising the ability of future generations to meet their own needs" (World Commission on Environment and Development, 1987). In the context of business, sustainable development involves adopting strategies and practices that fulfil the needs of the enterprise and its stakeholders in the present while also safeguarding, sustaining, and enhancing the human and natural resources that will be required in the future (Elkington, 1997; Andersson et al., 2022). This concept is commonly referred to as the triple bottom line (TBL) framework, which encompasses economic, environmental, and social dimensions (Elkington, 1997; Silvestri et al.,). Various factors drive the adoption of sustainable practices in the business sector, including regulatory pressures, consumer demand for ethical and environmentally friendly products, investor preferences for sustainable investments, and the intrinsic motivation to reduce environmental impacts (Porter and Kramer, 2006).

Technological advancements have also played a significant role in facilitating the development of green technologies, making sustainable practices more accessible and cost-effective (Schaltegger and Wagner, 2011). Despite the obvious benefits, businesses face numerous challenges when implementing sustainable practices. These challenges include the initial cost of transitioning, the complexity of accurately measuring sustainability metrics, and the need for cultural change within organizations (Bansal and Roth, 2000; Popescu, 2020). Furthermore, the lack of universally accepted standards for sustainability reporting can complicate efforts to communicate sustainability initiatives transparently (Gray, 2006; Mahmood and Uddin, 2021). Despite these challenges, several leading companies have successfully integrated sustainability into their core operations. For example, Unilever's Sustainable Living Plan aims to separate the company's growth from its environmental footprint while increasing its positive social impact (Siddique and Sultana, 2018). Similarly, Patagonia's commitment to environmental and social responsibility has been central to its business model, encompassing initiatives such as using recycled materials and donating a portion of its profits to environmental causes (Patagonia, 2020).

SIGNIFICANCE OF PROMOTING GREEN ORGANIZATIONAL CULTURE FOR AN ENVIRONMENTALLY RESPONSIBLE BEHAVIOUR IN BUSINESSES

A green organizational culture is characterized by several key features that collectively foster a sustainable workplace environment. Foremost among these is the unwavering commitment of leadership towards sustainability initiatives, which ensures that environmental considerations are at the forefront of the organization's agenda. Additionally, a strategic orientation towards minimizing environmental impacts, the adoption of green facilities, materials, and branding, as well as policies that incorporate environmental practices into everyday business operations, are indicative of a green culture. Furthermore, engaging employees through the formation of green teams and provision of sustainability-focused training, coupled with the diligent measurement and reporting of sustainability metrics, are critical components that contribute to the development and maintenance of a green organizational culture.

Developing a green organizational culture is a critical approach for businesses to promote sustainability (Lozano, 2013). It aligns the organization with shared environmental values and motivates employees to incorporate ecological considerations into daily decisions and tasks (Ones & Dilchert, 2012). A robust green culture improves consistency and coordination of sustainability efforts across departments

(Linnenluecke and Griffiths, 2010). Moreover, an organizational culture supporting sustainability can enhance a company's reputation among stakeholders. These issues highlight the necessity for sustainable development; economic growth that meets present needs without compromising the ability of future generations to meet their own needs (Emina, 2021). The sustainability concept demands that businesses pursue profits responsibly while supporting long-term ecological balance and social equity (Moldavska and Welo, 2019). For the corporate sector, environmental sustainability is no longer optional but imperative (Bansal and Song, 2017). Given that companies depend on functioning ecosystems and stable communities to produce goods, provide services, and maintain operations (Whiteman et al., 2013), they must adapt to climate change impacts and potential restrictions on resource use (Engert et al., 2016). A proactive environmental management can mitigate business risks and unlock opportunities in green technologies and sustainability consulting (Kiron et al., 2017).

Worthy of note is that the pursuit of sustainability transcends mere compliance with environmental and societal demands; it is a strategic initiative aimed at boosting an organization's reputation, fulfilling stakeholder expectations, and securing long-term viability Dhanda and Shrotryia (2021). The application of institutional theory combined with green organizational culture theory provides a robust framework for understanding and steering this transformative journey. These theories collectively offer insights into how organizations can align internal cultural practices with broader external pressures to achieve sustainable outcomes effectively.

To develop the exploration of how organizational culture influences sustainable development, this chapter explores the integration of Institutional Theory and theories surrounding Green Organizational Culture. This theoretical exploration provides a dual perspective on the external pressures and internal capabilities that organizations must navigate to encourage sustainability. Institutional Theory offers a foundational framework for understanding how organizations adapt to societal expectations. Grounded in the seminal works of DiMaggio and Powell (1983) on institutional isomorphism and expanded by Tribe (2022) through his model of the three pillars: regulative, normative, and cultural-cognitive. This theory explains how legal, social, and shared cognitive structures compel organizations to align with external standards, including those related to environmental sustainability. The mechanisms, categorised into coercive, mimetic, and normative pressures, help organizations gain legitimacy and operational stability by adhering to societal norms and regulations that govern sustainable practices.

Green Organizational Culture Theory complements this by focusing on the internal dynamics within organizations. It emphasizes the importance of cultivating an organizational culture that inherently supports sustainable practices through values, norms, and workplace behaviours. This theory stresses how a green organizational culture not only responds to but anticipates environmental challenges, integrating sustainability into the core strategic objectives of the organization. By encouraging a culture that prioritizes environmental stewardship, organizations can enhance their responsiveness to the institutional pressures outlined in Institutional Theory.

The interaction between these two theoretical frameworks is crucial for a comprehensive understanding of organizational sustainability. While Institutional Theory highlights the role of external pressures in shaping organizational practices towards sustainability, Green Organizational Culture Theory focuses on the internal transformation necessary to embed these practices deeply within the organizational ethos. For instance, Lewin's three-stage model of change; unfreezing, change, and refreezing, provides a methodological approach for organizations to transition towards greener practices, emphasizing the need for an adaptive and proactive organizational culture (Lewin, 1947).

Furthermore, Kotter's eight-step process for leading change reinforces this by detailing steps like creating a sense of urgency and forming powerful coalitions, which are vital for embedding sustainable development into the organizational culture (Kotter, 1996). These change models are instrumental in illustrating how organizations can systematically transition towards sustainability, not just in compliance with external demands but as a strategic enhancement of their internal capabilities.

In conclusion, by synthesizing Institutional Theory with Green Organizational Culture Theory, this chapter outlines how the pressures of the institutional environment interact with the strategic transformations within organizations to foster a robust green organizational culture. This integrated approach not only aligns with regulatory and societal expectations but also champions innovation and sustainability as core elements of organizational identity and strategy, thereby ensuring long-term viability and competitive advantage in a rapidly evolving global market.

IMPACT OF GREEN ORGANIZATIONAL CULTURE ON SUSTAINABLE DEVELOPMENT

The world increasingly recognizes the crucial role businesses play in achieving sustainable development. Organizations embracing a green organizational culture weave this commitment into the very fabric of their operations, actively contributing to the United Nations' 17 Sustainable Development Goals (SDGs). This section delves into the tangible impact of such a culture, highlighting its power to create positive change across environmental, social, and economic dimensions.

Environmental Benefits

Measuring and understanding the environmental impact is crucial for fostering a green organizational culture. By implementing sustainable strategies, businesses can achieve significant environmental benefits. Companies have the power to make a substantial difference in preserving natural resources and mitigating climate change. This can be achieved by reducing resource consumption, minimizing waste generation, and lowering greenhouse gas emissions. For instance, embracing energy-efficient technologies and renewable energy sources can effectively decrease carbon footprints and decrease reliance on fossil fuels. Additionally, implementing recycling and waste management programs can reduce landfill waste and promote a circular economy. Furthermore, prioritizing environmental sustainability, businesses not only fulfil their corporate social responsibility obligations but also gain a competitive advantage by appealing to environmentally conscious consumers.

It is essential to measure progress towards sustainability goals to hold individuals accountable, ensure transparency, and demonstrate the effectiveness of implemented practices. Green organizations focus on their environmental responsibilities and strive to make measurable progress in various Sustainable Development Goals (SDGs). Companies like Microsoft and Unilever employ life cycle assessment (LCA) to determine greenhouse gas emissions throughout their value chains. Microsoft is committed to achieving carbon negativity by 2030. To reach this goal, the company is investing heavily in renewable energy sources, carbon capture technologies, and projects that offset emissions, such as forestry initiatives (Microsoft, 2023). Through ambitious reduction targets, they actively contribute to SDG 13: Climate Action, effectively mitigating the impact of global warming. Since 2008, Unilever has made significant advancements in reducing water usage on a global scale. This has been accomplished through process improvements, rainwater harvesting, and the use of water-efficient technologies in agriculture. These efforts align with SDG 6: Clean Water and Sanitation (Unilever, 2023).

Monitoring energy consumption is a valuable tool for identifying areas in need of efficiency improvements. Tesla, for example, prioritizes energy efficiency in its manufacturing processes and products, enabling them to reduce their environmental impact and lower operational costs (Tesla, 2023). Similarly, industries with specific resource requirements can benefit from monitoring and analysing relevant metrics. Fashion brands, likewise, also monitor water usage for cotton cultivation and chemical usage in dyeing, while construction companies could track material waste and resource reuse. In 2022, H&M increased the use of recycled materials in its products to 24%. According to the H&M Group (2023), the company plans to exclusively use recycled or sustainably produced materials by 2030. This goal aligns with the principles of Responsible Consumption and Production (SDG 12) and aims to minimize the environmental and social impact of materials throughout their lifecycle. These examples illustrate how organizations can go beyond regulatory requirements by actively addressing climate change, conserving resources, and promoting responsible production practices.

Social Benefits

The implementation of sustainable strategies brings various social benefits. A green organizational culture not only prioritizes environmental responsibility but also places a high priority on social sustainability. This means that the organization's operations contribute positively to society and respect the well-being of individuals. When considering the long-term impact that an organization has on society, the social dimension is an extremely important factor. According to Wang et al., (2022), incorporating sustainability into corporate cultures present a good opportunity to link business operations with societal well-being and environmental health. A socially sustainable organization prioritizes the health and happiness of its workforce, recognizing that employees are significant assets (Wang et al., 2021). Investing in employee well-being means encouraging a healthy work-life balance, advocating for diversity and inclusion, and ensuring fair salaries and benefits. Google is well-known for its forward-thinking work culture, which goes beyond standard employee benefits. The company provides flexible work arrangements, wellness programs, and on-site healthcare services. This not only attracts top talent but also improves employee satisfaction, resulting in greater productivity and retention rates. Google also works toward achieving gender parity in its leadership and workforce, empowering women, and promoting equal opportunity in line with SDG 5 (Mirmotahari, 2022).

Furthermore, businesses can actively contribute to the socioeconomic progress of the areas where they operate by engaging in community development initiatives and helping local suppliers. This can be achieved through investments in local development, the creation of programs with shared value, and the empowerment of local communities. Patagonia is a prime example of a company that specializes in outdoor apparel and actively participates in local grassroots initiatives. They support environmental organizations by donating one percent of their sales through programs like the "1% for the Planet" project (Patagonia, 2023). By prioritising community engagement, Patagonia not only secures the future of the environment but also strengthens its bond with customers. Unilever collaborates with smallholder farmers in Africa to improve agricultural practices and livelihoods (Unilever, 2023). The company provides training and resources to these farmers during their partnership. Ben & Jerry's, an ice cream company, demonstrates its dedication to Fair Trade principles by sourcing certified cocoa and vanilla from farmers (Ben &

Jerry's, 2023). This not only ensures fair prices for farmers but also contributes to funding community development initiatives such as the establishment of schools and medical facilities. These examples illustrate how organizations can go beyond philanthropy and actively contribute to community development and the creation of opportunities for success.

Economic Benefits

It is important to note that sustainable strategies can bring substantial economic benefits to businesses, despite common misconceptions. While the upfront investment in sustainable practices may seem daunting, the long-term financial advantages and potential for revenue generation outweigh the initial costs. By adopting green practices like reducing energy consumption, water usage, and waste generation, organizations can achieve significant cost savings. Research indicates that when organizations implement environmentally friendly practices such as improving energy and water usage, reducing waste, and employing strategies to minimize their environmental impact, they contribute significantly to preserving ecosystems and natural resources. Studies conducted by Gürlek and Tuna (2018) and Roscoe (2019) suggest that when a company's leadership prioritizes environmental sustainability and integrates it into all aspects of the organization, it encourages widespread adoption of sustainable practices. An organization's commitment to eco-friendly principles can have a significant impact on sustainable development at the local, national, and global levels, both directly and indirectly. For example, Patagonia, a US-based outdoor apparel, and gear company widely recognized for its environmental, social, and governance (ESG) practices, was able to save on electricity costs and generate a $1 billion revenue increase in 2019 by switching to LED lighting throughout its stores and headquarters (Shaikh, 2023). This green initiative aligns with the goal of Responsible Consumption and Production (SDG 12).

Wang et al. (2022) discovered that when an organization's internal operations management reflects its commitment to environmental conservation and restoration, it inspires a larger network to strive for positive change. By promoting attitudes that prioritize waste reduction through recycling, composting, and lean manufacturing techniques, organizations can reduce disposal charges and decrease reliance on new materials, resulting in significant cost savings. Interface, a carpet company, exemplifies this approach by redesigning their products and implementing closed-loop recycling technologies to achieve zero waste to landfill (Interface, 2023). They specialize in sustainable flooring solutions made from recyclable materials, aligning with Goal 11 of the Sustainable Development Agenda to develop sustainable cities and communities. Moreover, fostering a green culture fosters a creative environment, motivating workers to generate innovative ways to enhance efficiency and minimize waste. Fatoki (2021) asserts that developing new sustainable products and services is key to improving efficiency and profitability. Tesla, on the other hand, contributes significantly to a sustainable energy future, particularly in the realm of Affordable and Clean Energy (SDG 7), by embracing electric vehicles and renewable energy, fuelling innovation, and revolutionizing the market (Tesla, 2023).

Empowering Employees for Environmental Responsibility

One of the most crucial aspects of running a sustainable business is educating and empowering employees to take action for the environment. By fostering a culture of environmental responsibility, organizations not only reduce their environmental impact but also enhance employee engagement, productivity, and innovation. Engaging employees in environmental initiatives is a strategy for instigating meaningful change and nurturing a pervasive culture of sustainability. Moreover, employees, armed with the right knowledge and motivation, can be powerful catalysts for change. In this section, we will explore how businesses can design and implement programs aimed at raising environmental awareness and fostering eco-friendly behaviours among employees.

Provide Training and Conduct Interactive Workshops

Organizations must ensure that all staff receive training, so they are fully aware of why certain changes are being made and their impact. All staff members should be trained in the company's environmental policies and procedures and understand what they mean in practice. Engaging workshops on topics such as energy conservation, waste reduction, and sustainable practices equip employees with the knowledge to make informed decisions. For example, companies like Patagonia regularly hold sustainability workshops for their employees, fostering a culture of environmental responsibility throughout the organization. These workshops are interactive, featuring guest speakers from environmental organizations, hands-on activities, and discussions on how environmental issues impact daily life (Singh, 2024). Patagonia also offers training programs on sustainable practices specific to their industry, such as responsible sourcing of materials and eco-friendly manufacturing processes.

Incentivizing Sustainable Actions: Aligning Personal and Corporate Goals

To truly integrate sustainability into a company's culture, training is essential, but incentives play a crucial role. When employees have tangible rewards tied to their sustainable efforts, they are more likely to participate. Companies can create incentive programmes that align personal and corporate goals, creating a win-win situation. Companies have implemented eco-friendly challenges to encourage employees to adopt sustainable practices in their daily lives. For example, PwC, a multinational professional services network, launched a global "Green Challenge" where employees competed in teams to reduce energy and water consumption, minimise waste, and travel sustainably. This challenge utilised a mobile app to track progress and featured a live leaderboard to foster friendly competition. PwC also offered incentives, such as extra vacation days or gift cards, to the winning teams, making the challenge even more engaging. According to Singh (2024), the results were impressive, with a 10% reduction in energy use and a 6% decrease in water usage.

Onboarding: Embracing Sustainability Superheroes

According to Avvio (2023), organizations should welcome new employees not just to the organisation, but to the mission of environmental responsibility. Onboarding presents a golden opportunity to transform new employees from enthusiastic newbies into sustainability superheroes. By equipping them with the knowledge and tools to make a difference, the organisation will foster a culture of environmental awareness that extends far beyond their first few weeks. Also, incorporating eco-challenges into onboarding activities, this could involve a friendly competition between teams to reduce paper waste during the first week, brainstorming sessions to identify areas for improved energy efficiency in the office, or even participating in a virtual tree-planting initiative. These challenges not only encourage active participation and engagement, but also spark creative thinking and problem-solving skills around sustainability.

SUCCESSFUL SUSTAINABILITY INITIATIVES BY LEADING COMPANIES

Companies who are at the forefront of their industry are beginning to understand the significance of incorporating environmental, social, and economic responsibility into their business operations. By doing so, they can enjoy rewards in terms of their brand reputation, employee engagement, and financial performance. This section examines some of the most effective sustainability projects that have been undertaken by top organisations, highlighting the impact that these programmes have had and providing sources of inspiration for future businesses.

Apple's Renewable Energy Commitment

Apple's global operations are powered by 100% renewable energy, a successful sustainability strategy that has significantly reduced its carbon footprint. By investing in renewable energy projects and working with suppliers to transition to sustainable energy sources, Apple exclusively uses renewable energy sources to power its data centers, offices, and retail stores (Unglesbee, 2023). Additionally, the company has made substantial progress in improving production processes to minimize waste and boost energy efficiency. These sustainability efforts by Apple have not only helped in environmental conservation but also resonated with consumers who value environmentally friendly businesses.

Unilever's Sustainable Sourcing

Unilever, a global consumer goods corporation, has integrated sustainability into its corporate strategy. One notable initiative they have taken is their commitment to sourcing sustainably produced palm oil. Unilever has acknowledged the significant environmental and social impacts associated with palm oil production. The company has committed to sourcing only palm oil that is certified as sustainable (Unilever, 2017 and Lawrence et al., 2019). Through partnerships with NGOs and suppliers, they have successfully reached their objective and are now establishing a benchmark for other companies in the industry (Unilever, 2023). Unilever's dedication to sustainable sourcing practices highlights their focus on responsible business conduct, ultimately contributing to the establishment of consumer trust.

Patagonia's Worn Wear Program

The outdoor gear manufacturer Patagonia launched a unique initiative called the Worn Wear program to advocate for sustainability. This campaign encourages repairing and reusing existing Patagonia gear instead of purchasing new items. Along with providing repair services, Patagonia sells second-hand gear and offers guidance on how to mend items at home. By promoting the principles of "reduce, reuse, and repair," Patagonia aims to minimize the environmental impact of their products and encourage more sustainable consumption practices (Patagonia, 2023). The Worn Wear initiative not only helps in waste reduction but also cultivates strong brand loyalty among customers who value the company's dedication to sustainability, contributing to its remarkable success (Batten, 2020).

Challenges and Strategies in Cultivating a Green Organizational Culture

Developing a green organizational culture can be challenging, despite the clear benefits it offers. Sin et al. (2021) suggest that cultivating a green organizational culture, which involves integrating sustainability and environmental responsibility into the organization's core values and operations, can be difficult for companies. The following are some common challenges encountered when trying to develop a green organizational culture and strategies for fostering a green organizational culture:

Resistance From Stakeholders

When businesses aim to implement sustainability efforts, they often encounter resistance from stakeholders such as employees, shareholders, or consumers who may be reluctant to change or sceptical about the benefits of sustainability. This reluctance to embrace sustainability can be fuelled by several factors, including fear of the unknown, a preference for familiar practices, or concerns about job security and performance implications (Sin et al., 2021). To overcome this obstacle, it is crucial to communicate effectively about the importance and advantages of sustainability initiatives. Providing compelling evidence, such as case studies or success stories from other companies, can help address doubts and generate support. A notable example is the case of Patagonia's "Don't Buy This Jacket" campaign, which initially faced criticism (Hwang et al., 2016). However, Patagonia managed to win over sceptics by clearly articulating their commitment to sustainability and highlighting their efforts to reduce waste and advocate for responsible consumption (Horsley-Summer, 2022).

Lack of Awareness and Knowledge

Organizations that lack awareness or have a limited understanding of sustainability initiatives may face challenges, especially in sectors that historically show little regard for environmental or societal issues. This lack of awareness can hinder the development of a culture of sustainability and the adoption of eco-friendly practices. For example, employees may not realize the environmental and financial benefits of saving energy and paper, overlooking the importance of these practices. One effective solution is to invest in training programs or seminars to educate employees about sustainability and its relevance to their industries (Faster Capital, 2023). Collaborating with external consultants or experts can provide valuable guidance and insights. Interactive seminars can spark interest and improve comprehension. By creating an environment that promotes continuous learning and knowledge sharing, businesses can overcome the challenge of insufficient awareness and establish a solid foundation for their sustainability initiatives.

Consumer Lack of Awareness

Consumer awareness influences green organizational culture. Many consumers are not fully aware of how their purchases impact the environment, and they are unsure of the sustainable options available. Additionally, some consumers question the credibility of companies' environmental claims, fearing greenwashing and misleading information about a product's eco-friendly features (de Freitas Netto, et al., 2020). Therefore, businesses should develop comprehensive marketing strategies to inform customers about the environmental impact of various products, utilizing diverse platforms like social media to reach a wider audience. It is crucial for companies to be transparent by providing clear, verifiable information about their sustainability efforts, certifications, and practices. For instance, Tesla not only excels in producing electric vehicles but also educates customers on the benefits of eco-friendly transportation. Through marketing campaigns, Tesla highlights the importance of moving away from fossil fuel-powered vehicles, emphasizing reduced carbon emissions and operational costs (Tesla, 2019).

Lack of Expertise

Transitioning to sustainable business practices requires expertise in various areas such as energy management, waste reduction, green supply chains, and sustainable procurement. Many organizations lack this expertise internally and struggle to identify significant green initiatives and efficiently implement them across their complex global operations. One effective way to overcome this obstacle is by hiring dedicated sustainability specialists. These experts can evaluate the company's environmental impact, identify key areas for improvement, and oversee implementation. Their knowledge can also influence investments in green technology, facility design, product development, and collaborations with vendors. For example, Coca-Cola appointed its first Chief Sustainability Officer to lead sustainability initiatives and help achieve the company's goal of net-zero emissions by 2040 (Coca Cola, 2021). The CSO brings substantial expertise in environmental science and policy. Similarly, Apple has a specialized team for supply chain innovation and environmental research to analyze products' life cycles and explore environmentally friendly alternatives. This approach has enabled Apple to shift to sustainable energy sources and boost the use of recycled rare metals in iPhones (2021). The expertise and experience that sustainability experts offer is invaluable in translating ambitious environmental goals into practical strategies. Understanding emerging technologies, managing stakeholders, and promoting sustainable operations are vital for organizations aiming to embrace eco-friendly practices.

Integration of technology and data-driven solutions: Technology will have a vital role in driving sustainability efforts within organizations. Developments in data analytics, artificial intelligence, and the internet of things will allow organizations to collect and analyze real-time data on resource consumption and environmental impacts. Artificial intelligence will be crucial in optimizing resource utilization, energy efficiency, and waste reduction. Smart technologies will facilitate real-time monitoring and data-driven decision-making for sustainable operations. The combination of real-time data, smart infrastructure, and automation will improve resource efficiency, streamline supply chains, and enhance environmental monitoring (World Economic Forum, 2023). Therefore, businesses should invest in technology to boost efficiency, mitigate environmental impact, and gain insights into sustainable practices. This investment will assist organizations in identifying areas for improvement, optimizing resource usage, and making informed decisions towards sustainability.

POTENTIAL SOLUTIONS AND IMPLEMENTATION TACTICS

While cultivating a green organizational culture presents various challenges, there are viable solutions that businesses can implement to overcome these barriers. The following strategies and tactics can assist organisations in fostering a sustainable culture:

1. **Establish clear sustainability goals and metrics**: The organisation can develop measurable and time-bound environmental targets aligned with the organization's vision. They can regularly track and communicate progress towards these goals to maintain accountability and transparency.
2. **Incentivize sustainable behaviours**: Implement reward systems that recognise and incentivise employees for adopting environmentally responsible practices. This can include monetary rewards, public recognition, or opportunities for professional development.
3. **Foster collaboration and knowledge sharing**: Encourage cross-functional collaboration and knowledge sharing among departments to identify and implement sustainable solutions. This can be achieved through regular meetings, workshops, or the formation of dedicated sustainability teams.
4. **Engage stakeholders and build partnerships**: Actively engage with stakeholders, such as customers, suppliers, and local communities, to understand their expectations and collaborate on sustainability initiatives. Also, the organisation should seek partnerships with environmental organizations, academic institu-

tions, or industry associations to leverage expertise and resources (Benn et al., 2014).

5. **Integrate sustainability into decision-making processes**: Ensuring that environmental considerations are integrated into all decision-making processes, from product design and procurement to operations and marketing. This can be facilitated by developing sustainability guidelines and conducting environmental impact assessments (Epstein & Buhovac, 2014).

By implementing these strategies and tactics, businesses can effectively address the challenges associated with nurturing a green organizational culture and create an environment that promotes sustainability and environmental responsibility.

CONCLUSION AND RECOMMENDATIONS

In summary, the exploration of Green Organizational Culture (GOC) within the context of Sustainable Development (SD) underscores a pivotal transformation in the corporate sector's approach towards environmental stewardship. This chapter has illuminated the essential role of integrating sustainability into organizational values, norms, and behaviors, advocating for a change in basic assumptions from profit-centric motives to fostering ecological balance and social equity. Through the lens of institutional theory and organizational change theory, the chapter dissected the mechanisms by which businesses can align with broader sustainability goals, leveraging both external pressures and internal change processes to embed environmental responsibility into the organizational fabric.

Conclusion: The critical examination of GOC and SD within businesses reveals that fostering an environmentally responsible organizational culture is not merely beneficial but imperative for navigating the complex landscape of global sustainability challenges. Organizations that successfully integrate green practices and values into their operational and strategic frameworks not only contribute positively to global environmental goals but also secure their long-term success and legitimacy in an increasingly eco-conscious market.

Recommendations: To further advance the integration of GOC and SD in businesses, the following recommendations are proposed:

1. **Enhanced Interdisciplinary Research:** Future studies should adopt an interdisciplinary approach, combining insights from environmental science, organizational behaviors, and strategic management to develop holistic strategies for embedding sustainability into corporate cultures.

2. **Empirical Case Studies:** There is a need for more empirical research focusing on case studies of companies that have successfully implemented green organizational cultures. These studies can provide valuable lessons and actionable strategies for other businesses seeking to embark on similar sustainability journeys.

3. **Innovation in Green Technologies:** Businesses should invest in research and development of green technologies and sustainable business models. This not only supports environmental goals but also opens new market opportunities and competitive advantages.

4. **Policy Advocacy:** Organizations should actively engage in policy discussions and advocacy efforts to shape regulatory frameworks that support sustainable business practices. Collaboration with governments, NGOs, and industry associations can amplify the impact of these efforts.

5. **Strengthening Stakeholder Engagement:** Companies should enhance their engagement with stakeholders, including employees, customers, suppliers, and local communities, to foster a shared commitment to sustainability. This can be achieved through transparent communication, participatory decision-making, and collaborative initiatives.

6. **Continuous Learning and Adaptation:** Organizations must commit to continuous learning and adaptation to keep pace with evolving sustainability standards and practices. Implementing mechanisms for regular review and feedback on sustainability initiatives can help in adjusting strategies to meet changing environmental and social needs.

By embracing these recommendations, businesses can not only navigate the complexities of integrating GOC and SD but also play a pivotal role in driving forward the global agenda for sustainability. The journey towards environmental responsibility and sustainable development is ongoing, and the corporate sector has both the opportunity and the obligation to lead by example, ensuring a prosperous future for both the planet and its inhabitants.

REFERENCES

Alvesson, M., & Sveningsson, S. (2015). *Changing organizational culture: Cultural change work in progress*. Routledge. 10.4324/9781315688404

Andersson, S., Svensson, G., Molina-Castillo, F. J., Otero-Neira, C., Lindgren, J., Karlsson, N. P., & Laurell, H. (2022). Sustainable development—Direct and indirect effects between economic, social, and environmental dimensions in business practices. *Corporate Social Responsibility and Environmental Management*, 29(5), 1158–1172. 10.1002/csr.2261

Apple. (2021). *Environmental Progress Report*. Apple._https://www.apple.com/environment/pdf/Apple_Environmental_Progress_Report_2021.pdf

Arum, E., Agyabeng-Mensah, Y., & Owusu, J. A. (2020). Translating environmental management practices into improved environmental performance via green organizational culture: Insight from Ghanaian manufacturing SMEs. *Journal of Supply Chain Management Systems*, 9(1), 31–49.

Baird, K., Harrison, G., & Reeve, R. (2007). The culture of Australian organizations and its relation with strategy. *International Journal of Business Studies: A Publication of the Faculty of Business Administration. Edith Cowan University*, 15(1), 15–41.

Bandoophanit, T. (2024). Holistic implementations of green supply chain management practices in Thai entrepreneurial ventures. *Journal of Small Business and Enterprise Development*. Emerald Publishing. https://www.emerald.com/insight/content/doi/10.1108/JSBED-01-2023-0001/full/html)

Batten. (2020). Patagonia's Worn Wear Collection Is Saving the Planet. *The Manual*. www.themanual.com/outdoors/oatagonia-worn-wear-collection-recycled-recommerce/

Baumgartner, R. J. (2009). *Organizational culture and leadership: Preconditions*.

Baumgartner, R. J., & Zielowski, C. (2007). Analysing zero emission strategies regarding impact on organizational culture and contribution to sustainable development. *Journal of Cleaner Production*, 15(13-14), 1321–1327. 10.1016/j.jclepro.2006.07.016

Belias, D., & Koustelios, A. (2014). Organizational culture and job satisfaction: A review. *International Review of Management and Marketing*, 4(2), 132–149.

Bocken, N. M. P., Short, S. W., Rana, P., & Evans, S. (2014). A literature and practice review to develop sustainable business model archetypes. *. Journal of Cleaner Production*, 65, 42–56. 10.1016/j.jclepro.2013.11.039

Cameron, K. S., & Quinn, R. E. (2006). *Diagnosing and changing organizational culture: Based on the competing values framework.* John Wiley & Sons.

Chatman, J. A., & Jehn, K. A. (1994). Assessing the relationship between industry characteristics and organizational culture: How different can you be? *Academy of Management Journal, 37*(3), 522–553. 10.2307/256699

Coca Cola. (2021). *Business & ESG Report: Refresh the World Make a Difference.* Coca Cola. https://www.coca-colacompany.com/content/dam/company/us/en/reports/coca-cola-business-environmental-social-governance-report-2021.pdf

de Freitas Netto, S. V., Sobral, M. F. F., Ribeiro, A. R. B., & Soares, G. R. L. (2020). Concepts and forms of greenwashing: A systematic review. *Environmental Sciences Europe, 32*(1), 19. 10.1186/s12302-020-0300-3

Deirmentzoglou, G. A., Agoraki, K. K., & Fousteris, A. E. (2020). Organizational culture and corporate sustainable development: Evidence from Greece. *International Journal of Business and Social Science, 11*(5), 92–98. 10.30845/ijbss.v11n5a9

Denison, D. R. (1990). *Corporate culture and organizational effectiveness.* John Wiley & Sons.

DiMaggio, P. J., & Powell, W. W. (1983). The iron cage revisited: Institutional isomorphism and collective rationality in organizational fields. *. American Sociological Review, 48*(2), 147–160. 10.2307/2095101

Elkington, J. (1997). The triple bottom line. *Environmental management: Readings and cases, 2*, 49-66.

Emina, K. A. (2021). Sustainable development and the future generations. *Social Sciences* [SHE Journal]. *Humanities and Education Journal, 2*(1), 57–71.

European Commission. (2021). *European Green Deal: Commission proposes transformation of EU economy and society to meet climate ambition.* EC. https://ec.europa.eu/commission/presscorner/detail/en/ip_21_3541

FasterCapital. (2023). *Sustainability Initiatives: Sustainable Strategies in the Competitive Landscape: A Winning Combination.* Faster Capital. https://fastercapital.com/content/Sustainability-Initiatives--Sustainable-Strategies-in-the-Competitive-Landscape--A-Winning-Combination.html

Fietz, B., & Günther, E. (2021). Changing organizational culture to establish sustainability. *Controlling & Management Review, 65*(3), 32–40. 10.1007/s12176-021-0379-4

Galpin, T., Whitttington, J. L., & Bell, G. (2015). Is your sustainability strategy sustainable? creating a culture of sustainability. *Corporate Governance (Bradford)*, 15(1), 1–17. 10.1108/CG-01-2013-0004

Ghorbani, M. (2023). Green Knowledge Management and Innovation for Sustainable Development: A Comprehensive Framework. In *ECKM 2023 24th European Conference on Knowledge Management*. Academic Conferences and publishing limited. 10.34190/eckm.24.1.1753

Gürlek, M., & Tuna, M. (2018). Reinforcing competitive advantage through green organizational culture and green innovation. *Service Industries Journal*, 38(7-8), 467–491. 10.1080/02642069.2017.1402889

Hakim, L. N. (2023). Green Manufacturing Practices and Green Innovation and Their Role in Sustainable Business Performance Through Culture Green Organization at Small Industrial Enterprises. In *International Conference on Economics Business Management and Accounting (ICOEMA)* (Vol. 2, pp. 366-376).

Hart, O. (1995). Corporate governance: Some theory and implications. *Economic Journal (London)*, 105(430), 678–689. 10.2307/2235027

H&M Group. (2023). *Sustainability: How we work with materials*. H&M. https://hmgroup.com/sustainability/circularity-and climate/materials/#: ~:text=How%20we%20work%20with%20materials,30%25%20recycled%20materials%20by%202025.

Hofstede, G. (2001). *Culture's consequences: Comparing values, behaviors, institutions, and organizations across nations*. Sage publications.

Horsley-Summer, B. (2022). *Better Brands 01: How Patagonia brought sustainability to the enlightened masses*. Avery and Brown. https://www.averyandbrown.com/outpost/better-brands-01-patagonia

Hwang, C., Lee, Y., Diddi, S., & Karpova, E. (2016). "Don't buy this jacket": Consumer reaction toward anti-consumption apparel advertisement. *Journal of Fashion Marketing and Management*, 20(4), 435–452. 10.1108/JFMM-12-2014-0087

Interface. (2023). *Lessons for the future: The Interface guide to changing your business to change the world*. Interface. https://www.interface.com/content/dam/interfaceinc/interface/sustainability/emea/25th-anniversary-report/Interface_MissionZeroCel_Booklet_EN.pdf

Jerónimo, H. M., Henriques, P. L., de Lacerda, T. C., da Silva, F. P., & Vieira, P. R. (2020). Going green and sustainable: The influence of green HR practices on the organizational rationale for sustainability. *Journal of Business Research*, 112, 413–421. 10.1016/j.jbusres.2019.11.036

Kotter, J. P. (1996). *Leading change*. Harvard Business School Press.

Lawrence, J., Rasche, A., & Kenny, K. (2019). Sustainability as Opportunity: Unilever's Sustainable Living Plan. In Lenssen, G. G., & Smith, N. C. (Eds.), *Managing Sustainable Business*. Springer., 10.1007/978-94-024-1144-7_21

Lewin, K. (1947). Frontiers in group dynamics: Concept, method, and reality in social science; social equilibria and social change. *Human Relations*, 1(1), 5–41. 10.1177/001872674700100103

Liu, X., & Lin, K. L. (2020). Green Organizational Culture, Corporate Social Responsibility Implementation, and Food Safety. *Frontiers in Psychology*, 11, 585435. 10.3389/fpsyg.2020.58543533240175

Luqman, R. A., Farhan, H. M., Shahzad, F., & Shaheen, S. (2012). 21st century challenges of educational leaders, way out and need of reflective practice. *International Journal of Learning and Development*, 2(1), 195–208. 10.5296/ijld.v2i1.1238

Mahmood, Z., & Uddin, S. (2021). Institutional logics and practice variations in sustainability reporting: Evidence from an emerging field. *Accounting, Auditing & Accountability Journal*, 34(5), 1163–1189. 10.1108/AAAJ-07-2019-4086

Majeed, M. A., Ahsan, T., & Gull, A. A. (2023). Does corruption sand the wheels of sustainable development? Evidence through green innovation. *Business Strategy and the Environment*. Wiley Online Library (https://onlinelibrary.wiley.com/doi/abs/10.1002/bse.3719)

Martela, F. (2019). What makes self-managing organizations novel? Comparing how Weberian bureaucracy, Mintzberg's adhocracy, and self-organizing solve six fundamental problems of organizing. *Journal of Organization Design*, 8(1), 1–23. 10.1186/s41469-019-0062-9

Microsoft. (2023). *2022 Environmental Sustainability Report Enabling sustainability for our company, our customers, and the world*. Microsoft. https://www.microsoft.com/en-us/corporate-responsibility/sustainability/report?ICID=SustainabilityReport22_MOI-ESblog

Mirmotahari, T. (2022). *Google Benefits and Perks for Employees - 11 ideas*. Perkup App. https://perkupapp.com/post/11-awesome-google-benefits-and-perks-for-employees

Mohezara, S., Nazria, M., Kaderb, M. A. R. A., Alib, R., & Yunusb, N. K. M. (2016). Corporate social responsibility in the Malaysian food retailing industry: An exploratory study. *Int. Acad. Res. J. Soc. Sci.*, 2, 66–72.

Nimfa, D. T., Latiff, A. S. A., Wahab, S. A., & Etheraj, P. (2021). Effect of organisational culture on sustainable growth of SMEs: Mediating role of innovation competitive advantage. *Journal of International Business and Management*, 4(2), 1–19. 10.37227/JIBM-2021-01-156

Noor Faezah, J., Yusliza, M. Y., & Ramayah, T. (2024). Mediating role of green culture and green commitment in implementing employee ecological behaviour. *Journal of Management Development*. Emerald Publishing. https://www.emerald.com/insight/content/doi/10.1108/JMD-08-2023-0258/full/html

Norton, T. A., Parker, S. L., Zacher, H., & Ashkanasy, N. M. (2015). Employee green behaviour: A theoretical framework, multilevel review, and future research agenda. *Organization & Environment*, 28(1), 103–125. 10.1177/1086026615575773

Olafsen, A. H., Nilsen, E. R., Smedsrud, S., & Kamaric, D. (2021). Sustainable development through commitment to organizational change: The implications of organizational culture and individual readiness for change. *Journal of Workplace Learning*, 33(3), 180–196. 10.1108/JWL-05-2020-0093

Pedersen, J. T. S., van Vuuren, D., Gupta, J., Santos, F. D., Edmonds, J., & Swart, R. (2022). IPCC emission scenarios: How did critiques affect their quality and relevance 1990–2022? *Global Environmental Change*, 75, 102538. 10.1016/j.gloenvcha.2022.102538

Popescu, C. R. G. (2020). Sustainability assessment: Does the OECD/G20 inclusive framework for BEPS (base erosion and profit shifting project) put an end to disputes over the recognition and measurement of intellectual capital? *Sustainability (Basel)*, 12(23), 10004. 10.3390/su122310004

Porter, M. E., & Kramer, M. R. (2006). The link between competitive advantage and corporate social responsibility. *Harvard Business Review*, 84(12), 78–92.17183795

Raza, M. Y., Saeed, A., Iqbal, N., & Faraz, N. A. (2021). Enabling digital transformation for sustainable change in organizations: An empirical study. *Sustainability*, 13(7), 3857.

Roscoe, S., Subramanian, N., Jabbour, C. J., & Chong, T. (2019). Green human resource management and the enablers of green organisational culture: Enhancing a firm's environmental performance for sustainable development. *Business Strategy and the Environment*, 28(5), 737–749. 10.1002/bse.2277

Sackmann, S. A. (1991). Uncovering culture in organizations. *The Journal of Applied Behavioral Science, 27*(3), 295–317. 10.1177/0021886391273005

Schillmann, C. (2020). *Patagonia Inc. under a sustainability perspective.*

Schneider, B., Ehrhart, M. G., & Macey, W. H. (2013). Organizational climate and culture. *Annual Review of Psychology, 64*(1), 361–388. 10.1146/annurev-psych-113011-14380922856467

Scott, W. R. (2008). *Institutions and organizations: Ideas and interests.* Sage Publications.

Shah, S. M. A., Jiang, Y., Wu, H., Ahmed, Z., Ullah, I., & Adebayo, T. S. (2021). Linking green human resource practices and environmental economics performance: The role of green economic organizational culture and green psychological climate. *International Journal of Environmental Research and Public Health, 18*(20), 10953. 10.3390/ijerph18201095334682698

Shahzad, F., & Luqman, A. (2012). Impact of Organizational Culture on Organizational Performance: An Overview. *Interdisciplinary Journal of Contemporary Research in Business, 3*(9), 975–985.

Shaikh, N. (2023). *Patagonia: $1B Revenue Surge through ESG Success.* https://www.linkedin.com/pulse/patagonia-1b-revenue-surge-through-esg-success-nabeel-shaikh/

Siddique, F. B., & Sultana, I. (2018). *Unilever Sustainable Living Plan: A Critical Analysis.*

Silvestri, C., Silvestri, L., Piccarozzi, M., & Ruggieri, A. (2022). Toward a framework for selecting indicators of measuring sustainability and circular economy in the agri-food sector: A systematic literature review. *The International Journal of Life Cycle Assessment*, 1–39.

Southey, F. (2019). *Nestlé talks challenges in sustainable soy: Complex supply chains and legal deforestation.* Food Navigator. https://www.foodnavigator.com/Article/2019/06/20/Nestle-talks-challenges-in-sustainable-soy-Complex-supply-chains-and-legal-deforestation?utm_source=copyright&utm_medium=OnSite&utm_campaign=copyright

Tesla. (2019). *Impact report.* Tesla. https://www.tesla.com/ns_videos/tesla-impact-report-2019.pdf

Tribe, H. (2022). *An exploratory investigation into how the implementation and internalization of processes within a MNC are affected by the regulatory, cognitive, and normative domains of institutionalism* [Doctoral dissertation, Brunel University London].

Umair, S., Waqas, U., Mrugalska, B. & Al Shamsi, I.R. (2023). *Environmental Corporate Social Responsibility, Green Talent Management, and Organization's Sustainable Performance in the Banking Sector of Oman: The Role of Innovative Work Behaviour and Green.*

Unglesbee, B. (2023). *More than 300 Apple suppliers have committed to clean energy.* Supply Chain Drive. https://www.supplychaindive.com/news/apple-suppliers -clean-energy-scope-3-carbon-neutral-products/693980/

Unilever. (2017). *Unilever responsible sourcing policy working in partnership with our suppliers.* Unilever. https://www.unilever.com/files/92ui5egz/production/ f51492642f57b314b05466b6194792e02d075d76.pdf

Unilever. (2023). *Annual Report and Accounts 2022. Delivering sustainable business performance.* Unilever. https://www.unilever.com/files/92ui5egz/production/ 257f12db9c95ffa2ed12d6f2e2b3ff67db49fd60.pdf

Wan, B. (2024). The Impact of Cultural Capital on Economic Growth Based on Green Low-Carbon Endogenous Economic Growth Model. *Sustainability.* MDPI. [Link](https://www.mdpi.com/2071-1050/16/5/1781)

Wang, C. H. (2019). How organizational green culture influences green performance and competitive advantage: The mediating role of green innovation. *Journal of Manufacturing Technology Management,* 30(4), 666–683. 10.1108/JMTM-09-2018-0314

Wang, S., Abbas, J., Sial, M.S., Álvarez-Otero, S. & Cioca, L.I. (2022). Achieving green innovation and sustainable development goals through green knowledge management: Moderating role of organizational green culture. *Journal of innovation & knowledge, 7*(4), p.100272.

WCED, S. W. S. (1987). World commission on environment and development. *Our common future, 17*(1), 1-91.

Weber, G., & Martensen, M. (2021). *Transforming organizational culture amidst a diverse workforce: A qualitative study in the service industry* (No. 1/2021). IUBH Discussion Papers-Human Resources.

Weiner, B. J. (2009). A theory of organizational readiness for change. *Implementation Science : IS,* 4(1), 67. 10.1186/1748-5908-4-6719840381

Whiteman, G., Walker, B., & Perego, P. (2013). Planetary boundaries: Ecological foundations for corporate sustainability. *Journal of Management Studies*, 50(2), 307–336. 10.1111/j.1467-6486.2012.01073.x

World Economic Forum. (2023). *The Global Risks Report 2023 18th Edition Insight Report*. WEF. https://www.weforum.org/publications/global-risks-report-2023/

Zammuto, R. F., Gifford, B., & Goodman, E. A. (2000). *Managerial ideologies, organization culture and the outcomes of innovation: A competing values perspective.*

Chapter 10
Local Governments' Roles in Sustainable Waste Management for a Green Economy

Olufemi Micheal Oladejo
University of KwaZulu-Natal, South Africa

Sybert Mutereko
Rhodes University, South Africa

Nyikiwa Agreement Mavunda
http://orcid.org/0000-0002-8161-7226
University of KwaZulu-Natal, South Africa

ABSTRACT

Waste management has become one of the most pressing challenges facing the world today, with the increasing population and urbanization leading to an exponential increase in waste generation. Hence, sustainable waste management which includes a hygienic environment, providing waste recycling and re-use logistics, and managing the consumption pattern of the local community are key components of a green economy. The study adopts a systemic and desktop review of the literature to explore the role of Local government (as the closest government to the people) in sustainable waste management for a green economy.

DOI: 10.4018/979-8-3693-2595-7.ch010

INTRODUCTION

Sustainable waste management (SWM) is a crucial aspect of building a green economy, and local governments play a vital role in achieving this goal (Zotos, Karagiannidis, Zampetoglou, Malamakis, Antonopoulos, Kontogianni, and Tchobanoglous, 2009). As the world is facing the consequences of unsustainable waste management, the need for a shift towards a green economy has become more urgent than ever. Waste management is not only a responsibility but also an opportunity for Local governments to contribute to the development of a sustainable and prosperous community. Waste management has been a major challenge facing governments globally. However, because local governments are the closest to the grassroots, they are at the center of the heat. With the increasing population and rapid urbanisation, the amount of waste generated has also increased significantly. A recent study puts global waste generation at 2.12 billion tons, while the World Bank projects a 70% increase by 2050 if the present unsustainable waste practices continue (Nandy, Fortunato, and Martins, 2022). Figure 1 below shows the global waste projection by 2050.

Figure 1. Waste increase projection

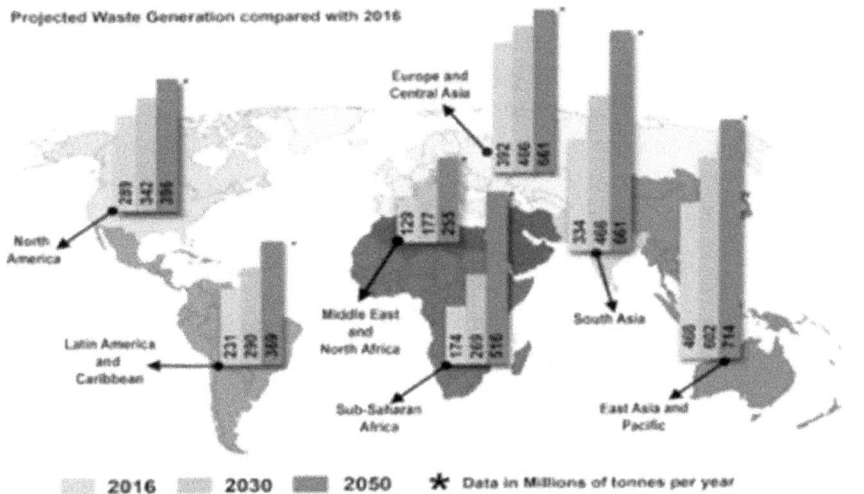

Source: (Nandy, Fortunato, and Martins, 2022).

Therefore, the concept of a green economy aims to achieve sustainable development by promoting economic growth while reducing environmental degradation and addressing social issues (Atiku, 2019; Elagroudy, Warith, and Zayat, 2016). The authors emphasized that SWM as a crucial component of a green economy contributes to reducing pollution, preserving resources, creating employment opportunities, and promoting innovation and technological development. Waste management involves a waste value chain that comprises various aspects which include generation, collection, transportation, treatment, and disposal (Wan, Shen, and Choi, 2019; Ugwu, Ozoegwu, Ozor, Agwu, and Mbahwa, 2021). From the generation stage, waste is separated into several categories such as biodegradable, biogenic, and recyclable for easy processing. Unfortunately, most countries do not follow this process. Hence, to promote collective practices that encourage responsible management methods such as source separation, Local government must engage waste generators (households and privates) in an awareness campaign towards waste recycling and reuse drive to achieve a green economy (Desa, Ba'yah Abd Kadir, & Yusooff, 2011; Tulebayeva, et al., 2020).

Understanding the process and procedure of recycling and reuse of waste is pivotal in decision-making regarding policies and planning. This chapter delves into the concepts surrounding recycling and reuse of waste with the aid of local government as a key player in the global move to achieve a green economy. Although, several studies have discussed the roles of various key players involved in waste management globally, however, this chapter focuses on the role of the Local government in sustainable waste management to achieve a green economy.

The chapter is divided into three sections, the first section conceptualizes sustainable waste management and a green economy with insight into the role of the Local government in achieving a green economy while the second section examines the implication of sustainable waste management to a green economy. This section looks at the impact and the benefits of sustainable waste management in a green economy with the support of the Local government. The last section concludes and summarizes the study.

STAINABLE WASTE MANAGEMENT AND GREEN ECONOMY

In recent years, there has been a growing concern about the impact of human activities on the environment (Nandy, Fortunato, and Martins, 2022). One of the major contributors to environmental degradation is the mismanagement of waste (Wan, Shen, and Choi, 2019). Also, evidence abounds that the increase in waste generation due to urbanization, industrialization, and population growth has posed a significant challenge to governments globally (Wan, Shen, and Choi, 2019; Mallak and Ishak, 2012; Ugwu, et al., 2021). Unfortunately, waste increase is inevitable according to the World Bank projection of 2012 and 2018 (Wan, Shen, and Choi, 2019; Zorpas, 2020).

Before now, countries managed waste by burning or burying it under the ground (landfill or incineration). These traditional methods of waste management are not only harmful to the environment but also unsustainable and are only a short-term waste management solution (Wan, Shen, and Choi, 2019; Nandy, Fortunato, and Martins, 2022). However, with the growing awareness of the impact of waste on climate change or global warming, there have been concerted efforts towards finding sustainable waste management methods that can help achieve a green economy. This is very important and urgently needed as the globe faces the devastating repercussions of several years of practice of unsustainable waste management methods that undermined both human and environmental protection (Atiku, 2020). These new methods such as waste-to-energy, waste-to-new product, etc, must, however, consider waste generation patterns, emissions data, impact characteristics, the risk associated with waste management practices, and the creation of new waste recycling technologies (Tulebayeva, et al, 2020; Mingaleva, et al., 2019; Ertz, et al., 2021),

Hence, SWM means waste management done or carried out in an environmentally friendly and responsible manner with both social and economic considerations (Nandy, Fortunato, and Martins, 2022). This involves reducing the amount of waste generated, reusing and recycling waste materials, and properly disposing of waste that cannot be reused or recycled. Also, it involves proper collection, sorting, treatment, and disposal of waste. This is the concern of every responsible and responsive government, especially the local governments as they coordinate all waste management activities within their jurisdiction through laws (Gorica, Kripa & Zenelaj, 2012). Studies have also shown that this approach aims to minimise the impact of waste on the environment and human health while promoting resource efficiency and economic growth (Wan, Shen, and Choi, 2019; Zorpas, 2020). One of the key strategies for SWM is the concept of '3Rs'- reduce, reuse, and recycle. This concept is discussed in detail below.

Waste Reduction Strategy

One of the best ways to manage waste is to produce or generate less waste (Wan, Shen, and Choi, 2019). This approach is plausible if sustainable consumption habits such as buying products with minimal packaging, using reusable containers, and avoiding single-use items can be adopted. Adopting this approach tends to reduce waste generation thereby reducing the negative impact of waste on the environment and the planet in the long run. Unfortunately, there is a low awareness of activities relating to waste reduction, reuse, and recycling (Zorpas, 2020). This awareness deficit has been attributed to a lack of willpower by stakeholders in waste management globally especially, the government in adhering to various treaties and accords reached on waste reduction. For instance, since the "Treaty of Rome 1957" which was amended in the Stockholm Agreement (UN, 1972), and the United Nations Sustainable Development Goals (SDGs) 2015, not much has been achieved in waste reduction (Zorpas, 2020).

In a study by Mallak and Ishak, (2012), on waste management strategy, waste reduction or waste minimization occupies the apex point in the hierarchy of waste management strategies as established by Environmental Protection Agency (EPA) in 1976. The idea of waste hierarchy was introduced over four decades in the United States and Europe as a strategy to prioritize the order of waste management processes (Pires & Martinho, 2019). According to Ewiijk and Stegemann, (2016), the practices range from prevention to disposal.

Waste minimization is a process of reducing waste from sources through waste prevention. This is also known as resource efficiency (Mallak and Ishak, 2012). Figure 2 below shows the waste management hierarchy. Accordingly, Mallak and Ishak (2012) revealed that waste minimization strategy plays a vital role in sustainable development. Unfortunately, waste minimization is sometimes misconstrued as waste recycling. While waste recycling involves converting or transforming waste into a new product, waste minimization is about reducing consumption or preventing waste at the point of purchase, reusing the product or repairing rather than replacing it (Ertz, et al., 2021). Deploying this strategy for domestic and industrial purposes has been recognized as an effective method to achieve sustainability in waste management due to its preventive and preservative impact on the environment (Atiku & Fapohunda, 2021; Fields & Atiku, 2017; Ewiijk & Stegemann, 2016).

Figure 2. Waste management hierarchy

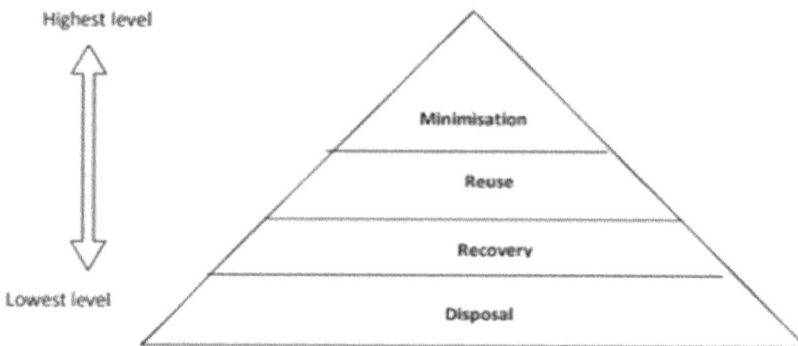

Source: Mallak and Ishak (2012).

Waste Reuse Strategy

Another key strategy in achieving SWM is the waste reuse option. Figure 2 above, shows reuse as the second-highest strategy in the SWM hierarchy. Waste reuse as a SWM strategy involves converting waste materials into other useful or usable materials. For instance, decomposing organic waste into nutrient-rich soil as a natural fertilizer. This has been confirmed to be strategic and very effective. On the one hand, it reduces the need for chemicals as in the case of fertilizers which can be harmful to the environment, and on the other hand, it reduces the amount of organic waste that ends up in landfills (Wan, Shen, and Choi, 2019). This is an effective and efficient way to manage waste sustainably and promote a circular economy (Lugo, Ail, and Castaldi, 2020). Drawing from Zhang et al., (2011), Wan, Shen, and Choi (2019: 4) posit 'that reuse of products extends products' life span and reduces the amount of waste directed to landfills and incineration'. Aside from the environmental benefits, waste reuse also brings about behavioral change in individual or organizational waste management orientation. The concept of reuse is an important component of a circular economy which is based on the principle of redesign, remanufacturing, and recovery (Lugo, Ail, and Castaldi, 2020). Arguably so, the authors believe this is a preferred method to limit or reduce waste generation. However, despite being an effective and potent strategy for SWM, the awareness of it is poor and low, especially in developing countries (Werff, et al., 2019; Lugo, Ail, and Castaldi, 2020).

Waste Recycling/Recovery Strategy

Recycling or recovery is the process of turning waste materials into new products, reducing the need for raw materials (Matinde et al., 2018). This process is recognised as one of the effective strategies of SWM. It helps to recover, reuse, and transform waste materials such as plastics, paper, polythene and organic waste into valuable and reusable materials and products. Also, it improves the economy and provides job opportunities (Lee, Kim, and Chong, 2016; Ugwu, et al., 2021). In addition, it conserves natural resources and reduces the amount of waste that ends up in landfills. Furthermore, there is preservation and protection of the environment from air pollution from hazardous waste which translates to the reduction of poisonous gas emissions generated from landfill sites (Elagroudy, Warith, and Zayat, 2016; Ugwu, et al., 2021).

Aside from the aforementioned, waste strategies, SWM also involves proper disposal of waste. Landfills and incinerators, the traditional methods of waste disposal, are not only unsustainable to the environment but also harmful. Landfills release methane, a potent greenhouse gas, into the atmosphere, contributing to climate change (Elagroudy, Warith, and Zayat, 2016). Incineration, on the other hand, releases pollutants such as dioxins and furans, which can have serious health and environmental impacts (Lee, Kim, and Chong, 2016). SWM therefore, promotes the use of environmentally friendly techniques such as anaerobic digestion, gasification, and pyrolysis, to convert waste into energy or fuel, without causing harm to the environment (Elagroudy, Warith, and Zayat, 2016).

The implementation of SWM strategies and practices contributes to achieving a green economy, an economic model that creates sustainable growth, mitigates environmental risks and enhances social well-being. Its core principles include low-carbon emission, resource efficiency, and social inclusion (Dogaru, 2020; Mikhn et al., 2021). Globally, nations are embracing the green economy model. For example, Russia, India, and China have implemented eco-transport systems, replacing fossil-fuel vehicles with solar and electric-powered ones. This has significantly reduced air pollution by mitigating greenhouse gas emissions (Mingaleva, et al., 2019). Additionally, the development of green technologies in various countries aims to safeguard human and environmental health (Mikho et al, 2021). In a green economy, economic prosperity and environmental protection are intertwined. To effectively address climate change and environmental degradation, governments at all levels must play a pivotal role in managing their territories and implementing sustainability principles (de Oliveira et al, 2013). Local governments particularly hold the responsibility to integrate environmental, social, and economic dimensions of sustainability (Gorica, Kripa and Zenelaj, 2012).

The Role of Local Government in Sustainable Waste Management

Local governments play a pivotal role in promoting SWM in the context of a green economy. They are responsible for waste management within their jurisdiction and have a direct impact on the environment and the health of their citizens. The role of Local governments is crucial to waste management policy and strategy implementation that promotes sustainability. As an arm of government closest to the grassroots, Local government is saddled with the responsibility of setting regulations and policies that encourage recycling and reuse of waste materials to achieve a green economy (Nyika, Onyari, Mishra, and Dinka, 2020; Viljoen, Schenck, Volschenk, Blaauw, and Grobler, 2021). Also, Local governments are expected to prepare Waste Management Plans (WMP) for waste monitoring and implementation and transform these plans into sustainable waste removal processes and services (DEFF, 2020a, DEFF, 2020b). In addition, Local governments are expected to organise and coordinate activities involved in waste removal, storage, and disposal. They are to educate the community on recycling and reuse and discourage environmental pollution through bylaws (DEFF, 2012; PMG, 2022). A few of the key roles of local governments in SWM are further discussed below.

Waste Collection and Disposals

Numerous as Local government functions and responsibilities are, waste collection and disposal is one of the most paramount. In the 1970s, waste management was primarily regarded as waste disposal while landfill was the only dominant disposal option (Davoudi and Evans, 2005). However, in recent times, waste management has become a complex issue that affects every community, and it is the responsibility of local governments to implement effective and innovative strategies and reforms to address it. One-way local government performs this responsibility is by providing waste collection bins or containers to households and businesses for waste disposal, which is regularly picked according to schedules. However, waste collection at the advanced level involves heavier equipment and manpower (Olukanni, Adeleke, and Aremu, 2016). Also, the pickup of wastes varies from household waste to commercial and industrial. Once these wastes are collected, the next is disposal. This is done through either landfill or incineration.

Waste Management Education and Planning

Effective waste management is essential for preserving a clean and healthy environment. This process encompasses the gathering, transportation, treatment, and disposal of waste materials in a secure and eco-conscious manner. (Longe, Ukpebor and Omole, 2009; Nyika, Onyari, Mishra, and Dinka, 2020). Aside from this, there is an important area of education, awareness and planning and the Local governments are in a better position to carry out this specific waste management task. The unique understanding of local waste generation patterns, the types of waste produced, and the existing waste management infrastructure is crucial in developing tailored waste management plans that cater to the specific needs of the community (Longe, Ukpebor and Omole, 2009). Based on this understanding, Local governments have a crucial role to play in educating and raising awareness among their communities about proper waste management practices. This includes promoting waste reduction and recycling, as well as educating the public on the importance of separating waste at the source. By doing so, individuals can be encouraged to take responsibility for their waste. Drawing from Oelofse and Godfrey (2008), Maluleke (2021) reveals that the general public often lacks knowledge about the waste lifecycle and the extent of pollution's impact on households. Furthermore, Deborah et al., (2021) emphasise that the gap in environmental awareness among citizens, particularly in developing countries, contributes to waste management challenges faced in those regions. Consequently, there is a pressing need for awareness and public education as a capacity intervention for waste management at the local government level.

Waste Partnerships and Collaborations

The significance of waste partnerships and collaborations cannot be overemphasised. Study has shown that, despite waste management being primarily the function of the public sector, inadequate service delivery has prompted the inclusion of the private sector through partnerships (Olukanni and Nwafor, 2019). These partnerships and collaborations on waste management necessitate cooperation and coordination between local governments and other stakeholders, including waste management companies, community organisations, and businesses. This arrangement is commonly known as a public-private partnership (PPP) (Forsy, 2005; Olukanni and Nwafor, 2019).

Public-private partnerships aim to address the challenges of waste management through joint efforts, resources, and expertise. Through collaborative efforts, stakeholders can enhance sustainable waste management practices that positively impact the community and the environment. A key benefit of waste partnerships lies in resource pooling (Atienza, 2020). Local governments often encounter limitations in waste management capabilities, such as funding, personnel, and equipment. Collaboration enables the sharing of these resources, reducing the financial and logistical burdens of joint projects. Additionally, these partnerships facilitate the exchange of expertise and knowledge (Atiku, 2020; Fields & Atiku, 2018), leading to innovative and efficient waste management solutions. By bringing together diverse perspectives and ideas, collaborations promote a comprehensive understanding of waste management challenges and opportunities.

Enforcement and Regulations

Local governments are responsible for establishing policies and regulations. One of the principal responsibilities of the Local government is the regulation of waste in their respective areas. Waste management is a crucial issue that affects not only the cleanliness and aesthetics of a locality but also has significant environmental and health implications. Hence, the enforcement of waste policies is done in several ways in different nations, which include inspections, penalties, and public awareness campaigns (Nwokike, 2020). For instance, in South Africa and Nigeria (West Africa), Local governments often have a dedicated department or team responsible for overseeing waste management, and their role involves inspecting waste facilities, and waste transport vehicles and conducting audits to ensure that all regulations are being adhered to (Oelofse & Godfrey, 2008; Ijaiya & Joseph, 2014). In cases where violations are found, the responsible party is issued a warning or fine, and in extreme cases. The primary objective of enforcing waste regulations is to ensure that waste is disposed of in an environmentally friendly manner. This means that waste must be properly segregated, and hazardous waste must be treated before disposal. These regulations also aim to promote recycling and composting, ultimately reducing the amount of waste that ends up in landfills (Oelofse & Godfrey, 2008; Ijaiya & Joseph, 2014). These regulations also address issues such as littering and illegal dumping, which contribute to environmental degradation. In recent years, the enforcement and regulations of waste have been strengthened and expanded to include electronic waste. With the rapid advancement of technology, electronic waste has become a major concern as it poses a significant threat to the environment (Van Yken, Boxall, Cheng, Nikoloski, Moheimani, and Kaksonen 2021). For instance, in Australia and China, Local governments are working towards implementing e-waste recycling programs, and have imposed regulations on the disposal of electronics to ensure they are recycled or disposed of properly (Davis & Herat, 2010; Wei & Liu, 2012).

Implication of Sustainable Waste Management to a Green Economy

Waste management has been a growing concern in recent years, as the ever-increasing global population continues to produce more and more waste. The traditional method of waste disposal, which involves incineration or landfill, is not sustainable in the long run and has negative consequences for the environment. As the world moves towards a more sustainable future, proper waste management has become a crucial component in building a green economy. SWM refers to the implementation of environmentally responsible and cost-effective practices to reduce, recycle, and manage waste (Nandy Fortunato, and Martins, 2022; Elsaid & Aghezzaf, 2015). This approach focuses on minimising the amount of waste that ends up in landfills or incinerators, and also, promotes the reuse, recycling, or recovery of materials. It also aims to reduce the environmental and health impacts of waste and promote the efficient use of resources (Lee, Kim, and Chong, 2016). SWM in a green economy has a few implications. These include greenhouse gas emission reduction, conservation of resources, job creation and economic growth, reduction of environmental pollution, and cost-saving/reduction. Hence, Local governments must implement and enforce comprehensive regulations and policies to ensure sustainable SWM and promote a green economy. These measures should prioritise environmentally friendly practices and ensure the realisation of the benefits associated with SWM. Moreover, a strategic and holistic approach to waste management, guided by green principles, is essential for long-term sustainability and economic vitality.

Greenhouse Gas Emission Reduction

Study has shown that one of the primary causes of change in the emission of greenhouse gases is the recent development of SWM strategies such as the 3Rs (reduction, reuse and recycle). Greenhouse gases, such as methane trap heat in the earth's atmosphere at landfill sites and this contributes to the warming of our planet (Lee, Kim, and Chong, 2016; Elagroudy, Warith and Zayat, 2016; Knickmeyer, 2020; Purmessur & Surroop, 2019; Un, 2023). Other greenhouse gas components include carbon dioxide and nitrous oxide. The devastating effects of these gases can only be managed by avoiding their release either through landfills or incineration. As a result, SWM plays a crucial role in reducing greenhouse gas emissions. In waste management, the traditional methods of waste management have been confirmed to contribute significantly to greenhouse gas emissions. Landfills are the largest human-caused source of methane emissions, with decomposing organic waste releasing large amounts of this potent greenhouse gas into the atmosphere. One approach Local government can use to manage greenhouse gas emissions from landfills is called a collection and control system (Purmessur & Surroop, 2019; Un, 2023). With this approach, the methane gas that is produced by the decomposition of organic waste in the landfill is processed to generate electricity or heat. This helps to reduce the amount of methane that is released into the atmosphere, which is a greenhouse gas that contributes to climate change. Landfill management is a complex issue, but it is essential to protect the health and safety of our communities and the environment. By working together, local governments can develop and implement effective landfill management strategies that meet the needs of their communities.

Incineration on the other hand is a waste management process that involves burning waste at high temperatures to reduce its volume and weight. It is a common method of waste disposal in many countries, but it can also be a controversial issue due to concerns about air pollution and the release of toxic gases such as carbon dioxide and other harmful gases thereby contributing to the greenhouse effect (Chen and Lin, 2007; Elagroudy, Warith and Zayat, 2016). Local governments must regulate by issuing permits to incineration facilities and set standards for their operation. These standards may include limits on emissions, requirements for monitoring and reporting, and safety protocols (Cui et al., 2024). Therefore, to achieve SWM, it is essential to reduce, reuse, and recycle waste materials.

Conservation of Resources

The world is facing a pressing issue of resource depletion and waste management. With the growing population and increasing consumption patterns, the need for conservation of resources has become more critical than ever. The current linear production-consumption-disposal model of the economy is not sustainable in the long run and is leading us towards an environmental crisis (Akimoto and Futagami, 2018). Therefore, the transition towards SWM is crucial to achieving a green economy, where resources are conserved, and waste is minimized. SWM is crucial to conserving resources by promoting the 3Rs - reduce, reuse, and recycle. According to Mallak and Ishak (2012), this transition is already taking place in many nations that have put in place policies to ensure compliance. The first step towards conserving resources is to reduce the need for them. By reducing waste generation, we are not only minimizing pollution but also minimizing the use of raw materials and energy. The second step is to reuse materials. Many items that we consider waste can be reused in different forms. By promoting the reuse of materials, we are extending their lifespan and reducing the need for new resources. The final step towards conserving resources is recycling. Recycling involves the process of converting waste materials into new products that can be used again. This reduces the demand for raw materials and minimizes the amount of waste sent to landfills. Local governments typically set up recycling programs that include curbside pickup, drop-off centers, and buyback programs. Local governments also enforce recycling regulations (Lee & Krieger, 2020; Agarwal, Werner, Lane & Lamborn, 2020). These regulations may require residents to recycle certain materials, such as paper, plastic, and metal. They may also prohibit residents from putting certain materials in their recycling bins, such as hazardous waste and food scraps. Therefore, the conservation of resources is a crucial implication of sustainable waste management towards a green economy. In addition, through technological innovations, waste can also be converted into renewable energy (gasification) (Elagroudy, Warith and Zayat, 2016).

Job Creation and Economic Growth

With rapid industrialization and population growth, the amount of waste being generated has increased exponentially. However, the concept of a green economy offers a solution to this problem, by promoting SWM as a means of achieving economic growth and creating job opportunities (Elagroudy, Warith and Zayat, 2016; Godfrey and Oelofse, 2017; Bala et al., 2021). A green economy aims to improve human well-being while reducing environmental risks and ecological scarcities. One of the key implications of SWM for a green economy is the creation of job opportunities, human development and innovative research (Elagroudy, Warith and Zayat, 2016; Forti et al., 2020). The transition from traditional waste disposal methods to SWM practices requires in-depth research and a skilled workforce. This requires the involvement of renowned researchers and professionals trained in waste management, renewable energy, and recycling technologies. This is possible through Local governments' funding, sponsorship, and partnership (Baud, Grafakos,

Hordijk, & Post, 2001). Also, by creating jobs in these areas, SWM contributes to the growth of the green economy. Moreover, the development of SWM systems also leads to the growth of new industries. For instance, the recycling industry has grown significantly in recent years, creating jobs in waste collection, sorting, and processing (Forti et al., 2020). This not only reduces the burden on the environment but also provides a significant boost to the economy. The tangible effects of this can be observed in countries such as India, China, the Philippines, and Peru (Baud, Grafakos, Hordijk, & Post, 2001).

Reduction of Environmental Pollution

Recent studies have shown that the amount of waste generated has reached alarming levels while the traditional methods of waste disposal continue to pose severe environmental consequences, such as air and water pollution, greenhouse gas emissions, and the depletion of natural resources (Lee, Kim, and Chong, 2016; Knickmeyer, 2020). Therefore, it is crucial to adopt SWM practices that reduce environmental pollution and promote a green economy. By this implementation such as composting and recycling, the amount of waste that needs to be burned or buried can be significantly reduced, resulting in fewer harmful emissions and a cleaner atmosphere. As such, the reduction of environmental pollution is a critical implication of SWM towards a green economy. Local governments can promote sustainable development by encouraging the use of renewable energy sources, such as solar and wind power, and supporting the development of green infrastructure, such as parks and green spaces. They can also promote sustainable transportation options, such as public transit, cycling, and walking, to reduce the number of vehicles on the road and decrease air pollution (Deakin, 2001; Patil, 2022). It not only helps us protect our planet but also has numerous economic benefits. Monitoring and evaluating the progress of environmental initiatives to ensure the effective achievement of intended goals is also critical. This can involve collecting data on pollution levels, tracking the implementation of environmental policies, and assessing the impact of these policies on the environment. This will help identify areas for improvement and make adjustments to strategies as needed (Pediairi, Donik & Moffar, 2010).

Cost-Saving/Reduction

Local governments are not profit-making entities, as such, their incomes only come from rates and taxes. More so, many Local governments have small revenue bases (DEA, 2012). Hence, with the ever-increasing population and the constant generation of waste, the need for sustainable waste management has become more critical than ever (Oelofse and Godfrey, 2008; Lee, Kim, and Chong, 2016). Failure to manage waste effectively not only leads to environmental degradation but also has economic consequences. In recent years, there has been a shift towards a green economy, with a focus on sustainable practices and cost-saving measures. Traditional waste management practices, such as landfilling and incineration, are harmful to the environment and also come at a high cost. Landfills require constant maintenance and monitoring, while incineration plants are expensive to build and operate. Hence, sustainable strategies such as waste buy-back and pay-as-you-throw, help to reduce waste generation to a great extent (Adeleke et al., 2021; Nahman, 2021). These practices reduce the amount of waste being sent to landfills and incinerators, resulting in reduced operating and maintenance costs. SWM has far-reaching implications for a green economy. By reducing costs, promoting resource efficiency, improving public health, and creating job opportunities, it is a win-win situation for both the environment and the economy. This is possible through Local governments' sustainability plans and strategic thinking.

CONCLUSION

This study set out to examine sustainable waste management in a green economy with a focus on the role of Local government. The study emphasized the 3Rs as part of the components of sustainable waste management strategic interventions needed for a green economy. Hence Local governments must develop strategic policies that will enhance and encourage the adoption of the reduction, reuse, and recycling of waste materials. Also, the study narrowed in on the specific roles of Local government in waste management. Globally a few of the identified roles include waste collection and disposal, waste management education and planning, waste partnership and collaborations, waste partnership and collaborations, and enforcement and regulations. These roles are essential to achieve sustainable waste management at the local government for a green economy. Furthermore, the study examined the implications of sustainable waste management for a green economy and some of the benefits captured include greenhouse gas emission reduction, conservation of resources, job creation and economic growth, reduction of environmental pollution, and cost-saving/reduction. The key finding in this study is that Local government has a crucial role to play in ensuring and enhancing sustainable waste management to achieve a green economy. One of the practical implications of this study is that it will help Local government leaders and Local government policymakers implement sustainable waste management practices with a holistic approach that drives sustainable and green initiatives, such as gasification, sustainable green education and awareness, sustainable waste management and green research. By embracing these practical implications, Local governments globally can make a meaningful impact on waste management and enhance a greener, more resilient economy. Hence, a concerted effort must be made through partnership and well-deployed incentive drives to encourage behavioral change and enhance sustainable waste management practices.

REFERENCES

Adeleke, O., Akinlabi, S., Jen, T. C., & Dunmade, I. (2021). Towards sustainability in municipal solid waste management in South Africa: A survey of challenges and prospects. *Transactions of the Royal Society of South Africa*, 76(1), 53–66. 10.1080/0035919X.2020.1858366

Akimoto, K., & Futagami, K. (2018). *Transition from a linear economy toward a circular economy in the Ramsey model* (No. 18-09).

Atiku, S. O. (2019). Institutionalizing Social Responsibility Through Workplace Green Behavior. In Atiku, S. (Ed.), *Contemporary Multicultural Orientations and Practices for Global Leadership* (pp. 183–199). IGI Global. 10.4018/978-1-5225-6286-3.ch010

Atiku, S. O. (2020). Knowledge Management for the Circular Economy. In Baporikar, N. (Ed.), *Handbook of Research on Entrepreneurship Development and Opportunities in Circular Economy* (pp. 520–537). IGI Global. 10.4018/978-1-7998-5116-5.ch027

Atiku, S. O. (Ed.). (2020). *Human Capital Formation for the Fourth Industrial Revolution*. IGI Global. 10.4018/978-1-5225-9810-7

Atiku, S. O., & Fapohunda, T. (Eds.). (2021). *Human Resource Management Practices for Promoting Sustainability*. IGI Global. 10.4018/978-1-7998-4522-5

Banerjee, S., & Sarkhel, P. (2020). Municipal solid waste management, household and local government participation: A cross country analysis. *Journal of Environmental Planning and Management*, 63(2), 210–235. 10.1080/09640568.2019.1576512

Baud, I. S. A., Grafakos, S., Hordijk, M., & Post, J. (2001). Quality of life and alliances in solid waste management: Contributions to urban sustainable development. *Cities (London, England)*, 18(1), 3–12. 10.1016/S0264-2751(00)00049-4

Cole, C., Quddus, M., Wheatley, A., Osmani, M., & Kay, K. (2014). The impact of Local Authorities' interventions on household waste collection: A case study approach using time series modelling. *Waste Management (New York, N.Y.)*, 34(2), 266–272. 10.1016/j.wasman.2013.10.01824256716

Davoudi, S., & Evans, N. (2005). The challenge of governance in regional waste planning. *Environment and Planning. C, Government & Policy*, 23(4), 493–517. 10.1068/c42m

de Oliveira, J. A. P., Doll, C. N., Balaban, O., Jiang, P., Dreyfus, M., Suwa, A., & Dirgahayani, P. (2013). Green economy and governance in cities: Assessing good governance in key urban economic processes. *Journal of Cleaner Production*, 58, 138–152. 10.1016/j.jclepro.2013.07.043

Deakin, E. (2001). *Sustainable development and sustainable transportation: strategies for economic prosperity, environmental quality, and equity.*

DEFF. (2012). *Addressing challenges with waste service provision in South Africa: Municipal waste sector plan.* Department of Environment, Forestry, and Fisheries.

DEFF. (2020a). *National Waste Management Strategy.* Department of Environment, Forestry, and Fisheries.

DEFF. (2020b). *National Waste Management Strategy.* Department of Environment, Forestry, and Fisheries.

Desa, A., Ba'yah Abd Kadir, N., & Yusooff, F. (2011). A study on the knowledge, attitudes, awareness status and behaviour concerning solid waste management. *Procedia: Social and Behavioral Sciences*, 18, 643–648. 10.1016/j.sbspro.2011.05.095

Dogaru, L. (2020). Eco-innovation and the contribution of companies to the sustainable development. *Procedia Manufacturing*, 46, 294–298. 10.1016/j.promfg.2020.03.043

Elagroudy, S., Warith, M. A., & El Zayat, M. (2016). *Municipal solid waste management and green economy.* Global Young Academy.

Elsaid, S., & Aghezzaf, E. H. (2015). A framework for sustainable waste management: Challenges and opportunities. *Management Research Review*, 38(10), 1086–1097. 10.1108/MRR-11-2014-0264

Ertz, M., Favier, R., Robinot, É., & Sun, S. (2021). To waste or not to waste? Empirical study of waste minimization behavior. *Waste Management (New York, N.Y.)*, 131, 443–452. 10.1016/j.wasman.2021.06.03234256344

Fields, Z., & Atiku, S. O. (2017). Collective Green Creativity and Eco-Innovation as Key Drivers of Sustainable Business Solutions in Organizations. In Fields, Z. (Ed.), *Collective Creativity for Responsible and Sustainable Business Practice* (pp. 1–25). IGI Global. 10.4018/978-1-5225-1823-5.ch001

Fields, Z., & Atiku, S. O. (2018). Collaborative Approaches for Communities of Practice Activities Enrichment. In Baporikar, N. (Ed.), *Knowledge Integration Strategies for Entrepreneurship and Sustainability* (pp. 304–333). IGI Global. 10.4018/978-1-5225-5115-7.ch015

Forti, V., Balde, C. P., Kuehr, R., & Bel, G. (2020). *The Global E-waste Monitor 2020: Quantities, flows and the circular economy potential.*

Godfrey, L. K., & Oelofse, S. H. (2008). *Systems approach to waste governance: Unpacking the challenges facing local government.*

Gorica, K., Kripa, D., & Zenelaj, E. (2012). The role of local government in sustainable development. *Acta Universitatis Danubius. Œconomica, 8*(2).

Ijaiya, H., & Joseph, O. T. (2014). Rethinking environmental law enforcement in Nigeria. *Beijing Law Review*, 5(4), 306–321. 10.4236/blr.2014.54029

Knickmeyer, D. (2020). Social factors influencing household waste separation: A literature review on good practices to improve the recycling performance of urban areas. *Journal of Cleaner Production*, 245, 118605. 10.1016/j.jclepro.2019.118605

Lee, S., Kim, J., & Chong, W. K. (2016). The causes of the municipal solid waste and the greenhouse gas emissions from the waste sector in the United States. *Waste Management (New York, N.Y.)*, 56, 593–599. 10.1016/j.wasman.2016.07.02227475865

Longe, E. O., Ukpebor, E. F., & Omole, D. O. (2009). HOUSEHOLD WASTE COLLECTION AND DISPOSAL: OJO LOCAL GOVERNMENT CASE STUDY, LAGOS STATE, NIGERIA. *Journal of Engineering Research*, 14(4).

Lugo, M., Ail, S. S., & Castaldi, M. J. (2020). Approaching a zero-waste strategy by reuse in New York City: Challenges and potential. *Waste Management & Research*, 38(7), 734–744. 10.1177/0734242X2091949632372709

Mallak, S. K., & Ishak, M. K. (2012). Waste minimization as sustainable waste management strategy for Malaysian industries. In *UMT 11th International Annual Symposium on Sustainability Science and Management*. UMT.

Maluleke, T.C. (2021). *A Review of Municipal Solid Waste Management Systems in Polokwane City Limpopo Province*. Northwest University, North West.

Matinde, E., Simate, G. S., & Ndlovu, S. (2018). Mining and metallurgical wastes: A review of recycling and re-use practices. *Journal of the Southern African Institute of Mining and Metallurgy*, 118(8), 825–844. 10.17159/2411-9717/2018/v118n8a5

Mikhno, I., Koval, V., Shvets, G., Garmatiuk, O., & Tamošiūnienė, R. (2021). *Green economy in sustainable development and improvement of resource efficiency*.

Mingaleva, Z., Vukovic, N., Volkova, I., & Salimova, T. (2019). Waste management in green and smart cities: A case study of Russia. *Sustainability (Basel)*, 12(1), 94. 10.3390/su12010094

Nahman, A. (2021). *Incentives for municipalities to divert waste from landfill in South Africa. Waste research development and innovation roadmap research report*. Council for Scientific and Industrial Research.

Nandy, S., Fortunato, E., & Martins, R. (2022). Green economy and waste management: An inevitable plan for materials science. *Progress in Natural Science*, 32(1), 1–9. 10.1016/j.pnsc.2022.01.001

Nwokike, L. I. (2020). Lagos Waste Management Authority Law 2007 and National Environmental Standards and Regulations Enforcement Agency (Establishment) Act 2007: A Comparative Appraisal. *AJLHR*, 4, 112.

Nyika, J. M., Onyari, E. K., Mishra, S., & Dinka, M. O. (2020). Waste Management in South Africa. In *Sustainable Waste Management Challenges in Developing Countries* (pp. 327–351). IGI Global. 10.4018/978-1-7998-0198-6.ch014

Oelofse, S. H., & Godfrey, L. (2008, November). Towards improved waste management services by local government–A waste governance perspective. In *Proceedings of Science: real and relevant Conference* (pp. 17-18).

Olukanni, D. O., Adeleke, J. O., & Aremu, D. D. (2016). A review of local factors affecting solid waste collection in Nigeria.

Pires, A., & Martinho, G. (2019). Waste hierarchy index for circular economy in waste management. *Waste Management (New York, N.Y.)*, 95, 298–305. 10.1016/j.wasman.2019.06.01431351615

PMG. (2022) *Status of Waste Management in South Africa. Cape Town, South Africa: Parliament of South Africa*. PMG.

Purmessur, B., & Surroop, D. (2019). Power generation using landfill gas generated from new cell at the existing landfill site. *Journal of Environmental Chemical Engineering*, 7(3), 103060. 10.1016/j.jece.2019.103060

Tulebayeva, N., Yergobek, D., Pestunova, G., Mottaeva, A., & Sapakova, Z. (2020). Green economy: Waste management and recycling methods. In *E3S Web of Conferences* (Vol. 159, p. 01012). EDP Sciences. 10.1051/e3sconf/202015901012

Ugwu, C. O., Ozoegwu, C. G., Ozor, P. A., Agwu, N., & Mbohwa, C. (2021). Waste reduction and utilization strategies to improve municipal solid waste management on Nigerian campuses. *Fuel Communications*, 9, 100025. 10.1016/j.jfueco.2021.100025

Un, C. (2023). A Sustainable approach to the conversion of waste into energy: Landfill gas-to-fuel technology. *Sustainability (Basel)*, 15(20), 14782. 10.3390/su152014782

Van Yken, J., Boxall, N. J., Cheng, K. Y., Nikoloski, A. N., Moheimani, N. R., & Kaksonen, A. H. (2021). E-waste.recycling and resource recovery: A review on technologies, barriers and enablers with a focus on oceania. *Metals*, 11(8), 1313. 10.3390/met11081313

Viljoen, J. M., Schenck, C. J., Volschenk, L., Blaauw, P. F., & Grobler, L. (2021). Household waste management practices and challenges in a rural remote town in the Hantam Municipality in the Northern Cape, South Africa. *Sustainability (Basel)*, 13(11), 5903. 10.3390/su13115903

Wan, C., Shen, G. Q., & Choi, S. (2019). Waste management strategies for sustainable development. In *Encyclopedia of sustainability in higher education* (pp. 2020–2028). Springer International Publishing. 10.1007/978-3-030-11352-0_194

Zorpas, A. A. (2020). Strategy development in the framework of waste management. *The Science of the Total Environment*, 716, 137088. 10.1016/j.scitotenv.2020.13708832059326

Zotos, G., Karagiannidis, A., Zampetoglou, S., Malamakis, A., Antonopoulos, I. S., Kontogianni, S., & Tchobanoglous, G. (2009). Developing a holistic strategy for integrated waste management within municipal planning: Challenges, policies, solutions and perspectives for Hellenic municipalities in the zero-waste, low-cost direction. *Waste Management (New York, N.Y.)*, 29(5), 1686–1692. 10.1016/j.wasman.2008.11.01619147341

Compilation of References

Ababneh, O. M. A. (2021). How do green HRM practices affect employees' green behaviours? The role of employee engagement and personality attributes. *Journal of Environmental Planning and Management*, 64(7), 1204–1226. 10.1080/09640568.2020.1814708

Abdelhay, A., & Abunaser, S. G. (2021). Modeling and Economic Analysis of Greywater Treatment in Rural Areas in Jordan Using a Novel Vertical-Flow Constructed Wetland. *Environmental Management*, 67(3), 477–488. 10.1007/s00267-020-01349-732856093

Abdul-Aziz H.M.et.al A. *Study of Baseline Data Regarding Solid Waste Management in the Holy City of Makkah during Hajj*. The Custodian of the Two Holy Mosques Institute of the Hajj Research; Medina, Saudi Arabia: 2007. Unpunished Report.

Abdulfatah, H. K., Stanley, O. I., Nzerem, P., & Jakada, K. (2019). Defining the optimal development strategy to maximize recovery and production rate from an integrated offshore water-flood project. *Society of Petroleum Engineers - SPE Nigeria Annual International Conference and Exhibition 2019, NAIC 2019*. ACM. 10.2118/198843-MS

Abubakar, I. R., & Aina, Y. A. Population Growth and Rapid Urbanization in the Developing World. IGI Global; Hershey, PA, USA: 2016. Achieving sustainable cities in Saudi Arabia: Juggling the competing urbanization challenges; pp. 234–255.

Abubakar, I. R. (2017). Household response to inadequate sewerage and garbage collection services in Abuja, Nigeria. *Journal of Environmental and Public Health*, 2017, 5314840. 10.1155/2017/531484028634496

Abul, S. (2010). Environmental and health impact of solid waste disposal at Manganin dumpsite in Manzini: Swaziland. *Journal of Sustainable Development in Africa*, 12, 64–78.

Addo, I. B., Adei, D., & Acheampong, E. O. (2015). Solid Waste Management and Its Health Implications on the Dwellers of Kumasi Metropolis, Ghana. *Curr. Res. J. Soc. Sci.*, 7(3), 81–93. 10.19026/crjss.7.5225

Adeleke, O., Akinlabi, S., Jen, T. C., & Dunmade, I. (2021). Towards sustainability in municipal solid waste management in South Africa: A survey of challenges and prospects. *Transactions of the Royal Society of South Africa*, 76(1), 53–66. 10.1080/0035919X.2020.1858366

Adel, H. M. (2021). Mapping and assessing green entrepreneurial performance: Evidence from a vertically integrated organic beverages supply chain. *Journal of Entrepreneurship and Innovation in Emerging Economies*, 7(1), 78–98. 10.1177/2393957520983722

Adewumi, S. A., Ajadi, T., & Ntshangase, B. (2022). Green human resource management and green environmental workplace behaviour in the Thekwini municipality of South Africa. *International Journal of Research in Business and Social Science, 2147-4478, 11*(4), 159–170.

Adisa, T. A., Aiyenitaju, O., & Adekoya, O. D. (2021). The work–family balance of British working women during the COVID-19 pandemic. *Journal of Work-Applied Management*, 13(2), 241–260. 10.1108/JWAM-07-2020-0036

Adisa, T., & Abdulkareem, I. (2017). Long working hours and the challenges of work-life balance: The case of Nigerian medical doctors. *British Academy of Management*, 2017(October), 1–12.

Adisa, T., Abdulkareem, I., & Isiaka, S. (2019). Patriarchal hegemony: Investigating the impact of patriarchy on women's work-life balance. *Gender in Management*, 34(1), 19–33. 10.1108/GM-07-2018-0095

Agiakloglou, C., & Gkouvakis, M. (2022). Policy implications and welfare analysis under the possibility of default for the Euro zone area. *Journal of Economic Asymmetries, 25*(November 2020), e00246. 10.1016/j.jeca.2022.e00246

Ahmad, S., 2015. Green human resource management: Policies and practices. *Cogent business & management, 2*(1), p.1030817.

Ahmed, R. R., & Zhang, X. (2021). Multi-stage network-based two-type cost minimization for the reverse logistics management of inert construction waste. *Waste Management (New York, N.Y.)*, 120, 805–819. 10.1016/j.wasman.2020.11.00433279346

Akimoto, K., & Futagami, K. (2018). *Transition from a linear economy toward a circular economy in the Ramsey model* (No. 18-09).

Akmal T., Jamil F. Health impact of Solid Waste Management Practices on Household: The case of Metropolitans of Islamabad-Rawalpindi, Pakistan. *Heliyon*. 2021. doi:. 2021.e07327.10.1016/j. heliyon

AL-agele, H. A., Nackley, L., & Higgins, C. W.AL-agele. (2021). A Pathway for Sustainable Agriculture. *Sustainability (Basel)*, 13(8), 4328. 10.3390/su13084328

Allen, Y. (2007). *Innovation pushes Edmonton to the leading edge of waste management*. FCM. https://www.fcm.ca/Documents/presentations/2007/mission/Innovation_pushes_Edmonto to the leading edge of waste management EN.pdf

Allenby, B., & Fullerton, A. (1991). Design for Environment: A new strategy for environmental management. *Pollution Prevention Review*, 2(1), 51–61.

Al-Romeedy, B. S. (2019). Green human resource management in Egyptian travel agencies: Constraints of implementation and requirements for success. *Journal of Human Resources in Hospitality & Tourism*, 18(4), 529–548. 10.1080/15332845.2019.1626969

Alvesson, M., & Sveningsson, S. (2015). *Changing organizational culture: Cultural change work in progress*. Routledge. 10.4324/9781315688404

Amrutha, V. N., & Geetha, S. N. (2020). A systematic review on green human resource management: Implications for social sustainability. *Journal of Cleaner Production*, 247, 119131. 10.1016/j.jclepro.2019.119131

Andersson, S., Svensson, G., Molina-Castillo, F. J., Otero-Neira, C., Lindgren, J., Karlsson, N. P., & Laurell, H. (2022). Sustainable development—Direct and indirect effects between economic, social, and environmental dimensions in business practices. *Corporate Social Responsibility and Environmental Management*, 29(5), 1158–1172. 10.1002/csr.2261

Ansari, N. Y., Farrukh, M., & Raza, A. (2021). Green human resource management and employees' pro-environmental behaviours: Examining the underlying mechanism. *Corporate Social Responsibility and Environmental Management*, 28(1), 229–238. 10.1002/csr.2044

Anton, A. (2016). Green human resource management practices: A review. *Sri Lankan Journal of Human Resource Management*, 5(1), 1–16.

Appelbaum, E., Bailey, T., Berg, P., & Kalleberg, A. (2000). *Manufacturing Advantage: Why High-Performance Work Systems Pay off*. Cornell University Press.

Apple. (2021). *Environmental Progress Report*. Apple. https://www.apple.com/environment/pdf/Apple_Environmental_Progress_Report_2021.pdf

Aprianti, V., Hurriyati, R., Gaffar, V., & Wibowo, L. A. (2021). The effect of green trust and attitude toward purchasing intention of green products: A case study of the green apparel industry in Indonesia. *The Journal of Asian Finance. Economics and Business*, 8(7), 235–244.

Aragon-Correa, J. A., Martin-Tapia, I., & Hurtado-Torres, N. E. (2013). Proactive environmental strategies and employee inclusion: The positive effects of information sharing and promoting collaboration and the influence of uncertainty. *Organization & Environment*, 26(2), 139–161. 10.1177/1086026613489034

Aragon-Correa, J. A., & Sharma, S. (2003). A contingent resource based view of proactive corporate environmental strategy. *Academy of Management Review*, 28(1), 71–88. 10.5465/amr.2003.8925233

Ardiza, F., Nawangsari, L. C., & Sutawidjaya, A. H. (2021). The influence of green performance appraisal and green compensation to improve employee performance through OCBE. *International Review of Management and Marketing*, 11(4), 13–22. 10.32479/irmm.11632

Arum, E., Agyabeng-Mensah, Y., & Owusu, J. A. (2020). Translating environmental management practices into improved environmental performance via green organizational culture: Insight from Ghanaian manufacturing SMEs. *Journal of Supply Chain Management Systems*, 9(1), 31–49.

Ashu, A. B., & Lee, S. (2021). The Effects of Climate Change on the Reuse of Agricultural Drainage Water in Irrigation. *KSCE Journal of Civil Engineering*, 25(3), 1116–1129. 10.1007/s12205-021-0004-2

Atiku, S. O. (2019). Institutionalizing Social Responsibility Through Workplace Green Behavior. In Atiku, S. (Ed.), *Contemporary Multicultural Orientations and Practices for Global Leadership* (pp. 183–199). IGI Global. 10.4018/978-1-5225-6286-3.ch010

Atiku, S. O. (2020). Knowledge Management for the Circular Economy. In Baporikar, N. (Ed.), *Handbook of Research on Entrepreneurship Development and Opportunities in Circular Economy* (pp. 520–537). IGI Global. 10.4018/978-1-7998-5116-5.ch027

Atiku, S. O. (Ed.). (2020). *Human Capital Formation for the Fourth Industrial Revolution*. IGI Global. 10.4018/978-1-5225-9810-7

Atiku, S. O., & Abatan, A. A. (2021). Strategic Capabilities for the Sustainability of Small, Medium, and Micro Enterprises. In Ayandibu, A. (Ed.), *Reshaping Entrepreneurship Education With Strategy and Innovation* (pp. 17–44). IGI Global. 10.4018/978-1-7998-3171-6.ch002

Atiku, S. O., & Anane-Simon, R. (2022). Stimulating Creativity and Innovation Through Apt Educational Policy. In Fields, Z. (Ed.), *Achieving Sustainability Using Creativity, Innovation, and Education: A Multidisciplinary Approach* (pp. 113–133). IGI Global. 10.4018/978-1-7998-7963-3.ch006

Atiku, S. O., & Fapohunda, T. (Eds.). (2021). *Human Resource Management Practices for Promoting Sustainability*. IGI Global. 10.4018/978-1-7998-4522-5

Baggio, G., Qadir, M., & Smakhtin, V. (2021). Freshwater availability status across countries for human and ecosystem needs. *The Science of the Total Environment*, 792, 148230. 10.1016/j.scitotenv.2021.14823034147805

Baird, K., Harrison, G., & Reeve, R. (2007). The culture of Australian organizations and its relation with strategy. *International Journal of Business Studies: A Publication of the Faculty of Business Administration. Edith Cowan University*, 15(1), 15–41.

Balogun, A. L., Adebisi, N., Abubakar, I. R., Dano, U. L., & Tella, A. (2022). Digitalization for transformative urbanization, climate change adaptation, and sustainable farming in Africa: Trend, opportunities, and challenges. *Journal of Integrative Environmental Sciences*, 19(1), 17–37. 10.1080/1943815X.2022.2033791

Bandoophanit, T. (2024). Holistic implementations of green supply chain management practices in Thai entrepreneurial ventures. *Journal of Small Business and Enterprise Development*. Emerald Publishing. https://www.emerald.com/insight/content/doi/10.1108/JSBED-01-2023-0001/full/html)

Banerjee, S. (2001). Managerial perceptions of corporate environmentalism: Interpretation from industry and strategic implications for organizations. *Journal of Management Studies*, 38(4), 489–513. 10.1111/1467-6486.00246

Banerjee, S., & Sarkhel, P. (2020). Municipal solid waste management, household and local government participation: A cross country analysis. *Journal of Environmental Planning and Management, 63*(2), 210–235. 10.1080/09640568.2019.1576512

Bangwal, D., & Tiwari, P. (2015). Green HRM–A way to greening the environment. *IOSR Journal of Business and Management, 17*(12), 45–53.

Bangwal, D., Tiwari, P., & Chamola, P. (2017). Green HRM, work-life and environment performance. *International Journal of Environment. Workplace and Employment, 4*(3), 244–268. 10.1504/IJEWE.2017.087808

Batool, S. A., & Chaudhry, M. N. (2009). The impact of municipal solid waste treatment methods on greenhouse gas emissions in Lahore, Pakistan. *Waste Management (New York, N.Y.), 29*(1), 63–69. 10.1016/j.wasman.2008.01.01318387288

Batten. (2020). Patagonia's Worn Wear Collection Is Saving the Planet. *The Manual.* www.themanual.com/outdoors/oatagonia-worn-wear-collection-recycled-recommerce/

Batwara, A., Sharma, V., Makkar, M., & Giallanza, A. (2022). An Empirical Investigation of Green Product Design and Development Strategies for Eco Industries Using Kano Model and Fuzzy AHP. *Sustainability (Basel), 14*(14), 8735. 10.3390/su14148735

Baud, I. S. A., Grafakos, S., Hordijk, M., & Post, J. (2001). Quality of life and alliances in solid waste management: Contributions to urban sustainable development. *Cities (London, England), 18*(1), 3–12. 10.1016/S0264-2751(00)00049-4

Baumgartner, R. J. (2009). *Organizational culture and leadership: Preconditions.*

Baumgartner, R. J., & Zielowski, C. (2007). Analysing zero emission strategies regarding impact on organizational culture and contribution to sustainable development. *Journal of Cleaner Production, 15*(13-14), 1321–1327. 10.1016/j.jclepro.2006.07.016

Becker, B. (1998). High performance work systems and firm performance: A synthesis of research and managerial implications. *Research in Personnel and Human Resources Management, 16,* 53.

Belias, D., & Koustelios, A. (2014). Organizational culture and job satisfaction: A review. *International Review of Management and Marketing, 4*(2), 132–149.

Bocken, N. M. P., Short, S. W., Rana, P., & Evans, S. (2014). A literature and practice review to develop sustainable business model archetypes. *. Journal of Cleaner Production, 65,* 42–56. 10.1016/j.jclepro.2013.11.039

Boiral, O., Paillé, P., & Raineri, N. (2015). The nature of employees' pro-environmental behaviors. *The Psychology of Green Organizations, 1*(March), 12–32. 10.1093/acprof:oso/9780199997480.003.0002

Bombiak, E. (2020). Barierrs to implementing the concept of green human resource management the case of Poland. *European Research Studies Journal, 23*(4).

Boztepe, A. (2012). Green marketing and its impact on consumer buying behavior. *European Journal of Economic & Political Studies*, 5(1).

Braungart, M., & McDonough, W. (2009). *Cradle to cradle: Remaking the way we make things.* Vintage Books.

Brennan, M., Rondón-Sulbarán, J., Sabogal-Paz, L. P., Fernandez-Ibañez, P., & Galdos-Balzategui, A. (2021). Conceptualising global water challenges: A transdisciplinary approach for understanding different discourses in sustainable development. *Journal of Environmental Management, 298*. https://doi.org/10.1016/j.jenvman.2021.113361

Brockett, J. (2007). Prepare now for big rise in 'green' jobs'. *People Management*, 13(10), 9.

Budak, A. (2020). Sustainable reverse logistics optimization with triple bottom line approach: An integration of disassembly line balancing. *Journal of Cleaner Production*, 270, 122475. 10.1016/j.jclepro.2020.122475

Bunn, S. E. (2016). Grand challenge for the future of freshwater ecosystems. *Frontiers in Environmental Science*, 4(MAR), 1–4. 10.3389/fenvs.2016.00021

Cameron, K. S., & Quinn, R. E. (2006). *Diagnosing and changing organizational culture: Based on the competing values framework.* John Wiley & Sons.

Cantor, D. E., Morrow, P. C., & Montabon, F. (2012). Engagement in environmental behaviors among supply chain management employees: An organizational support theoretical perspective. *The Journal of Supply Chain Management*, 48(3), 33–51. 10.1111/j.1745-493X.2011.03257.x

Cao, Z., Zhou, L., Gao, Z., Huang, Z., Jiao, X., Zhang, Z., Ma, K., Di, Z., & Bai, Y. (2021). Comprehensive benefits assessment of using recycled concrete aggregates as the substrate in constructed wetland polishing effluent from wastewater treatment plant. *Journal of Cleaner Production*, 288, 125551. 10.1016/j.jclepro.2020.125551

Cascardo, A. (2007, October 9-13). Indoor air pollution: An ever-growing threat to our society. *Executive Housekeeping Today.*

Castillo-Díaz, F. J., Belmonte-Ureña, L. J., Batlles-delaFuente, A., & Camacho-Ferre, F. (2023). Strategic evaluation of the sustainability of the Spanish primary sector within the framework of the circular economy. *Sustainable Development (Bradford)*, 1–16. 10.1002/sd.2837

Chams, N., & García-Blandón, J. (2019). On the importance of sustainable human resource management for the adoption of sustainable development goals. *Resources, Conservation and Recycling*, 141(February), 109–122. 10.1016/j.resconrec.2018.10.006

Chan, E. S., & Hawkins, R. (2010). Attitude towards EMSs in an international hotel: An exploratory case study. *International Journal of Hospitality Management*, 29(4), 641–651. 10.1016/j.ijhm.2009.12.002

Compilation of References

Chatman, J. A., & Jehn, K. A. (1994). Assessing the relationship between industry characteristics and organizational culture: How different can you be? *Academy of Management Journal*, 37(3), 522–553. 10.2307/256699

Chaturvedi, P., Kulshreshtha, K., Tripathi, V., & Agnihotri, D. (2024). Investigating the impact of restaurants' sustainable practices on consumers' satisfaction and revisit intentions: A study on leading green restaurants. *Asia-Pacific Journal of Business Administration*, 16(1), 41–62. 10.1108/APJBA-09-2021-0456

Chaudhary, R. (2018). Can green human resource management attract young talent? An empirical analysis. *Evidenced-Based HRM*, 6(3), 305–319. 10.1108/EBHRM-11-2017-0058

Chaudhary, R. (2019). Green human resource management in Indian automobile industry. *Journal of Global Responsibility*, 10(2), 161–175. 10.1108/JGR-12-2018-0084

Chekima, B., Wafa, S. A. W. S. K., Igau, O. A., Chekima, S., & Sondoh, S. L.Jr. (2016). Examining green consumerism motivational drivers: Does premium price and demographics matter to green purchasing? *Journal of Cleaner Production*, 112, 3436–3450. 10.1016/j.jclepro.2015.09.102

Cherian, J., & Jacob, J. (2012). A study of Green HR practices and its effective implementation in the organization: A review. *International Journal of Business and Management*, 7(21), 25–33. 10.5539/ijbm.v7n21p25

Chuah, S. C., Mohd, I. H., Kamaruddin, J. N. B., & Md Noh, N. (2021). Impact of Green Human Resource Management Practices Towards Green Lifestyle and Job Performance. *Global Business and Management Research*, 13(4).

Ciccullo, F., Pero, M., Caridi, M., Gosling, J., & Purvis, L. (2018). Integrating the environmental and social sustainability pillars into the lean and agile supply chain management paradigms: A literature review and future research directions. *Journal of Cleaner Production*, 172, 2336–2350. 10.1016/j.jclepro.2017.11.176

Ciemleja, G., & Lace, N. (2011). The model of sustainable performance of small and medium-sized enterprises. *The Engineering Economist*, 22(5), 501–509.

Clair, J. A., Milliman, J., & Whelan, K. S. (1996). Toward an environmentally sensitive eco-philosophy for business management. *Industrial & Environmental Crisis Quarterly*, 9(3), 289–326. 10.1177/108602669600900302

Coca Cola. (2021). *Business & ESG Report: Refresh the World Make a Difference*. Coca Cola. https://www.coca-colacompany.com/content/dam/company/us/en/reports/coca-cola-business-environmental-social-governance-report-2021.pdf

Cole, C., Quddus, M., Wheatley, A., Osmani, M., & Kay, K. (2014). The impact of Local Authorities' interventions on household waste collection: A case study approach using time series modelling. *Waste Management (New York, N.Y.)*, 34(2), 266–272. 10.1016/j.wasman.2013.10.01824256716

Dahlmann, F., Branicki, L., & Brammer, S. (2019). Managing carbon aspirations: The influence of corporate climate change targets on environmental performance. *Journal of Business Ethics*, 158(1), 1–24. 10.1007/s10551-017-3731-z

Daily, B. F., Bishop, J. W., & Govindarajulu, N. (2009). A conceptual model for organizational citizenship behavior directed toward the environment. *Business & Society*, 48(2), 243–256. 10.1177/0007650308315439

Daily, B. F., & Huang, S. C. (2001). Achieving sustainability through attention to human resource factors in environmental management. *International Journal of Operations & Production Management*, 21(12), 1539–1552. 10.1108/01443570110410892

Damarayudha, T. R., Sadat, A. M., & Febrillia, I. (2023). Pengaruh Green Marketing Mix Terhadap Purchase Intention Dengan Environmental Knowledge Sebagai Variabel Moderator: Survei Pada Toko Furniture Modern. *Indonesian Journal of Economy, Business. Entrepreneurship and Finance*, 3(2), 306–322.

Dangelico, R &Vocalelli, D. (2017). "Green Marketing": an analysis of definitions, strategy steps, and tools through a systematic review of the literature. *Journal of Cleaner Production.* S0959652617316372–. 10.1016/j.jclepro.2017.07.184

Dangelico, R. M. (2016). Green Product Innovation: Where we are and Where we are Going. *Business Strategy and the Environment*, 25(8), 560–576. 10.1002/bse.1886

Daniel, O. Okanigbe et al IOP Conf. Series: Materials Science and Engineering 391 (2018) 012006 10.1088/1757-899X/391/1/012006

Danirmala, L., & Prajogo, W. (2022). The mediating role of green training to the influence of green organizational culture to green organizational citizenship behavior and green employee involvement. *International Journal of Human Capital Management*, 6(1), 66–75. 10.21009/IJHCM.06.01.6

Davoudi, S., & Evans, N. (2005). The challenge of governance in regional waste planning. *Environment and Planning. C, Government & Policy*, 23(4), 493–517. 10.1068/c42m

de Freitas Netto, S. V., Sobral, M. F. F., Ribeiro, A. R. B., & Soares, G. R. L. (2020). Concepts and forms of greenwashing: A systematic review. *Environmental Sciences Europe*, 32(1), 19. 10.1186/s12302-020-0300-3

De Luca, A. I., Iofrida, N., Leskinen, P., Stillitano, T., Falcone, G., Strano, A., & Gulisano, G. (2017). Life cycle tools combined with multi-criteria and participatory methods for agricultural sustainability: Insights from a systematic and critical review. *The Science of the Total Environment*, 595, 352–370. 10.1016/j.scitotenv.2017.03.28428395257

de Oliveira, J. A. P., Doll, C. N., Balaban, O., Jiang, P., Dreyfus, M., Suwa, A., & Dirgahayani, P. (2013). Green economy and governance in cities: Assessing good governance in key urban economic processes. *Journal of Cleaner Production*, 58, 138–152. 10.1016/j.jclepro.2013.07.043

Compilation of References

Deakin, E. (2001). *Sustainable development and sustainable transportation: strategies for economic prosperity, environmental quality, and equity.*

Debrah, Y. A., Oseghale, R. O., & Adams, K. (2018). Human capital development, innovation and international competitiveness in Sub-Saharan Africa. In Adeleye, I., & Esposito, M. (Eds.), *Africa's Competitiveness in the Global Economy* (pp. 219–248). Palgrave Macmillan. 10.1007/978-3-319-67014-0_9

Declercq, R., Loubier, S., Condom, N., & Molle, B. (2020). Socio-Economic Interest of Treated Wastewater Reuse in Agricultural Irrigation and Indirect Potable Water Reuse: Clermont-Ferrand and Cannes Case Studies' Cost–Benefit Analysis. *Irrigation and Drainage*, 69(S1), 194–208. 10.1002/ird.2205

DEFF. (2012). *Addressing challenges with waste service provision in South Africa: Municipal waste sector plan.* Department of Environment, Forestry, and Fisheries.

DEFF. (2020a). *National Waste Management Strategy.* Department of Environment, Forestry, and Fisheries.

Deirmentzoglou, G. A., Agoraki, K. K., & Fousteris, A. E. (2020). Organizational culture and corporate sustainable development: Evidence from Greece. *International Journal of Business and Social Science*, 11(5), 92–98. 10.30845/ijbss.v11n5a9

Dell. (2022). *Concept Luna.* https://www.dell.com/en-us/blog/concept-luna-whats-next/

Denison, D. R. (1990). *Corporate culture and organizational effectiveness.* John Wiley & Sons.

Deri, M. N. (2022). Green practices among hotels in the Sunyani Municipality of Ghana. *Journal of Business and Environmental Management*, 1(1), 1–22. 10.59075/jbem.v1i1.147

Desa, A., Ba'yah Abd Kadir, N., & Yusooff, F. (2011). A study on the knowledge, attitudes, awareness status and behaviour concerning solid waste management. *Procedia: Social and Behavioral Sciences*, 18, 643–648. 10.1016/j.sbspro.2011.05.095

Deshwal, P. (2015). Green HRM: An organizational strategy of greening people. *International Journal of Applied Research*, 1(13), 176–181.

Di Vaio, A., Trujillo, L., D'Amore, G., & Palladino, R. (2021). Water governance models for meeting sustainable development Goals:A structured literature review. *Utilities Policy, 72*, 101255. 10.1016/j.jup.2021.101255

Diaz-Elsayed, N., Rezaei, N., Ndiaye, A., & Zhang, Q. (2020). Trends in the environmental and economic sustainability of wastewater-based resource recovery: A review. *Journal of Cleaner Production*, 265, 121598. 10.1016/j.jclepro.2020.121598

Diaz-Farina, E., Díaz-Hernández, J. J., & Padrón-Fumero, N. (2021). Analysis of hospitality waste generation: Impacts of services and mitigation strategies. *Annals of Tourism Research Empirical Insights*, 4(1), 12–24.

DiMaggio, P. J., & Powell, W. W. (1983). The iron cage revisited: Institutional isomorphism and collective rationality in organizational fields. *. *American Sociological Review*, 48(2), 147–160. 10.2307/2095101

Ding, L., Wang, T., & Chan, P. (2023). Forward and reverse logistics for circular economy in construction: A systematic literature review. *Journal of Cleaner Production*, 388, 135981. 10.1016/j.jclepro.2023.135981

Dogaru, L. (2020). Eco-innovation and the contribution of companies to the sustainable development. *Procedia Manufacturing*, 46, 294–298. 10.1016/j.promfg.2020.03.043

DuBois, C. L., & Dubois, D. A. (2012). Strategic HRM as social design for environmental sustainability in organization. *Human Resource Management*, 51(6), 799–826. 10.1002/hrm.21504

Dumont, J. (2015). *Green human resource management and employee workplace outcomes* [Thesis, University of South Australia].

Dumont, J., Shen, J., & Deng, X. (2017). Effects of green HRM practices on employee workplace green behavior: The role of psychological green climate and employee green values. *Human Resource Management*, 56(4), 613–627. 10.1002/hrm.21792

Durif, F., Boivin, C., & Julien, C. (2010). In search of a green product definition. *Innovative Marketing, 6*(1).

Dzikriansyah, M., Ni, N., Shah, S. Z., & Soomro, B. A. (2023). Transparent reporting and communication strategies for enhancing business reputation and credibility in sustainability.

Edwards, R. J., & Rothbard, P. N. (2000). Mechanisms linking work and family: Clarifying the relationship between work and family constructs. Academy of Management. *Academy of Management Review*, 25(1), 178–199. 10.2307/259269

Elagroudy, S., Warith, M. A., & El Zayat, M. (2016). *Municipal solid waste management and green economy*. Global Young Academy.

Elkington, J. (1997). The triple bottom line. *Environmental management: Readings and cases,* 2, 49-66.

Ellen MacArthur Foundation. (2015). *Schools of thought – industrial ecology.* Ellen MacArthur Foundation. https://www.ellenmacarthurfoundation.org/circular-economy/schools-of-thought/cradle2cradle

Elsaid, S., & Aghezzaf, E. H. (2015). A framework for sustainable waste management: Challenges and opportunities. *Management Research Review*, 38(10), 1086–1097. 10.1108/MRR-11-2014-0264

Emina, K. A. (2021). Sustainable development and the future generations. *Social Sciences* [SHE Journal]. *Humanities and Education Journal*, 2(1), 57–71.

Compilation of References

Ertz, M., Favier, R., Robinot, É., & Sun, S. (2021). To waste or not to waste? Empirical study of waste minimization behavior. *Waste Management (New York, N.Y.)*, 131, 443–452. 10.1016/j. wasman.2021.06.03234256344

European Centre for the Development of Vocational Training. (2012). *A Strategy for Green Skills? A Study on Skill Needs and Training Has Wider Lessons for Successful Transition to a Green Economy: Briefing Report*. European Centre for the Development of Vocational Training.

European Commission. (2021). *European Green Deal: Commission proposes transformation of EU economy and society to meet climate ambition*. EC. https://ec.europa.eu/commission/presscorner/detail/en/ip_21_3541

Eyring, G. (1992). *Green Products by Design: Choices for a Cleaner Environment*. Diane Pub Co.

Fairphone. (2023). https://www.fairphone.com/en/

Faisal, S. (2023). Green human resource management—A synthesis. *Sustainability (Basel)*, 15(3), 2259. 10.3390/su15032259

Fang, B., Yu, J., Chen, Z., Osman, A. I., Farghali, M., Ihara, I., Hamza, E. H., Rooney, D. W., & Yap, P.-S. (2023). Artificial intelligence for waste management in smart cities: A review. *Environmental Chemistry Letters*, 21(4), 1959–1989. 10.1007/s10311-023-01604-3

FasterCapital. (2023). *Sustainability Initiatives: Sustainable Strategies in the Competitive Landscape: A Winning Combination*. Faster Capital. https://fastercapital.com/content/Sustainability -Initiatives--Sustainable-Strategies-in-the-Competitive-Landscape--A-Winning-Combination .html

Fayyazi, M., Shahbazmoradi, S., Afshar, Z. and Shahbazmoradi, M., 2015. Investigating the barriers of the green human resource management implementation in oil industry. *Management science letters, 5*(1), pp.101-108.

Fields, Z., & Atiku, S. O. (2017). Collective Green Creativity and Eco-Innovation as Key Drivers of Sustainable Business Solutions in Organizations. In Fields, Z. (Ed.), *Collective Creativity for Responsible and Sustainable Business Practice* (pp. 1–25). IGI Global. 10.4018/978-1-5225-1823-5. ch001

Fields, Z., & Atiku, S. O. (2018). Collaborative Approaches for Communities of Practice Activities Enrichment. In Baporikar, N. (Ed.), *Knowledge Integration Strategies for Entrepreneurship and Sustainability* (pp. 304–333). IGI Global. 10.4018/978-1-5225-5115-7.ch015

Fietz, B., & Günther, E. (2021). Changing organizational culture to establish sustainability. *Controlling & Management Review*, 65(3), 32–40. 10.1007/s12176-021-0379-4

Fitrianingrum, A. (2020). *Greenwashing, does it Work Well for Indonesian Millennials Buyers.*

Forkan, M., Rizvi, M. M., & Chowdhury, M. A. M. (2022). Multiobjective reverse logistics model for inventory management with environmental impacts: An application in industry. *Intelligent Systems with Applications*, 14, 200078. 10.1016/j.iswa.2022.200078

Forti, V., Balde, C. P., Kuehr, R., & Bel, G. (2020). *The Global E-waste Monitor 2020: Quantities, flows and the circular economy potential.*

Fryxell, G. E., & Lo, C. W. (2003). The influence of environmental knowledge and values on managerial behaviours on behalf of the environment: An empirical examination of managers in China. *Journal of Business Ethics*, 46(1), 45–69. 10.1023/A:1024773012398

Gadenne, D. L., Kennedy, J., & McKeiver, C. (2009). An empirical study of environmental awareness and practices in SMEs. *Journal of Business Ethics*, 84(1), 45–63. 10.1007/s10551-008-9672-9

Gaffar, V., & Koeswandi, T. (2021). Climate Change And The Sustainable Small And Medium-Sized Enterprises. In *Handbook Of Research On Climate Change And The Sustainable Financial Sector* (pp. 171–189). IGI Global. 10.4018/978-1-7998-7967-1.ch011

Gaffar, V., Rahayu, A., Adi Wibowo, L., & Tjahjono, B. (2021). The adoption of circular economy principles in the hotel industry. *Journals and Gaffar, Vanessa and Rahayu, Agus and Adi Wibowo, Lili and Tjahjono, Benny, The Adoption of Circular Economy Principles in the Hotel Industry (June 30, 2021). Reference to this paper should be made as follows. Gaffar*, V, 92–97.

Galpin, T., Whitttington, J. L., & Bell, G. (2015). Is your sustainability strategy sustainable? creating a culture of sustainability. *Corporate Governance (Bradford)*, 15(1), 1–17. 10.1108/CG-01-2013-0004

Gao, X., & Cao, C. (2020). A novel multi-objective scenario-based optimization model for sustainable reverse logistics supply chain network redesign considering facility reconstruction. *Journal of Cleaner Production*, 270, 122405. 10.1016/j.jclepro.2020.122405

Gazzeh, K., Abubakar, I. R., & Hammad, E. (2022). Impacts of COVID-19 Pandemic on the Global Flows of People and Goods: Implications on the Dynamics of Urban Systems. *Land (Basel)*, 11(3), 429. 10.3390/land11030429

Ghanbarzadeh-Shams, M., Yaghin, R. G., & Sadeghi, A. H. (2022). A hybrid fuzzy multi-objective model for carpet production planning with reverse logistics under uncertainty. *Socio-Economic Planning Sciences*, 83, 101344. 10.1016/j.seps.2022.101344

Ghazali, I., Abdul-Rashid, S. H., Dawal, S. Z. M., Irianto, I., Herawan, S. G., Ho, F.-H., Abdullah, R., Abdul Rasib, A. H., & Padzil, N. W. S. (2023). Embedding Green Product Attributes Preferences and Cultural Consideration for Product Design Development: A Conceptual Framework. *Sustainability (Basel)*, 15(5), 4542. 10.3390/su15054542

Gholizadeh, H., Goh, M., Fazlollahtabar, H., & Mamashli, Z. (2022). Modelling uncertainty in sustainable-green integrated reverse logistics network using metaheuristics optimization. *Computers & Industrial Engineering*, 163, 107828. 10.1016/j.cie.2021.107828

Ghorbani, M. (2023). Green Knowledge Management and Innovation for Sustainable Development: A Comprehensive Framework. In *ECKM 2023 24th European Conference on Knowledge Management*. Academic Conferences and publishing limited. 10.34190/eckm.24.1.1753

Gise, (2009).Improving Operations Performance in a small Company: A Case Study.*International Journal of Operations & Production Management,20*(3).

Giusti L. A review of waste management practices and their impact on human health. *Waste Manag.* 2009; **29:2227**–2239.n .10.1016/j.wasman.2009.03.028

Godfrey, L. K., & Oelofse, S. H. (2008). *Systems approach to waste governance: Unpacking the challenges facing local government.*

Goedkoop, M., Heijungs, R., Huijbregts, M., De Schryver, A., Struijs, J., & Van Zelm, R. (2009). ReCiPe 2008. *Potentials, May 2014*, 1–44. https://www.pre-sustainability.com/download/misc/ReCiPe_main_report_final_27-02-2009_web.pdf

Gorica, K., Kripa, D., & Zenelaj, E. (2012). The role of local government in sustainable development. *Acta Universitatis Danubius. Œconomica, 8*(2).

Govindarajulu, N., & Daily, B. F. (2004). Motivating employees for environmental improvement. *Industrial Management & Data Systems*, 104(4), 364–372. 10.1108/02635570410530775

Graedel, T. E., Comrie, P. R., & Sekutowski, J. C. (1995). Green Product Design. *AT & T Technical Journal*, 74(6), 17–25. 10.1002/j.1538-7305.1995.tb00262.x

Guerci, M., Montanari, F., Scapolan, A., & Epifanio, A. (2016). Green and nongreen recruitment practices for attracting job applicants: Exploring independent and interactive effects. *International Journal of Human Resource Management*, 27(2), 129–150. 10.1080/09585192.2015.1062040

Guerrero, L. A., Maas, G., & Hogland, W. (2012). Solid waste management challenges for cities in developing countries. *Waste Management (New York, N.Y.)*, 33(1), 220–232. 10.1016/j.wasman.2012.09.00823098815

Guggeri, E. M., Ham, C., Silveyra, P., Rossit, D. A., & Piñeyro, P. (2023). Goal programming and multi-criteria methods in remanufacturing and reverse logistics: Systematic literature review and survey. *Computers & Industrial Engineering*, 185, 109587. 10.1016/j.cie.2023.109587

Gunasekara, L., Robb, D. J., & Zhang, A. (2023). Used product acquisition, sorting and disposition for circular supply chains: Literature review and research directions. *International Journal of Production Economics*, 260, 108844. 10.1016/j.ijpe.2023.108844

Gunawan, A. I., Amalia, F. A., Ramadhan, M., & Bansah, P. F. (2024). Exploring The Reasons Of Indonesian Young Adult Consumers Toward Sustainably Packaged Food & Beverages Product. [JMI]. *Journal of Marketing Innovation*, 4(1). 10.35313/jmi.v4i1.106

Gürlek, M., & Tuna, M. (2018). Reinforcing competitive advantage through green organizational culture and green innovation. *Service Industries Journal*, 38(7-8), 467–491. 10.1080/02642069.2017.1402889

Gutowski, T., Murphy, C., Allen, D., Bauer, D., Bras, B., Piwonka, T., Sheng, P., Sutherland, J., Thurston, D., & Wolff, E. (2005). Environmentally benign manufacturing: Observations from Japan, Europe and the United States. *Journal of Cleaner Production*, 13(1), 1–17. 10.1016/j.jclepro.2003.10.004

H&M Group. (2023). *Sustainability: How we work with materials*. H&M. https://hmgroup.com/sustainability/circularity-and climate/materials/#: ~:text=How%20we%20work%20with%20materials,30%25%20recycled%20materials%20by%202025.

Hakim, L. N. (2023). Green Manufacturing Practices and Green Innovation and Their Role in Sustainable Business Performance Through Culture Green Organization at Small Industrial Enterprises. In *International Conference on Economics Business Management and Accounting (ICOEMA)* (Vol. 2, pp. 366-376).

Hallak, B. K., Nasr, W. W., & Jaber, M. Y. (2021). Re-ordering policies for inventory systems with recyclable items and stochastic demand–Outsourcing vs. in-house recycling. *Omega*, 105, 102514. 10.1016/j.omega.2021.102514

Hameed, Z., Khan, I. U., Islam, T., Sheikh, Z., & Naeem, R. M. (2020). Do green HRM practices influence employees' environmental performance? *International Journal of Manpower*, 41(7), 1061–1079. 10.1108/IJM-08-2019-0407

Handajani, L., Husnan, L. H., & Rifai, A. (2019). Kajian Tentang Inisiasi Praktik Green Banking Pada Bank BUMN di Indonesia. *Jurnal Economia Review of Business and Economics*, 15(1), 1–16.

Han, H., Hsu, J., & Sheu, C. (2010). Application of the theory of planned behavior to green hotel choice: Testing the effect of environmental friendly activities. *Tourism Management*, 31(3), 325–334. 10.1016/j.tourman.2009.03.013

Hanna, M. D., Rocky Newman, W., & Johnson, P. (2000). Linking operational and environmental improvement through employee involvement. *International Journal of Operations & Production Management*, 20(2), 148–165. 10.1108/01443570010304233

Hariyanto, O. I. B. (2019). *Customer Green Awareness and Eco-Label for Organic Products*. In: 2019 International Conference of Organizational Innovation (2019 ICOI).

Hart, O. (1995). Corporate governance: Some theory and implications. *Economic Journal (London)*, 105(430), 678–689. 10.2307/2235027

Hashem, T. N., & Al-Rifai, N. A. (2011). The influence of applying green marketing mix by chemical industries companies in three Arab States in West Asia on consumer's mental image. *International Journal of Business and Social Science*, 2(3).

Heong, Y. M., Sern, L. C., Kiong, T. T., & Mohamad, M. M. B. (2016). The role of higher order thinking skills in green skill development. *MATEC Web of Conferences, 70*, 05001. EDP Sciences.

Hernu, R. (2022). *Sustainable performance: How to make success last?* [Post]. LinkedIn. https://www.linkedin.com/pulse/sustainable-performance-how-make-success-last-raphaelle-hernu

Hettiarachchi, H., Meegoda, J., & Ryu, S. (2018). Organic Waste Buyback as a Viable Method to Enhance Sustainable Municipal Solid Waste Management in Developing Countries. *International Journal of Environmental Research and Public Health*, 15(11), 2483. 10.3390/ijerph1511248330405058

Hignett, S., & McDermott, H. (2015). Qualitative methodology. *Evaluation of human work*, 119-138.

Hofstede, G. (2001). *Culture's consequences: Comparing values, behaviors, institutions, and organizations across nations.* Sage publications.

Hong, Z., Wang, H., & Gong, Y. (2019). Green product design considering functional-product reference. *International Journal of Production Economics*, 210, 155–168. 10.1016/j.ijpe.2019.01.008

Hong, Z., Wang, H., & Yu, Y. (2018). Green product pricing with non-green product reference. *Transportation Research Part E, Logistics and Transportation Review*, 115, 1–15. 10.1016/j.tre.2018.03.013

Hoorn Weg, D., & Giannelli, N. (2007). *Managing Municipal Solid Waste in Latin America and the Caribbean: Integrating the Private Sector, Harnessing Incentives.* World Bank.

Hoornweg, D., & Bhada-Tata, P. What a Waste: A Global Review of Solid Waste Management. Urban Development Series. World Bank; Washington, DC, USA: 2012. Knowledge Papers No. 15.

Horsley-Summer, B. (2022). *Better Brands 01: How Patagonia brought sustainability to the enlightened masses.* Avery and Brown. https://www.averyandbrown.com/outpost/better-brands-01-patagonia

Hristov, J., Barreiro-Hurle, J., Salputra, G., Blanco, M., & Witzke, P. (2021). Reuse of treated water in European agriculture: Potential to address water scarcity under climate change. *Agricultural Water Management*, 251(March), 106872. 10.1016/j.agwat.2021.10687234079159

Hsieh, Y. (2012). Hotel companies' environmental policies & practices: A content analysis of web pages. *International Journal of Contemporary Hospitality Management*, 24(1), 97–121. 10.1108/095961112

Hu, J., Li, X., Wang, N., & Jiang, B. (2022). Green Product Design. In N. Wang, Q. Jiang, B. Jiang, & Z. He, *Enterprises' Green Growth Model and Value Chain Reconstruction* (pp. 155–183). Springer Nature Singapore. 10.1007/978-981-19-3991-4_7

Hwang, C., Lee, Y., Diddi, S., & Karpova, E. (2016). "Don't buy this jacket": Consumer reaction toward anti-consumption apparel advertisement. *Journal of Fashion Marketing and Management*, 20(4), 435–452. 10.1108/JFMM-12-2014-0087

Iannuzzi, A. (2016). *Greener Products.* CRC Press. 10.1201/b11276

Ijaiya, H., & Joseph, O. T. (2014). Rethinking environmental law enforcement in Nigeria. *Beijing Law Review*, 5(4), 306–321. 10.4236/blr.2014.54029

IKEA. (2023). *KUNGSBACKA white kitchen guide*. IKEA. https://www.ikea.com/au/en/rooms/kitchen/

Interface. (2023). *Lessons for the future: The Interface guide to changing your business to change the world*. Interface. https://www.interface.com/content/dam/interfaceinc/interface/sustainability/emea/25th-anniversary-report/Interface_MissionZeroCel_Booklet_EN.pdf

Islam, M. A., Hunt, A., Jantan, A. H., Hashim, H., & Chong, C. W. (2020). Exploring challenges and solutions in applying green human resource management practices for the sustainable workplace in the ready-made garment industry in Bangladesh. *Business Strategy & Development*, 3(3), 332–343. 10.1002/bsd2.99

Islam, M. A., Jantan, A. H., Yusoff, Y. M., Chong, C. W., & Hossain, M. S. (2023). Green Human Resource Management (GHRM) practices and millennial employees' turnover intentions in tourism industry in malaysia: Moderating role of work environment. *Global Business Review*, 24(4), 642–662. 10.1177/0972150920907000

Ismail, I. J. (2023). The role of technological absorption capacity, enviropreneurial orientation, and green marketing in enhancing business' sustainability: Evidence from fast-moving consumer goods in Tanzania. *Technological Sustainability*, 2(2), 121–141. 10.1108/TECHS-04-2022-0018

Jabbour, C. J. C. (2015). Environmental training and environmental management maturity of Brazilian companies with ISO14001: Empirical evidence. *Journal of Cleaner Production*, 96, 331–338. 10.1016/j.jclepro.2013.10.039

Jabbour, C. J. C., & Santos, F. C. A. (2008). Relationships between human resource dimensions and environmental management in companies: Proposal of a model. *Journal of Cleaner Production*, 16(1), 51–58. 10.1016/j.jclepro.2006.07.025

Jabbour, C. J. C., Santos, F. C. A., & Nagano, M. S. (2010). Contributions of HRM throughout the stages of environmental management: Methodological triangulation applied to companies in Brazil. *International Journal of Human Resource Management*, 21(7), 1049–1089. 10.1080/09585191003783512

Jackson, S. E., Renwick, D. W., Jabbour, C. J., & Muller-Camen, M. (2011). State-of-the-art and future directions for green human resource management: Introduction to the special issue. *German Journal of Human Resource Management*, 25(2), 99–116. 10.1177/239700221102500203

Jafri, S. (2012). Green HR practices: An empirical study of certain automobile organizations of India. *Human Resource Management*, 42(4), 6193–6198.

Jahangiri, A., Asadi-Gangraj, E., & Nemati, A. (2022). Designing a reverse logistics network to manage construction and demolition wastes: A robust bi-level approach. *Journal of Cleaner Production*, 380, 134809. 10.1016/j.jclepro.2022.134809

Jennah, H., & Ismail, A. (2023). Pengaruh Green Marketing Mix Terhadap Purchase Decision Dalam Menggunakan Eco Friendly Product. *Journal of Trends Economics and Accounting Research*, 3(4), 390–398. 10.47065/jtear.v3i4.636

Jerónimo, H. M., Henriques, P. L., de Lacerda, T. C., da Silva, F. P., & Vieira, P. R. (2020). Going green and sustainable: The influence of green HR practices on the organizational rationale for sustainability. *Journal of Business Research*, 112, 413–421. 10.1016/j.jbusres.2019.11.036

Joshi, G., & Dhar, R. L. (2020). Green training in enhancing green creativity via green dynamic capabilities in the Indian handicraft sector: The moderating effect of resource commitment. *Journal of Cleaner Production*, 267, 121948. 10.1016/j.jclepro.2020.121948

Kabera, T., & Nishimwe, H. (2019). *Systems Analysis of Municipal Solid Waste Management and Recycling System in East Africa: Benchmarking Performance in Kigali City, Rwanda*. EDP Sciences.

Kannan, D., Solanki, R., Darbari, J. D., Govindan, K., & Jha, P. C. (2023). A novel bi-objective optimization model for an eco-efficient reverse logistics network design configuration. *Journal of Cleaner Production*, 394, 136357. 10.1016/j.jclepro.2023.136357

Kartawinata, B. R., Maharani, D., Pradana, M., & Amani, H. M. (2020, August). The role of customer attitude in mediating the effect of green marketing mix on green product purchase intention in love beauty and planet products in indonesia. In *Proceedings of the International Conference on Industrial Engineering and Operations Management* (Vol. 1, pp. 3023-3033).

Kasim, A. (2017). Corporate environmentalism in the hotel sector: Evidence of drivers and barriers in Penang, Malaysia. *Journal of Sustainable Tourism*, 15(6), 680–699. 10.2167/jost575.0

Katadata Insight Center. (2024). *Digitalisasi UMKM*. KIC.

Khan, M. S. M., & Kaneesamkandi, Z. (2013). Biodegradable waste to biogas: Renewable energy option for the Kingdom of Saudi Arabia. *International Journal of Innovation and Applied Studies*, 4, 101–113.

Khan, S. J., Dhir, A., Parida, V., & Papa, A. (2021). Past, present, and future of green product innovation. *Business Strategy and the Environment*, 30(8), 4081–4106. 10.1002/bse.2858

Khaskhely, M. K., Qazi, S. W., Khan, N. R., Hashmi, T., & Chang, A. A. R. (2022). Understanding the impact of green human resource management practices and dynamic sustainable capabilities on corporate sustainable performance: Evidence from the manufacturing sector. *Frontiers in Psychology*, 13, 1–17. 10.3389/fpsyg.2022.84448835846624

Khateeb, F.R. and Nabi, T., 2023. Green Human Resource Management: A Review of Two Decades of Research. *Management Research & Practice, 15*(2).

Kilic, H. S., Kalender, Z. T., Solmaz, B., & Iseri, D. (2023). A two-stage MCDM model for reverse logistics network design of waste batteries in Turkey. *Applied Soft Computing*, 143, 110373. 10.1016/j.asoc.2023.110373

Kim, Y. J., Kim, W. G., Choi, H. M., & Phetvaroon, K. (2019). The effect of green human resource management on hotel employees' eco-friendly behavior and environmental performance. *International Journal of Hospitality Management*, 76, 83–93. 10.1016/j.ijhm.2018.04.007

Kitazawa, S., & Sarkis, J. (2000). The relationship between ISO 14001 and continuous source reduction programmes. *International Journal of Operations & Production Management*, 20(2), 225–248. 10.1108/01443570010304279

Knickmeyer, D. (2020). Social factors influencing household waste separation: A literature review on good practices to improve the recycling performance of urban areas. *Journal of Cleaner Production*, 245, 118605. 10.1016/j.jclepro.2019.118605

Kodua, L. T., Xiao, Y., Adjei, N. O., Asante, D., Ofosu, B. O., & Amankona, D. (2022). Barriers to green human resources management (GHRM) implementation in developing countries. Evidence from Ghana. *Journal of Cleaner Production*, 340, 130671. 10.1016/j.jclepro.2022.130671

Kosatica, M. (2024). Semiotic landscape in a green capital: The political economy of sustainability and environment. *Linguistic Landscape*, 10(2), 136–165. 10.1075/ll.23016.kos

Kotter, J. P. (1996). *Leading change*. Harvard Business School Press.

Krimpas, N. A., Salamaliki, P. K., & Venetis, I. A. (2021). Factor decomposition of disaggregate inflation: The case of Greece. *International Journal of Computational Economics and Econometrics*, 11(1), 84–104. 10.1504/IJCEE.2021.111713

Kuo, T.-C., Huang, S. H., & Zhang, H.-C. (2001). Design for manufacture and design for 'X': Concepts, applications, and perspectives. *Computers & Industrial Engineering*, 41(3), 241–260. 10.1016/S0360-8352(01)00045-6

Lakho, F. H., Qureshi, A., Igodt, W., Le, H. Q., Depuydt, V., Rousseau, D. P. L., & Van Hulle, S. W. H. (2022). Life cycle assessment of two decentralized water treatment systems combining a constructed wetland and a membrane based drinking water production system. *Resources, Conservation and Recycling, 178*, 106104. 10.1016/j.resconrec.2021.106104

Laroche, M., Bergeron, J., & Barbaro-Forleo, G. (2001). Targeting consumers who are willing to pay more for environmentally friendly products. *Journal of Consumer Marketing*, 18(6), 503–520. 10.1108/EUM0000000006155

Laschinger, H. K. S., Finegan, J. E., Shamian, J., & Wilk, P. (2004). A longitudinal analysis of the impact of workplace empowerment on work satisfaction. *Journal of Organizational Behavior: The International Journal of Industrial. Journal of Organizational Behavior*, 25(4), 527–545. 10.1002/job.256

Lashari, I. A., Li, Q., Maitlo, Q., Bughio, F. A., Jhatial, A. A., & Rashidi Syed, O. (2022). Environmental sustainability through green HRM: Measuring the perception of university managers. *Frontiers in Psychology*, 13, 1007710. 10.3389/fpsyg.2022.100771036467149

Lawrence, J., Rasche, A., & Kenny, K. (2019). Sustainability as Opportunity: Unilever's Sustainable Living Plan. In Lenssen, G. G., & Smith, N. C. (Eds.), *Managing Sustainable Business*. Springer., 10.1007/978-94-024-1144-7_21

Lee, H. L., & Holweg, M. (2010). Supply chain integration: The role of product and process modularity. *International Journal of Production Research*, 48(1), 169–190.

Lee, K. H. (2009). Why and how to adopt green management into business organizations: The case study of Korean SMEs in manufacturing industry. *Management Decision*, 47(7), 1101–1121. 10.1108/00251740910978322

Lee, K. H., & Ball, R. (2003). Achieving Sustainable Corporate Competitiveness: Strategic Link between Top Management"s (Green) Commitment and Corporate Environmental Strategy. *Greener Management International*, 2003(44), 89–104. 10.9774/GLEAF.3062.2003.wi.00009

Lee, S., Kim, J., & Chong, W. K. (2016). The causes of the municipal solid waste and the green-house gas emissions from the waste sector in the United States. *Waste Management (New York, N.Y.)*, 56, 593–599. 10.1016/j.wasman.2016.07.02227475865

Leigh, M., & Li, X. (2015). Industrial ecology, industrial symbiosis and supply chain environmental sustainability: A case study of a large UK distributor. *Journal of Cleaner Production*, 106, 632–643. 10.1016/j.jclepro.2014.09.022

Lei, J., Che, A., & Van Woensel, T. (2023). Collection-disassembly-delivery problem of disassembly centers in a reverse logistics network. *European Journal of Operational Research*.

Leonidou, L. C., Leonidou, C. N., Fotiadis, T. A., & Zeriti, A. (2013). Resources and Capabilities asDrivers of Hotel Environmental Marketing Strategy: Implications for Competitive Advantage andPerformance. *Tourism Management*, 35, 94–110. 10.1016/j.tourman.2012.06.003

Levente Szász, Ottó Csíki, Béla-Gergely Rácz, Sustainability management in the global automotive industry: A theoretical model and survey study, International Journal of Production Economics, Volume 235,2021

Lewin, K. (1947). Frontiers in group dynamics: Concept, method, and reality in social science; social equilibria and social change. *Human Relations*, 1(1), 5–41. 10.1177/001872674700100103

Liao, G. H. W., & Luo, X. (2022). Collaborative reverse logistics network for electric vehicle batteries management from sustainable perspective. *Journal of Environmental Management*, 324, 116352. 10.1016/j.jenvman.2022.11635236208516

Likhitkar, P. & Verma, P. (2017). Impact of green HRM practices on organization sustainability and employee retention. *International journal for innovative research in multidisciplinary field*, 3(5), 152-157.

Likhitkar, P., & Verma, P. (2017). Impact of green HRM practices on organization sustainability and employee retention. *International Journal for Innovative Research in Multidisciplinary Field*, 3(5), 152–157.

Lin, J., Li, X., Zhao, Y., Chen, W., & Wang, M. (2023). Design a reverse logistics network for end-of-life power batteries: A case study of Chengdu in China. *Sustainable Cities and Society*, 98, 104807. 10.1016/j.scs.2023.104807

Liu, J., Gao, X., Cao, Y., Mushtaq, N., Chen, J., & Wan, L. (2022). Catalytic effect of green human resource practices on sustainable development goals: Can individual values moderate an empirical validation in a developing economy? *Sustainability (Basel)*, 14(21), 14502. 10.3390/su142114502

Liu, X., & Lin, K. L. (2020). Green Organizational Culture, Corporate Social Responsibility Implementation, and Food Safety. *Frontiers in Psychology*, 11, 585435. 10.3389/fpsyg.2020.58543533240175

Longe, E. O., Ukpebor, E. F., & Omole, D. O. (2009). HOUSEHOLD WASTE COLLECTION AND DISPOSAL: OJO LOCAL GOVERNMENT CASE STUDY, LAGOS STATE, NIGERIA. *Journal of Engineering Research*, 14(4).

López-Serrano, M. J., Lakho, F. H., Van Hulle, S. W. H., & Batlles-delaFuente, A. (2023). Life cycle cost assessment and economic analysis of a decentralized wastewater treatment to achieve water sustainability within the framework of circular economy. *Oeconomia Copernicana*, 14(1), 103–133. 10.24136/oc.2023.003

López-Serrano, M. J., Velasco-Muñoz, J. F., Aznar-Sánchez, J. A., & Román-Sánchez, I. M. (2021). Financial evaluation of the use of reclaimed water in agriculture in Southeastern Spain, a mediterranean region. *Agronomy (Basel)*, 11(11), 2218. 10.3390/agronomy11112218

Lubbe, W., ten Ham-Baloyi, W., & Smit, K. (2020). The integrative literature review as a research method: A demonstration review of research on neurodevelopmental supportive care in preterm infants. *Journal of Neonatal Nursing*, 26(6), 308–315. 10.1016/j.jnn.2020.04.006

Lugo, M., Ail, S. S., & Castaldi, M. J. (2020). Approaching a zero-waste strategy by reuse in New York City: Challenges and potential. *Waste Management & Research*, 38(7), 734–744. 10.1177/0734242X2091949632372709

Luo, J. M., Chau, K. Y., Fan, Y., & Chen, H. (2021). Barriers to the implementation of Green Practices in the Integrated Resort Sector. *SAGE Open*, 11(3), 1–15. 10.1177/21582440211030277

Luqman, R. A., Farhan, H. M., Shahzad, F., & Shaheen, S. (2012). 21st century challenges of educational leaders, way out and need of reflective practice. *International Journal of Learning and Development*, 2(1), 195–208. 10.5296/ijld.v2i1.1238

Lush Retail. (2024). *Lish shampoo bars*. Lush. https://www.lush.com/us/en_us/c/shampoo-bars

Lutterbeck, C. A., Kist, L. T., Lopez, D. R., Zerwes, F. V., & Machado, E. L. (2017). Life cycle assessment of integrated wastewater treatment systems with constructed wetlands in rural areas. *Journal of Cleaner Production*, 148, 527–536. 10.1016/j.jclepro.2017.02.024

Mahmood, Z., & Uddin, S. (2021). Institutional logics and practice variations in sustainability reporting: Evidence from an emerging field. *Accounting, Auditing & Accountability Journal*, 34(5), 1163–1189. 10.1108/AAAJ-07-2019-4086

Majeed, M. A., Ahsan, T., & Gull, A. A. (2023). Does corruption sand the wheels of sustainable development? Evidence through green innovation. *Business Strategy and the Environment*. Wiley Online Library (https://onlinelibrary.wiley.com/doi/abs/10.1002/bse.3719)

Malik, S., Kumar, S. (2012). Management of Hotel Waste: A Case Study of Small Hotels of Haryana State. *ArthPrabandh: A Journal of Economics and Management, 1*(09) 43-55.

Mallak, S. K., & Ishak, M. K. (2012). Waste minimization as sustainable waste management strategy for Malaysian industries. In *UMT 11th International Annual Symposium on Sustainability Science and Management*. UMT.

Maluleke, T.C. (2021). *A Review of Municipal Solid Waste Management Systems in Polokwane City Limpopo Province*. Northwest University, North West.

Manaf, L. A., Samah, M. A. A., & Zukki, N. I. M. (2009). Municipal solid waste management in Malaysia: Practices and challenges. *Waste Management (New York, N.Y.)*, 29(11), 2902–2906. 10.1016/j.wasman.2008.07.01519540745

Manaktola, K., & Jauhari, V. (2007). Exploring consumer attitude and behaviour towards green practices in the lodging industry in India. *International Journal of Contemporary Hospitality Management*, 19(5), 364–377. 10.1108/09596110710757534

Mandip, G. (2012). Green HRM: People management commitment to environmental sustainability. *Research Journal of Recent Sciences, ISSN*, 2277, 2502.

Maniam, G., Zakaria, N. A., Leo, C. P., Vassilev, V., Blay, K. B., Behzadian, K., & Poh, P. E. (2022). An assessment of technological development and applications of decentralized water reuse: A critical review and conceptual framework. *WIREs. Water*, 9(3), 1–31. 10.1002/wat2.1588

Margaretha, M., & Saragih, S. (2013). *Developing New Corporate Culture through Green Human Resource Practice*, International Conference on Business, Economics, and Accounting 20 – 23 March 2013, Bangkok – Thailands.

Marhatta, S., & Adhikari, S. (2013). Green HRM and sustainability. *International eJournal of Ongoing Research in Management & IT*. www.asmgroup.edu.in/incon/publication/incon13-hr -006pdf

Martela, F. (2019). What makes self-managing organizations novel? Comparing how Weberian bureaucracy, Mintzberg's adhocracy, and self-organizing solve six fundamental problems of organizing. *Journal of Organization Design*, 8(1), 1–23. 10.1186/s41469-019-0062-9

Martins, J. M., Aftab, H., Mata, M. N., Majeed, M. U., Aslam, S., Correia, A. B., & Mata, P. N. (2021). Assessing the impact of green hiring on sustainable performance: Mediating role of green performance management and compensation. *International Journal of Environmental Research and Public Health*, 18(11), 5654. 10.3390/ijerph1811565434070535

Masri, H. A., & Jaaron, A. A. (2017). Assessing green human resources management practices in Palestinian manufacturing context: An empirical study. *Journal of Cleaner Production*, 143, 474–489. 10.1016/j.jclepro.2016.12.087

Matinde, E., Simate, G. S., & Ndlovu, S. (2018). Mining and metallurgical wastes: A review of recycling and re-use practices. *Journal of the Southern African Institute of Mining and Metallurgy*, 118(8), 825–844. 10.17159/2411-9717/2018/v118n8a5

Mbasera, M., Du Plessis, E., Saayman, M. & Kruger, M. (2016). Environmentally-friendly practices in hotels. *ActaCommercii* 16(1)

McAllister, J. Factors Influencing Solid-Waste Management in the Developing World. All Graduate Plan B and Other Reports. 528. 2015. ((accessed on 9 November 2021)).https://digitalcommons.usu.edu/grad reports/528

McDougal, F., White, P., Franke, M., & Hindle, P. (2001). *Integrated Solid Waste Management: A Life Cycle Inventory* (2nd ed.). Blackwell Science. 10.1002/9780470999677

Melián-Navarro, A., & Ruiz-Canales, A. (2020). Evaluation in carbon dioxide equivalent and chg emissions for water and energy management in water users associations. A case study in the southeast of spain. *Water (Basel)*, 12(12), 3536. 10.3390/w12123536

Menezes, E., Filho, S., & Drigo, E. (2017). Analysis of organisational and human factors in the local production arrangement of the hotel chain to avoid social and environmental impacts, case study of Maragogi, Alagoas, Brazil. In *Advances in Social & Occupational Ergonomics: Proceedings of the AHFE 2016 International Conference on Social and Occupational Ergonomics*, (pp. 421-433). Springer International Publishing.

Menikpura, S. N. M., Gheewala, S. H., & Bonnet, S. (2012). Sustainability assessment of municipal solid waste management in Sri Lanka: Problems and prospects. *Journal of Material Cycles and Waste Management*, 14(3), 181–192. 10.1007/s10163-012-0055-z

Mensah, I. (2006). Environmental management practices among hotels in the greater Accra region. *International Journal of Hospitality Management*, 25(3), 0–431.

Microsoft. (2023). *2022 Environmental Sustainability Report Enabling sustainability for our company, our customers, and the world.* Microsoft. https://www.microsoft.com/en-us/corporate-responsibility/sustainability/report?ICID=SustainabilityReport22_MOI-ESblog

Miezah, K., Obiri-Danso, K., Kádár, Z., Fei-Baffoe, B., & Mensah, M. Y. (2015). Obiri-DansoK.,KádárZ.,Fei-Baffoe B.,Mensah M.Y. Municipal solid waste characterization and quantification as a measure towards effective waste management Ghana. *Waste Management (New York, N.Y.)*, 46, 15–27. 10.1016/j.wasman.2015.09.00926421480

Mikhno, I., Koval, V., Shvets, G., Garmatiuk, O., & Tamošiūnienė, R. (2021). *Green economy in sustainable development and improvement of resource efficiency.*

Millar, M., Mayer, K. J., & Baloglu, S. (2012). Importance of green hotel attributes to business and leisure travellers. *Journal of Hospitality Marketing & Management*, 21(4), 395–413. 10.1080/19368623.2012.624294

Mingaleva, Z., Vukovic, N., Volkova, I., & Salimova, T. (2019). Waste management in green and smart cities: A case study of Russia. *Sustainability (Basel)*, 12(1), 94. 10.3390/su12010094

Compilation of References

Mirmotahari, T. (2022). *Google Benefits and Perks for Employees - 11 ideas.* Perkup App. https://perkupapp.com/post/11-awesome-google-benefits-and-perks-for-employees

Mohan, V., Deepak, B., & Mona, S. (2017). Reduction and Management of Waste in Hotel Industries. *International Journal of Engineering Research and Applications*, 7(7), 34–37. 10.9790/9622-0707103437

Mohezara, S., Nazria, M., Kaderb, M. A. R. A., Alib, R., & Yunusb, N. K. M. (2016). Corporate social responsibility in the Malaysian food retailing industry: An exploratory study. *Int. Acad. Res. J. Soc. Sci.*, 2, 66–72.

Montiel, I., Cuervo-Cazurra, A., Park, J., Antolín-López, R., & Husted, B. W. (2021). Implementing the United Nations' sustainable development goals in international business. *Journal of International Business Studies*, 52(5), 999–1030. 10.1057/s41267-021-00445-y34054154

Moreo, A. (2008). Green Consumption in hotel Industry an examination of consumer attitudes. Google scholar accessed 27 March 2019.

Morioka, S. N., & de Carvalho, M. M. (2016). A systematic literature review towards a conceptual framework for integrating sustainability performance into business. *Journal of Cleaner Production*, 136, 134–146. 10.1016/j.jclepro.2016.01.104

Mousa, S. K., & Othman, M. (2020). The impact of green human resource management practices on sustainable performance in healthcare organisations: A conceptual framework. *Journal of Cleaner Production*, 243, 118595. 10.1016/j.jclepro.2019.118595

Muhammad, N. (2023, October 13). *Usaha Mikro Tetap Merajai UMKM, Berapa Jumlahnya?* [webpage]. Diakses pada https://databoks.katadata.co.id/datapublish/2023/10/13/usaha-mikro-tetap-merajai-umkm-berapa-jumlahnya

Mungai, M. &Urungu, R. (2013).An assessment of management commitment to application of green practices in 4-5-star hotels in Mombasa, Kenya.*Information and knowledge management*, 3(6), 40-47.

Muogbo, U. (2013). The impact of employee motivation on organisational performance (A study of some selected firms in Anambra state Nigeria). *The International Journal of Engineering and Science*, 2, 70–80.

Muster, V., & Schrader, U. (2011). Green work-life balance: A new perspective for Green HRM. *Zeitschrift Fur Personalforschung*, 25(2), 140–156. 10.1177/239700221102500205

Myers, M. D. (2019). *Qualitative research in business and management.*

Nahman, A. (2021). *Incentives for municipalities to divert waste from landfill in South Africa. Waste research development and innovation roadmap research report.* Council for Scientific and Industrial Research.

Naini, S. R., Mekapothula, R. R., Jain, R., & Manohar, S. (2024). Redefining green consumerism: A diminutive approach to market segmentation for sustainability. *Environmental Science and Pollution Research International*, 1–17. 10.1007/s11356-023-31717-938180668

Nanayakkara, P. R., Jayalath, M. M., Thibbotuwawa, A., & Perera, H. N. (2022). A circular reverse logistics framework for handling e-commerce returns. *Cleaner Logistics and Supply Chain*, 5, 100080. 10.1016/j.clscn.2022.100080

Nandy, S., Fortunato, E., & Martins, R. (2022). Green economy and waste management: An inevitable plan for materials science. *Progress in Natural Science*, 32(1), 1–9. 10.1016/j.pnsc.2022.01.001

Natakoesoemah, S., & Adiarsi, G. R. (2020). The Indonesian Millenials Consumer Behaviour on Buying Eco-Friendly Products: The Relationship Between Environmental Knowledge and Perceived Consumer Effectiveness. *International Journal of Multicultural and Multireligious Understanding*, 7(9), 292–302.

Ni, N., Shah, S. Z., & Soomro, B. A. (2023). *The impact of transparent reporting on corporate sustainability performance.*

Nimfa, D. T., Latiff, A. S. A., Wahab, S. A., & Etheraj, P. (2021). Effect of organisational culture on sustainable growth of SMEs: Mediating role of innovation competitive advantage. *Journal of International Business and Management*, 4(2), 1–19. 10.37227/JIBM-2021-01-156

Noor Faezah, J., Yusliza, M. Y., & Ramayah, T. (2024). Mediating role of green culture and green commitment in implementing employee ecological behaviour. *Journal of Management Development*. Emerald Publishing. https://www.emerald.com/insight/content/doi/10.1108/JMD-08-2023-0258/full/html

Noordiatmoko, D., & Riyadi, B. S. (2023). The Analysis of Sustainable Performance Management of Government Institution in Indonesia: A Public Policy Perspective. *International Journal of Membrane Science and Technology*, 10(3), 1146–1157. 10.15379/ijmst.v10i3.1684

Norton, T. A., Parker, S. L., Zacher, H., & Ashkanasy, N. M. (2015). Employee green behaviour: A theoretical framework, multilevel review, and future research agenda. *Organization & Environment*, 28(1), 103–125. 10.1177/1086026615575773

Nosrati-Abarghooee, S., Sheikhalishahi, M., Nasiri, M. M., & Gholami-Zanjani, S. M. (2023). Designing reverse logistics network for healthcare waste management considering epidemic disruptions under uncertainty. *Applied Soft Computing*, 142, 110372. 10.1016/j.asoc.2023.11037237168874

Notpla. (2023). *Notpla Disappearing Packaging*. Notpla. https://www.notpla.com/

Novela, S., & Hansopaheluwakan, S. (2018). Analysis of Green Marketing Mix Effect on Customer Satisfaction using 7p Approach. *Pertanika Journal of Social Sciences & Humanities*.

Nurfitriya, M., Fauziyah, A., Koeswandi, T. A. L., Yusuf, I., & Rachmani, N. N. (2022). Peningkatan Literasi Digital Marketing UMKM Kota Tasikmalaya. *Acitya Bhakti*, 2(1), 57. 10.32493/acb.v2i1.14618

Nwokike, L. I. (2020). Lagos Waste Management Authority Law 2007 and National Environmental Standards and Regulations Enforcement Agency (Establishment) Act 2007: A Comparative Appraisal. *AJLHR*, 4, 112.

Nyika, J. M., Onyari, E. K., Mishra, S., & Dinka, M. O. (2020). Waste Management in South Africa. In *Sustainable Waste Management Challenges in Developing Countries* (pp. 327–351). IGI Global. 10.4018/978-1-7998-0198-6.ch014

Odeku, K. O. (2018). Proactive responses to mitigate climate change impacts by the hospitality sector in South Africa. *African Journal of Hospitality, Tourism and Leisure*, 7, 1–13.

Oelofse, S. H., & Godfrey, L. (2008, November). Towards improved waste management services by local government–A waste governance perspective. In *Proceedings of Science: real and relevant Conference* (pp. 17-18).

Ogbeibu, S., Emelifeonwu, J., Pereira, V., Oseghale, R., Gaskin, J., Sivarajah, U., & Gunasekaran, A. (2023). Demystifying the roles of organisational smart technology, artificial intelligence, robotics, and algorithms capability: A strategy for green human resource management and environmental sustainability. *Business Strategy and the Environment*, 33, 369–388. 10.1002/bse.3495

Ogbeibu, S., Emelifeonwu, J., Senadjki, A., Gaskin, J., & Kaivo-oja, J. (2020). Technological turbulence and greening of team creativity, product innovation, and human resource management: Implications for sustainability. *Journal of Cleaner Production*, 244, 118703. 10.1016/j.jclepro.2019.118703

Ojo, A. O., Tan, C. N. L., & Alias, M. (2022). Linking green HRM practices to environmental performance through pro-environment behaviour in the information technology sector. *Social Responsibility Journal*, 18(1), 1–18. 10.1108/SRJ-12-2019-0403

Ojok, J. (2013). Rate and quantities of household solid waste generated in Kampala City, Uganda. *Sci.J.Environ.Eng. Res.*, 2013. Advance online publication. 10.7237/sjeer/237

Olafsen, A. H., Nilsen, E. R., Smedsrud, S., & Kamaric, D. (2021). Sustainable development through commitment to organizational change: The implications of organizational culture and individual readiness for change. *Journal of Workplace Learning*, 33(3), 180–196. 10.1108/JWL-05-2020-0093

Olanipekun, A. O., Xia, B., Hon, C., & Hu, Y. (2017). Project owners' motivation for delivering green building projects. *Journal of Construction Engineering and Management*, 143(9), 04017068. 10.1061/(ASCE)CO.1943-7862.0001363

Olukanni, D. O., Adeleke, J. O., & Aremu, D. D. (2016). A review of local factors affecting solid waste collection in Nigeria.

Omole, D. O., Jim-George, T., & Akpan, V. E. (2019). Economic Analysis of Wastewater Reuse in Covenant University. *Journal of Physics: Conference Series*, 1299(1), 012125. 10.1088/1742-6596/1299/1/012125

Opatha, H. H., & Arulrajah, A. A. (2014). Green Human Resource Management: Simplified general reflections. *International Business Research*, 7(8), 101–112. 10.5539/ibr.v7n8p101

Oseghale, O. R., Mulyata, J., & Debrah, Y. A. (2018). Global talent management. In Manchando, C., & Davim, J. P. (Eds.), *Organizational Behaviour and Human Resource Management* (pp. 139–155). Springer. 10.1007/978-3-319-66864-2_6

Otter, P., Sattler, W., Grischek, T., Jaskolski, M., Mey, E., Ulmer, N., Grossmann, P., Matthias, F., Malakar, P., Goldmaier, A., Benz, F., & Ndumwa, C. (2020). Economic evaluation of water supply systems operated with solar-driven electro-chlorination in rural regions in Nepal, Egypt and Tanzania. *Water Research*, 187, 116384. 10.1016/j.watres.2020.11638432980605

Paauwe, J. (2009). HRM and performance: Achievements, methodological issues and prospects. *Journal of Management Studies*, 46(1), 129–142. 10.1111/j.1467-6486.2008.00809.x

Paauwe, J., Boon, C., Boselie, P., & Den Hartog, D. (2013). *Reconceptualizing fit in strategic human resource management: 'Lost in translation?' In Human Resource Management and Performance: Achievements and Challenges.* John Wiley & Sons Ltd.

Paillé, P., Chen, Y., Boiral, O., & Jin, J. (2014). The impact of human resource management on environmental performance: An employee-level study. *Journal of Business Ethics*, 121(3), 451–466. 10.1007/s10551-013-1732-0

Park, A. Y., & Krause, R. M. (2021). Exploring the landscape of sustainability performance management systems in US local governments. *Journal of Environmental Management*, 279, 111764. 10.1016/j.jenvman.2020.11176433360650

Patagonia. (2023). *Gear for a good time and a long time.* Patagonia. https://wornwear.patagonia.com/

Patagonia. (2024). *Black Hole® Bags.* Patagonia. https://www.patagonia.com/shop/gear/bags/black-hole

Pedersen, J. T. S., van Vuuren, D., Gupta, J., Santos, F. D., Edmonds, J., & Swart, R. (2022). IPCC emission scenarios: How did critiques affect their quality and relevance 1990–2022? *Global Environmental Change*, 75, 102538. 10.1016/j.gloenvcha.2022.102538

Perron, G. M., Côté, R. P., & Duffy, J. F. (2006). Improving environmental awareness training in business. *Journal of Cleaner Production*, 14(6–7), 551–562. 10.1016/j.jclepro.2005.07.006

Pham, N. T., Tučková, Z., & Jabbour, C. J. C. (2019). Greening the hospitality industry: How do green human resource management practices influence organizational citizenship behavior in hotels? A mixed-methods study. *Tourism Management*, 72, 386–399. 10.1016/j.tourman.2018.12.008

Philips. (2024). *Philips Hue, smart home lighting made brilliant.* Phillips. https://www.philips-hue.com/en-in

Phillips, L. (2007). Go green to gain the edge over rivals. *People Management*, 23(August), 9.

Pinzone, M., Guerci, M., Lettieri, E., & Redman, T. (2016). Progressing in the change journey towards sustainability in healthcare: The role of 'Green'HRM. *Journal of Cleaner Production*, 122, 201–211. 10.1016/j.jclepro.2016.02.031

Pires, A., & Martinho, G. (2019). Waste hierarchy index for circular economy in waste management. *Waste Management (New York, N.Y.)*, 95, 298–305. 10.1016/j.wasman.2019.06.01431351615

PMG. (2022) *Status of Waste Management in South Africa. Cape Town, South Africa: Parliament of South Africa*. PMG.

Pokhrel, D., & Viraraghavan, T. (2005). Municipal solid waste management in Nepal: Practices and challenges. *Waste Management (New York, N.Y.)*, 25(5), 555–562. 10.1016/j.wasman.2005.01.02015925764

Popescu, C. R. G. (2020). Sustainability assessment: Does the OECD/G20 inclusive framework for BEPS (base erosion and profit shifting project) put an end to disputes over the recognition and measurement of intellectual capital? *Sustainability (Basel)*, 12(23), 10004. 10.3390/su122310004

Porter, M. E., & Kramer, M. R. (2006). The link between competitive advantage and corporate social responsibility. *Harvard Business Review*, 84(12), 78–92.17183795

Porter, M. E., & Kramer, M. R. (2019). Creating shared value: How to reinvent capitalism—and unleash a wave of innovation and growth. *Harvard Business Review*, 323–346. 10.1007/978-94-024-1144-7_16

Priansa, D. J. (2016). The influence of E-WOM and perceived value on consumer decisions to shop online at Lazada. *Ecodemic Journal of Management and Business Economics*, 4(1), 117–124.

Pucik, V., Bjorkman, I., Evans, P., & Stahl, G. K. (2023). *The global challenge: Managing people across borders*. Edward Elgar Publishing. 10.4337/9781035300723

Purmessur, B., & Surroop, D. (2019). Power generation using landfill gas generated from new cell at the existing landfill site. *Journal of Environmental Chemical Engineering*, 7(3), 103060. 10.1016/j.jece.2019.103060

Raharja, S. U. J., & Chan, A. (2021). Youth's Green Consumer Behavior: A Study In Citarum Watersehd West Java Indonesia. *AdBispreneur: Jurnal Pemikiran dan Penelitian Administrasi Bisnis dan Kewirausahaan, 6*(3).

Rahmawati, M., Pratiwi, S. R., Devi, C., & Nainggolan, Y. T. (2022). Penerapan Strategi Green Marketing Di Tengah Pandemi Covid-19. *Jurnal Ekonomika*, 13(01), 1–18. 10.35334/jek.v13i0.2410

Ramachandra, T. V., Bharath, H. A., Kulkarni, G., & Han, S. S. (2018). Municipal solid waste: Generation, composition and GHG emissions in Bangalore, India. *Renewable & Sustainable Energy Reviews*, 82, 1122–1136. 10.1016/j.rser.2017.09.085

Ramasamy, A. (2017). *A study on implications of implementing green HRM in the corporate bodies with special reference to developing nations*.

Ramlee, M. (2015). Green and sustainable development for TVET in Asia. *The International Journal of Technical and Vocational Education*, 11(2), 133–142.

Randa, I. O., & Atiku, S. O. (2021). SME Financial Inclusivity for Sustainable Entrepreneurship in Namibia During COVID-19. In Baporikar, N. (Ed.), *Handbook of Research on Sustaining SMEs and Entrepreneurial Innovation in the Post-COVID-19 Era* (pp. 373–396). IGI Global. 10.4018/978-1-7998-6632-9.ch018

Rani, S., & Mishra, K. (2014). Green HRM: Practices and strategic implementation in the organisations. *International Journal on Recent and Innovation Trends in Computing and Communication*, 2(11), 3633–3639.

Rau, H., Wu, J.-J., & Procopio, K. M. (2023). Exploring green product design through TRIZ methodology and the use of green features. *Computers & Industrial Engineering*, 180, 109252. 10.1016/j.cie.2023.109252

Raut, R. D., Narkhede, B., & Gardas, B. B. (2017). To identify the critical success factors of sustainable supply chain management practices in the context of oil and gas industries: ISM approach. *Renewable & Sustainable Energy Reviews*, 68, 33–47. 10.1016/j.rser.2016.09.067

Rayner, J., & Morgan, D. (2018). An empirical study of 'green' workplace behaviours: Ability, motivation, and opportunity. *Asia Pacific Journal of Human Resources*, 56(1), 56–78. 10.1111/1744-7941.12151

Raza, M. Y., Saeed, A., Iqbal, N., & Faraz, N. A. (2021). Enabling digital transformation for sustainable change in organizations: An empirical study. *Sustainability*, 13(7), 3857.

Reinholt, M., Pedersen, T., & Foss, N. J. (2011). Why a central network position isn't enough: The role of motivation and ability for knowledge sharing in employee networks. *Academy of Management Journal*, 54(6), 1277–1297. 10.5465/amj.2009.0007

Ren, S., Tang, G., & Jackson, E, S. (. (2018). Green human resource management research in emergence: A review and future directions. *Asia Pacific Journal of Management*, 35, 769–803. 10.1007/s10490-017-9532-1

Renwick, D., Redman, T., & Maguire, S. (2008). Green HRM: A review, process model, and research agenda. *University of Sheffield Management School Discussion Paper, 1*(1), 1-46.

Renwick, D. W., Redman, T., & Maguire, S. (2013). Green human resource management: A review and research agenda. *International Journal of Management Reviews*, 15(1), 1–14. 10.1111/j.14 68-2370.2011.00328.x

Reynolds, M., Salter, N., Muranko, Ż., Nolan, R., & Charnley, F. (2024). Product life extension behaviours for electrical appliances in UK households: Can consumer education help extend product life amid the cost-of-living crisis? *Resources, Conservation and Recycling*, 205, 107527. 10.1016/j.resconrec.2024.107527

Rodríguez de Sá Silva, A. C.., Bimbato, A. M., Balestieri, J. A. P., & Vilanova, M. R. N. (2022). Exploring environmental, economic and social aspects of rainwater harvesting systems: A review. *Sustainable Cities and Society, 76.* https://doi.org/10.1016/j.scs.2021.103475

Rogerson, J. M., & Sims, S. R. (2012). The greening of urban hotels in South Africa: Evidence from Gauteng. *Urban Forum23*(3), 391–407.

Rogerson, J. M. (2012). The Boutique hotel industry in South Africa: Definition, scope and organisation. *Symposium on Motivation*, 27, 65-116.

Romli, N. A., Safitri, D., & Yustitia, P. (2023). Strategi Komunikasi Pemasaran Hijau Dalam Pemberdayaan Kewirausahaan Masyarakat Mat Peci. *IKRA-ITH HUMANIORA: Jurnal Sosial dan Humaniora, 7*(3), 59-71.

Roscoe, S., Subramanian, N., Jabbour, C. J., & Chong, T. (2019). Green human resource management and the enablers of green organisational culture: Enhancing a firm's environmental performance for sustainable development. *Business Strategy and the Environment*, 28(5), 737–749. 10.1002/bse.2277

Roudbari, E. S., Ghomi, S. F., & Sajadieh, M. S. (2021). Reverse logistics network design for product reuse, remanufacturing, recycling and refurbishing under uncertainty. *Journal of Manufacturing Systems*, 60, 473–486. 10.1016/j.jmsy.2021.06.012

Sackmann, S. A. (1991). Uncovering culture in organizations. *The Journal of Applied Behavioral Science*, 27(3), 295–317. 10.1177/0021886391273005

Saeed, B. B., Afsar, B., Hafeez, S., Khan, I., Tahir, M., & Afridi, M. A. (2019). Promoting employee's proenvironmental behavior through green human resource management practices. *Corporate Social Responsibility and Environmental Management*, 26(2), 424–438. 10.1002/csr.1694

Saeed, M. O., Hassan, M. N., & Mujeebu, M. A. (2009). Assessment of municipal solid waste generation and recyclable materials potential in Kuala Lumpur, Malaysia. *Waste Management (New York, N.Y.)*, 29(7), 2209–2213. 10.1016/j.wasman.2009.02.01719369061

Sakwa, S. M. (2018). *Factors affecting implementation of green human resource practices in the civil service in Kenya* [Doctoral dissertation, University of Nairobi].

Sana, S. S. (2020). Price competition between green and non green products under corporate social responsible firm. *Journal of retailing and consumer services, 55*, 102118.

Sánchez Pérez, J. A., Arzate, S., Soriano-Molina, P., García Sánchez, J. L., Casas López, J. L., & Plaza-Bolaños, P. (2020). Neutral or acidic pH for the removal of contaminants of emerging concern in wastewater by solar photo-Fenton? A techno-economic assessment of continuous raceway pond reactors. *The Science of the Total Environment*, 736(May), 139681. 10.1016/j.scitotenv.2020.13968132479960

Sanyal, C., & Haddock-Millar, J. (2018). Employee engagement in managing environmental performance: A case study of the planet champion initiative, McDonalds UK and Sweden. In Renwick, D. W. S. (Ed.), *Contemporary Developments in Green Human Resource Management Research: Towards Sustainability in Action* (pp. 39–56). Taylor and Francis. 10.4324/9781315768953-3

Sar, K., & Ghadimi, P. (2023). A systematic literature review of the vehicle routing problem in reverse logistics operations. *Computers & Industrial Engineering*, 177, 109011. 10.1016/j.cie.2023.109011

Sarkis, J., Gonzalez-Torre, P., & Adenso-Diaz, B. (2010). Stakeholder pressure and the adoption of environmental practices: The mediating effect of training. *Journal of Operations Management*, 28(2), 163–176. 10.1016/j.jom.2009.10.001

Saud, J. S. (2013). Solid waste management utilizing microbial consortia and its comparative effectiveness study with vermicomposting. *International Journal of Engineering Research & Technology (Ahmedabad)*, 2(10), 2870–2885.

Saxena, N., Sarkar, B., Wee, H. M., Reong, S., Singh, S. R., & Hsiao, Y. L. (2023). A reverse logistics model with eco-design under the Stackelberg-Nash equilibrium and centralized framework. *Journal of Cleaner Production*, 387, 135789. 10.1016/j.jclepro.2022.135789

Scarlat, N., Motola, V., Dallemand, J. F., Monforti-Ferrario, F., & Mofor, L. (2015). Evaluation of energy potential of municipal solid waste from African urban areas. *Renewable & Sustainable Energy Reviews*, 50, 1269–1286. 10.1016/j.rser.2015.05.067

Schillmann, C. (2020). *Patagonia Inc. under a sustainability perspective.*

Schneider, B., Ehrhart, M. G., & Macey, W. H. (2013). Organizational climate and culture. *Annual Review of Psychology*, 64(1), 361–388. 10.1146/annurev-psych-113011-14380922856467

Scott, W. R. (2008). *Institutions and organizations: Ideas and interests.* Sage Publications.

Shafaei, A., Nejati, M., & Yusoff, Y. M. (2020). Green human resource management: A two-study investigation of antecedents and outcomes. *International Journal of Manpower*, 41(7), 1041–1060. 10.1108/IJM-08-2019-0406

Shah, S. Z., & Soomro, B. A. (2023). *Sustainable business practices: Strategies for green human resource management.*

Shah, M. (2019). Green human resource management: Development of a valid measurement scale. *Business Strategy and the Environment*, 28(5), 771–785. 10.1002/bse.2279

Shah, S. M. A., Jiang, Y., Wu, H., Ahmed, Z., Ullah, I., & Adebayo, T. S. (2021). Linking green human resource practices and environmental economics performance: The role of green economic organizational culture and green psychological climate. *International Journal of Environmental Research and Public Health*, 18(20), 10953. 10.3390/ijerph18201095334682698

Shahzad, F., & Luqman, A. (2012). Impact of Organizational Culture on Organizational Performance: An Overview. *Interdisciplinary Journal of Contemporary Research in Business*, 3(9), 975–985.

Shaikh, N. (2023). *Patagonia: $1B Revenue Surge through ESG Success*. https://www.linkedin.com/pulse/patagonia-1b-revenue-surge-through-esg-success-nabeel-shaikh/

Sharma, A., & Iyer, G. R. (2012). Resource-constrained product development: Implications for green marketing and green supply chains. *Industrial Marketing Management*, 41(4), 599–608. 10.1016/j.indmarman.2012.04.007

Sharma, G., & Bansal, P. (2017). The role of sustainability in business decision-making. *Academy of Management Journal*, 60(4), 1352–1380.

Sharma, S., Gururani, S., & Sarkar, P. (2023). Measuring ideation effectiveness in bioinspired design. *Artificial Intelligence for Engineering Design, Analysis and Manufacturing*, 37, e14. 10.1017/S0890060423000070

Sharma, S., & Sarkar, P. (2019). Biomimicry: Exploring Research, Challenges, Gaps, and Tools. In Chakrabarti, A. (Ed.), *Research into Design for a Connected World* (Vol. 134, pp. 87–97). Springer Singapore. http://link.springer.com/10.1007/978-981-13-5974-3_810.1007/978-981-13-5974-3_8

Shehawy, Y. M., & Khan, S. M. F. A. (2024). Consumer readiness for green consumption: The role of green awareness as a moderator of the relationship between green attitudes and purchase intentions. *Journal of Retailing and Consumer Services*, 78, 103739. 10.1016/j.jretconser.2024.103739

Shoda, N. (2013). *Barriers to Sustainable Beverage Packaging*. California Polytechnic State University.

Siddique, F. B., & Sultana, I. (2018). *Unilever Sustainable Living Plan: A Critical Analysis*.

Silvestri, C., Silvestri, L., Piccarozzi, M., & Ruggieri, A. (2022). Toward a framework for selecting indicators of measuring sustainability and circular economy in the agri-food sector: A systematic literature review. *The International Journal of Life Cycle Assessment*, 1–39.

Sinambela, E. A., Azizah, E. I., & Putra, A. R. (2022). The Effect of Green Product, Green Price, and Distribution Channel on The Intention to Repurchasing Simple Face Wash. *Journal of Business and Economics Research (JBE)*, 3(2), 156-162.

Singh, A., Tyagi, P. K., & Garg, A. (Eds.). (2024). *Sustainable Disposal Methods of Food Wastes in Hospitality Operations*. IGI Global. 10.4018/979-8-3693-2181-2

Southey, F. (2019). *Nestlé talks challenges in sustainable soy: Complex supply chains and legal deforestation*. Food Navigator. https://www.foodnavigator.com/Article/2019/06/20/Nestle-talks-challenges-in-sustainable-soy-Complex-supply-chains-and-legal-deforestation?utm_source=copyright&utm_medium=OnSite&utm_campaign=copyright

Spenceley, A. (2005). Tourism certification initiatives in Africa. The International Ecotourism Society (TIES), Washington, DC.

Steinfeld, H., & Gerber, P. (2010). Livestock production and the global environment: Consume less or produce better? *Proceedings of the National Academy of Sciences of the United States of America*, 107(43), 18237–18238. 10.1073/pnas.101254110720935253

Suarna, I. W. (2018). Bali dalam Tarikan Pembangunan Berkelanjutan. *Jurnal Bali Membangun Bali*, 1(3), 199–206. 10.51172/jbmb.v1i3.31

Suasana, I. G. A. K. G., & Ekawati, N. W. (2018). Environmental commitment and green innovation reaching success new products of creative industry in Bali. *The Journal of Business and Retail Management Research*, 12(4). 10.24052/JBRMR/V12IS04/ART-25

Subramanian, P. & Jaganathan, A. (2022). Promoting Environment Sustainability Through Green HRM: The Socially Responsible Organizations. *Ushus Journal of Business Management, 21*(2), 01-13.

Suhartanto, D., Dean, D., Amalia, F. A., & Triyuni, N. N. (2024). Attitude formation towards green products evidence in Indonesia: Integrating environment, culture, and religion. *Asia Pacific Business Review*, 30(1), 94–114. 10.1080/13602381.2022.2082715

Suharti, L., & Sugiarto, A. (2020). A qualitative study of green HRM practices and their benefits in the organization: An Indonesian company experience. *Business: Theory and Practice*, 21(1), 200–211. 10.3846/btp.2020.11386

Sutikno, A. N. (2020). Bonus demografi di indonesia. *VISIONER: Jurnal Pemerintahan Daerah Di Indonesia*, 12(2), 421–439.

Suttell, R. (2005). Hospitality and IAQ. *Buildings*, (November), 62–74.

Tamim, M. S., & Akter, L. (2024). Green marketing impact on youth purchasing: Bangladesh district-wise study on consumer intentions. *Annals of Management and Organization Research*, 5(3), 205–217. 10.35912/amor.v5i3.1818

Tang, F. E. (2012). A study of water consumption in two Malaysian resorts.*International journal of environmental, Ecological and geophysical engineering*, 6(8), 88-93.

Tang, G., Chen, Y., Jiang, Y., Paillé, P., & Jia, J. (2018). Green human resource management practices: Scale development and validity. *Asia Pacific Journal of Human Resources*, 56(1), 31–55. 10.1111/1744-7941.12147

Teixeira, A. A., Jabbour, C. J. C., de Sousa Jabbour, A. B. L., Latan, H., & De Oliveira, J. H. C. (2016). Green training and green supply chain management: Evidence from Brazilian firms. *Journal of Cleaner Production*, 116, 170–176. 10.1016/j.jclepro.2015.12.061

Tesla. (2019). *Impact report*. Tesla. https://www.tesla.com/ns_videos/tesla-impact-report-2019.pdf

Thoibah, W., Arif, M., & Harahap, R. D. (2022). Implementasi Green Marketing Pada UMKM Upaya Memasuki Pasar Internasional (Studi Kasus pada Creabrush Indonesia). *Jurnal Ekonomika Dan Bisnis*, 2(3), 798–805.

Timothy, D. J. &Teye, V. B. (2009). *Tourism & Lodging sector*. UK-Oxford Elsevier INC.

Tociu, C., Ciobotaru, I. E., Maria, C., Déak, G., Ivanov, A. A., Marcu, E., Marinescu, F., Savin, I., & Noor, N. M. (2019). Exhaustive approach to livestock wastewater treatment in irrigation purposes for a better acceptability by the public. *AIP Conference Proceedings*, 2129(July), 020066. Advance online publication. 10.1063/1.5118074

Tribe, H. (2022). *An exploratory investigation into how the implementation and internalization of processes within a MNC are affected by the regulatory, cognitive, and normative domains of institutionalism* [Doctoral dissertation, Brunel University London].

Truchado, P., Gil, M. I., López, C., Garre, A., López-Aragón, R. F., Böhme, K., & Allende, A. (2021). New standards at European Union level on water reuse for agricultural irrigation: Are the Spanish wastewater treatment plants ready to produce and distribute reclaimed water within the minimum quality requirements? *International Journal of Food Microbiology*, 356(June), 109352. 10.1016/j.ijfoodmicro.2021.10935234385095

Tseng, M.-L., Chiu, A. S. F., Tan, R. R., & Siriban-Manalang, A. B. (2013). Sustainable consumption and production for Asia: Sustainability through green design and practice. *Journal of Cleaner Production*, 40, 1–5. 10.1016/j.jclepro.2012.07.015

Tulebayeva, N., Yergobek, D., Pestunova, G., Mottaeva, A., & Sapakova, Z. (2020). Green economy: Waste management and recycling methods. In *E3S Web of Conferences* (Vol. 159, p. 01012). EDP Sciences. 10.1051/e3sconf/202015901012

Uang, S.-T., & Liu, C.-L. (2013). The Development of an Innovative Design Process for Eco-efficient Green Products. In Kurosu, M. (Ed.), *Human-Computer Interaction. Users and Contexts of Use* (Vol. 8006, pp. 475–483). Springer Berlin Heidelberg. 10.1007/978-3-642-39265-8_53

Ugwu, C. O., Ozoegwu, C. G., Ozor, P. A., Agwu, N., & Mbohwa, C. (2021). Waste reduction and utilization strategies to improve municipal solid waste management on Nigerian campuses. *Fuel Communications*, 9, 100025. 10.1016/j.jfueco.2021.100025

Umair, S., Waqas, U., Mrugalska, B. & Al Shamsi, I.R. (2023). *Environmental Corporate Social Responsibility, Green Talent Management, and Organization's Sustainable Performance in the Banking Sector of Oman: The Role of Innovative Work Behaviour and Green.*

Un, C. (2023). A Sustainable approach to the conversion of waste into energy: Landfill gas-to-fuel technology. *Sustainability (Basel)*, 15(20), 14782. 10.3390/su152014782

UNEP (United Nations Environment Programme). (2018). *Emissions Gap Report 2018*. United Nations Environment Programme.

UNEP. (2019). *Global Environment Outlook – GEO-6: Healthy Planet*. Healthy People. 10.1017/9781108627146

Unglesbee, B. (2023). *More than 300 Apple suppliers have committed to clean energy.* Supply Chain Drive. https://www.supplychaindive.com/news/apple-suppliers-clean-energy-scope-3 -carbon-neutral-products/693980/

Unilever. (2017). *Unilever responsible sourcing policy working in partnership with our suppliers.* Unilever. https://www.unilever.com/files/92ui5egz/production/f51492642f57b314b054 66b6194792e02d075d76.pdf

Unilever. (2023). *Annual Report and Accounts 2022. Delivering sustainable business performance.* Unilever. https://www.unilever.com/files/92ui5egz/production/257f12db9c95ffa2ed12 d6f2e2b3ff67db49fd60.pdf

United Nations Industrial Development Organisation (UNIDO). (2022) *Sustainable development goals.* UN. https://www.unido.org/unido-sdgs

United Nations. (2021). The United Nations World Water Development Report: Va*luing Water.* UNESCO, Paris.

Van Yken, J., Boxall, N. J., Cheng, K. Y., Nikoloski, A. N., Moheimani, N. R., & Kaksonen, A. H. (2021). E-waste.recycling and resource recovery: A review on technologies, barriers and enablers with a focus on oceania. *Metals*, 11(8), 1313. 10.3390/met11081313

Viljoen, J. M., Schenck, C. J., Volschenk, L., Blaauw, P. F., & Grobler, L. (2021). Household waste management practices and challenges in a rural remote town in the Hantam Municipality in the Northern Cape, South Africa. *Sustainability (Basel)*, 13(11), 5903. 10.3390/su13115903

Vohra, N. D., & Arora, H. (2021). *Quantitative techniques in management.* McGraw Hill.

Waage, S. A. (2007). Re-considering product design: A practical "road-map" for integration of sustainability issues. *Journal of Cleaner Production*, 15(7), 638–649. 10.1016/j.jclepro.2005.11.026

Wagner, S. A. (2002). *Understanding green consumer behaviour: A qualitative cognitive approach.* Routledge. 10.4324/9780203444030

Waldegrave, W., & Davis, S. C. (1987). POLLUTION ABATEMENT TECHNOLOGY AWARD CEREMONY. *Journal of the Royal Society of Arts*, 135(5372), 603–608.

Wan, B. (2024). The Impact of Cultural Capital on Economic Growth Based on Green Low-Carbon Endogenous Economic Growth Model. *Sustainability*. MDPI. [Link](https://www.mdpi.com/ 2071-1050/16/5/1781)

Wan, C., Shen, G. Q., & Choi, S. (2019). Waste management strategies for sustainable development. In *Encyclopedia of sustainability in higher education* (pp. 2020–2028). Springer International Publishing. 10.1007/978-3-030-11352-0_194

Wang, M.-C. (2016). Development of An Innovative Design Process for Green Products. *Proceedings of International Conference on Artificial Life and Robotics,* (vol. *21*, 108–111). IEEE. 10.5954/ICAROB.2016.OS1-5

Compilation of References

Wang, S., Abbas, J., Sial, M.S., Álvarez-Otero, S. & Cioca, L.I. (2022). Achieving green innovation and sustainable development goals through green knowledge management: Moderating role of organizational green culture. *Journal of innovation & knowledge, 7*(4), p.100272.

Wang, C. H. (2019). How organizational green culture influences green performance and competitive advantage: The mediating role of green innovation. *Journal of Manufacturing Technology Management*, 30(4), 666–683. 10.1108/JMTM-09-2018-0314

WCED, S. W. S. (1987). World commission on environment and development. *Our common future, 17*(1), 1-91.

Weber, G., & Martensen, M. (2021). *Transforming organizational culture amidst a diverse workforce: A qualitative study in the service industry* (No. 1/2021). IUBH Discussion Papers-Human Resources.

Weerasooriya, R. R., Liyanage, L. P. K., Rathnappriya, R. H. K., Bandara, W. B. M. A. C., Perera, T. A. N. T., Gunarathna, M. H. J. P., & Jayasinghe, G. Y. (2021). Industrial water conservation by water footprint and sustainable development goals: a review. In *Environment, Development and Sustainability, 23*(9). 10.1007/s10668-020-01184-0

Weiner, B. J. (2009). A theory of organizational readiness for change. *Implementation Science : IS*, 4(1), 67. 10.1186/1748-5908-4-6719840381

Whiteman, G., Walker, B., & Perego, P. (2013). Planetary boundaries: Ecological foundations for corporate sustainability. *Journal of Management Studies*, 50(2), 307–336. 10.1111/j.1467-6486.2012.01073.x

Woo, E. J. (2021). The necessity of environmental education for employee green behaviour. *East Asian Journal of Business Economics*, 9(4), 29–41.

World Economic Forum. (2023). *The Global Risks Report 2023 18th Edition Insight Report.* WEF. https://www.weforum.org/publications/global-risks-report-2023/

Wulandari, N. P. D. (2018). Between Eco-Education And Critical Thinking: The Application of Emancipatory Learning on Gaining The Awareness of Environmental Problems. In *Bali. In Proceding-International Seminar Culture Change and Sustainable Development in Multidisciplinary Approach: Education, Environment, Art, Politic, Economic, Law, and Tourism* (pp. 126–132). Udayana University.

Wuni, I. Y. (2022). Mapping the barriers to circular economy adoption in the construction industry: A systematic review, Pareto analysis, and mitigation strategy map. *Building and Environment*, 223, 109453. 10.1016/j.buildenv.2022.109453

Xie, H., & Lau, T. C. (2023). Evidence-Based Green Human Resource Management: A Systematic Literature Review. *Sustainability (Basel)*, 15(14), 10941. 10.3390/su151410941

Xu, X., Wang, F., Chen, Y., Yang, B., Zhang, S., Song, X., & Shen, L. (2023). Design of urban medical waste recycling network considering loading reliability under uncertain conditions. *Computers & Industrial Engineering*, 183, 109471. 10.1016/j.cie.2023.109471

Yacob, S., Erida, E., Machpuddin, A., & Alamsyah, D. J. M. S. L. (2021). A model for the business performance of micro, small and medium enterprises: Perspective of social commerce and the uniqueness of resource capability in Indonesia. *Management Science Letters*, 11(1), 101–110. 10.5267/j.msl.2020.8.025

Yang, C. L., Lin, S. P., Chan, Y. H., & Sheu, C. (2010). Mediated effect of environmental management on manufacturing competitiveness: An empirical study. *International Journal of Production Economics*, 123(1), 210–220. 10.1016/j.ijpe.2009.08.017

Yang, Q., Fu, L., Liu, X., & Cheng, M. (2018). Evaluating the Efficiency of Municipal Solid Waste Management in China. *International Journal of Environmental Research and Public Health*, 15(11), 2448. 10.3390/ijerph1511244830400237

Yaputra, H., Risqiani, R., Lukito, N., & Sukarno, K. P. (2023). Pengaruh Green Marketing, Sustainable Advertising, Eco Packaging/Labeling Terhadap Green Purchasing Behavior (Studi Pada Kendaraan Listrik). [IMA]. *Journal of Indonesia Marketing Association*, 2(1), 71–90.

Yi-Fei, G. (2017). Green Innovation Design of Products under the Perspective of Sustainable Development. *IOP Conference Series. Earth and Environmental Science*, 51, 012011. 10.1088/1742-6596/51/1/012011

Yong, J. Y., Yusliza, M. Y., Ramayah, T., Chiappetta Jabbour, C. J., Sehnem, S., & Mani, V. (2020). Pathways towards sustainability in manufacturing organisations: Empirical evidence on the role of green human resource management. *Business Strategy and the Environment*, 29(1), 212–228. 10.1002/bse.2359

Yousif, D. F., & Scott, S. (2007). Governing solid waste management in Mazatenango, Guatemala: Problems and prospects. *International Development Planning Review*, 29(4), 433–450. 10.3828/idpr.29.4.2

Yusiana, R., Widodo, A., & Sumarsih, U. (2021). Integration Consumer Response during the Pandemic Covid-19 on Advertising: Perception Study on Eco Labeling and Eco Brand Products Eco Care. *Inclusive Society and Sustainability Studies*, 1(2), 45–56. 10.31098/issues.v1i2.708

Zadeh, S. M., Hunt, D. V. L., Lombardi, D. R., & Rogers, C. D. F. (2013). Shared urban greywater recycling systems: Water resource savings and economic investment. *Sustainability (Basel)*, 5(7), 2887–2912. 10.3390/su5072887

Zafar, S. Solid Waste Management in Saudi Arabia. EcoMENA. 2015. ((accessed on 24 February 2021)). Available online: https://www.ecomena.org/tag/dammam/

Zaid, A. A., Jaaron, A. A., & Bon, A. T. (2018). The impact of green human resource management and green supply chain management practices on sustainable performance: An empirical study. *Journal of Cleaner Production*, 204, 965–979. 10.1016/j.jclepro.2018.09.062

Zammuto, R. F., Gifford, B., & Goodman, E. A. (2000). *Managerial ideologies, organization culture and the outcomes of innovation: A competing values perspective.*

Compilation of References

Zhang, D. Q., Tan, S. K., & Gersberg, R. M. (2010). Municipal solid waste management in China: Status, problems and challenges. *Journal of Environmental Management*, 91(8), 1623–1633. 10.1016/j.jenvman.2010.03.01220413209

Zhang, X., Zhu, S., Dai, S., Jiang, Z., Gong, Q., & Wang, Y. (2024). Optimization of third party take-back enterprise collection strategy based on blockchain and remanufacturing reverse logistics. *Computers & Industrial Engineering*, 187, 109846. 10.1016/j.cie.2023.109846

Zhuo, L., & Shengxue, Y. (2010). A Research on Green Product Design Process and Evaluation. *2010 3rd International Conference on Information Management, Innovation Management and Industrial Engineering*, 612–614. 10.1109/ICIII.2010.466

Zoogah, D. (2011). The dynamics of Green HRM behaviors: A cognitive social information processing approach. *Zeitschrift fur Personalforschung*, 25(2), 117–139. 10.1177/239700221102500204

Zorpas, A. A. (2020). Strategy development in the framework of waste management. *The Science of the Total Environment*, 716, 137088. 10.1016/j.scitotenv.2020.13708832059326

Zotos, G., Karagiannidis, A., Zampetoglou, S., Malamakis, A., Antonopoulos, I. S., Kontogianni, S., & Tchobanoglous, G. (2009). Developing a holistic strategy for integrated waste management within municipal planning: Challenges, policies, solutions and perspectives for Hellenic municipalities in the zero-waste, low-cost direction. *Waste Management (New York, N.Y.)*, 29(5), 1686–1692. 10.1016/j.wasman.2008.11.01619147341

Related References

To continue our tradition of advancing academic research, we have compiled a list of recommended IGI Global readings. These references will provide additional information and guidance to further enrich your knowledge and assist you with your own research and future publications.

Ajiboye, O. E., & Yusuff, O. S. (2017). Foreign Land Acquisition: Food Security and Food Chains – The Nigerian Experience. In I. Management Association (Ed.), *Natural Resources Management: Concepts, Methodologies, Tools, and Applications* (pp. 1524-1545). Hershey, PA: IGI Global. https://doi.org/10.4018/978-1-5225 -0803-8.ch072

Alapiki, H. E., & Amadi, L. A. (2018). Sustainable Food Consumption in the Neoliberal Order: Challenges and Policy Implications. In Obayelu, A. (Ed.), *Food Systems Sustainability and Environmental Policies in Modern Economies* (pp. 90–123). Hershey, PA: IGI Global. 10.4018/978-1-5225-3631-4.ch005

Altaş, A. (2018). Geographical Information System Applications Utilized in Museums in Turkey Within the Scope of the Cultural Heritage Tourism: A Case Study of Mobile Application of Müze Asist. In Chaudhuri, S., & Ray, N. (Eds.), *GIS Applications in the Tourism and Hospitality Industry* (pp. 42–60). Hershey, PA: IGI Global. 10.4018/978-1-5225-5088-4.ch002

Andreea, I. R. (2018). Beyond Macroeconomics of Food and Nutrition Security. *International Journal of Sustainable Economies Management*, 7(1), 13–22. 10.4018/ IJSEM.2018010102

Anwar, J. (2017). Reproductive and Mental Health during Natural Disaster: Implications and Issues for Women in Developing Nations – A Case Example. In I. Management Association (Ed.), *Gaming and Technology Addiction: Breakthroughs in Research and Practice* (pp. 446-472). Hershey, PA: IGI Global. https://doi.org/ 10.4018/978-1-5225-0778-9.ch021

Awadh, H., Aksissou, M., Benhardouze, W., Darasi, F., & Snaiki, J. (2018). Socio-economic Status of Artisanal Fishers in the West Part of Moroccan Mediterranean. *International Journal of Social Ecology and Sustainable Development*, 9(1), 40–52. 10.4018/IJSESD.2018010104

Aye, G. C., & Haruna, R. F. (2018). Effect of Climate Change on Crop Productivity and Prices in Benue State, Nigeria: Implications for Food Security. In Erokhin, V. (Ed.), *Establishing Food Security and Alternatives to International Trade in Emerging Economies* (pp. 244–268). Hershey, PA: IGI Global. 10.4018/978-1-5225-2733-6. ch012

Azizan, S. A., & Suki, N. M. (2017). Consumers' Intentions to Purchase Organic Food Products. In Esakki, T. (Ed.), *Green Marketing and Environmental Responsibility in Modern Corporations* (pp. 68–82). Hershey, PA: IGI Global. 10.4018/978-1-5225-2331-4.ch005

Barakabitze, A. A., Fue, K. G., Kitindi, E. J., & Sanga, C. A. (2017). Developing a Framework for Next Generation Integrated Agro Food-Advisory Systems in Developing Countries. In I. Management Association (Ed.), *Agri-Food Supply Chain Management: Breakthroughs in Research and Practice* (pp. 47-67). Hershey, PA: IGI Global. 10.4018/978-1-5225-1629-3.ch004

Beachcroft-Shaw, H., & Ellis, D. (2018). Using Successful Cases to Promote Environmental Sustainability: A Social Marketing Approach. In I. Management Association (Ed.), *Sustainable Development: Concepts, Methodologies, Tools, and Applications* (pp. 936-953). Hershey, PA: IGI Global. 10.4018/978-1-5225-3817-2.ch042

Behnassi, M., Kahime, K., Boussaa, S., Boumezzough, A., & Messouli, M. (2017). Infectious Diseases and Climate Vulnerability in Morocco: Governance and Adaptation Options. In I. Management Association (Ed.), *Public Health and Welfare: Concepts, Methodologies, Tools, and Applications* (pp. 91-109). Hershey, PA: IGI Global. https://doi.org/10.4018/978-1-5225-1674-3.ch005

Bekele, F., & Bekele, I. (2017). Social and Environmental Impacts on Agricultural Development. In Ganpat, W., Dyer, R., & Isaac, W. (Eds.), *Agricultural Development and Food Security in Developing Nations* (pp. 21–56). Hershey, PA: IGI Global. 10.4018/978-1-5225-0942-4.ch002

Benaouda, A., & García-Peñalvo, F. J. (2018). Towards an Intelligent System for the Territorial Planning: Agricultural Case. In García-Peñalvo, F. (Ed.), *Global Implications of Emerging Technology Trends* (pp. 158–178). Hershey, PA: IGI Global. 10.4018/978-1-5225-4944-4.ch010

Bhaskar, A., Rao, G. B., & Vencatesan, J. (2017). Characterization and Management Concerns of Water Resources around Pallikaranai Marsh, South Chennai. In Rao, P., & Patil, Y. (Eds.), *Reconsidering the Impact of Climate Change on Global Water Supply, Use, and Management* (pp. 102–121). Hershey, PA: IGI Global. 10.4018/978-1-5225-1046-8.ch007

Bhyan, P., Shrivastava, B., & Kumar, N. (2022). Requisite Sustainable Development Contemplating Buildings: Economic and Environmental Sustainability. In Hussain, A., Tiwari, K., & Gupta, A. (Eds.), *Addressing Environmental Challenges Through Spatial Planning* (pp. 269–288). IGI Global. https://doi.org/10.4018/978-1-7998-8331-9.ch014

Bogataj, D., & Drobne, D. (2017). Control of Perishable Goods in Cold Logistic Chains by Bionanosensors. In Joo, S. (Ed.), *Applying Nanotechnology for Environmental Sustainability* (pp. 376–402). Hershey, PA: IGI Global. 10.4018/978-1-5225-0585-3.ch016

Bogataj, D., & Drobne, D. (2017). Control of Perishable Goods in Cold Logistic Chains by Bionanosensors. In I. Management Association (Ed.), *Materials Science and Engineering: Concepts, Methodologies, Tools, and Applications* (pp. 471-497). Hershey, PA: IGI Global. https://doi.org/10.4018/978-1-5225-1798-6.ch019

Bogueva, D., & Marinova, D. (2018). What Is More Important: Perception of Masculinity or Personal Health and the Environment? In Bogueva, D., Marinova, D., & Raphaely, T. (Eds.), *Handbook of Research on Social Marketing and Its Influence on Animal Origin Food Product Consumption* (pp. 148–162). Hershey, PA: IGI Global. 10.4018/978-1-5225-4757-0.ch010

Bouzid, M. (2017). Waterborne Diseases and Climate Change: Impact and Implications. In Bouzid, M. (Ed.), *Examining the Role of Environmental Change on Emerging Infectious Diseases and Pandemics* (pp. 89–108). Hershey, PA: IGI Global. 10.4018/978-1-5225-0553-2.ch004

Bowles, D. C. (2017). Climate Change-Associated Conflict and Infectious Disease. In Bouzid, M. (Ed.), *Examining the Role of Environmental Change on Emerging Infectious Diseases and Pandemics* (pp. 68–88). Hershey, PA: IGI Global. 10.4018/978-1-5225-0553-2.ch003

Buck, J. J., & Lowry, R. K. (2017). Oceanographic Data Management: Quills and Free Text to the Digital Age and "Big Data". In Diviacco, P., Leadbetter, A., & Glaves, H. (Eds.), *Oceanographic and Marine Cross-Domain Data Management for Sustainable Development* (pp. 1–22). Hershey, PA: IGI Global. 10.4018/978-1-5225-0700-0.ch001

Buse, C. G. (2017). Are Climate Change Adaptation Policies a Game Changer?: A Case Study of Perspectives from Public Health Officials in Ontario, Canada. In Bouzid, M. (Ed.), *Examining the Role of Environmental Change on Emerging Infectious Diseases and Pandemics* (pp. 230–257). Hershey, PA: IGI Global. 10.4018/978-1-5225-0553-2.ch010

Calderon, F. A., Giolo, E. G., Frau, C. D., Rengel, M. G., Rodriguez, H., Tornello, M., & Gallucci, R. (2018). Seismic Microzonation and Site Effects Detection Through Microtremors Measures: A Review. In Ceryan, N. (Ed.), *Handbook of Research on Trends and Digital Advances in Engineering Geology* (pp. 326–349). Hershey, PA: IGI Global. 10.4018/978-1-5225-2709-1.ch009

Carfi, D., Donato, A., & Panuccio, D. (2018). A Game Theory Coopetitive Perspective for Sustainability of Global Feeding: Agreements Among Vegan and Non-Vegan Food Firms. In I. Management Association (Ed.), *Game Theory: Breakthroughs in Research and Practice* (pp. 71-104). Hershey, PA: IGI Global. https://doi.org/10.4018/978-1-5225-2594-3.ch004

Castagnolo, V. (2018). Analyzing, Classifying, Safeguarding: Drawing for the Borgo Murattiano Neighbourhood of Bari. In Carlone, G., Martinelli, N., & Rotondo, F. (Eds.), *Designing Grid Cities for Optimized Urban Development and Planning* (pp. 93–108). Hershey, PA: IGI Global. 10.4018/978-1-5225-3613-0.ch006

Chekima, B. (2018). The Dilemma of Purchase Intention: A Conceptual Framework for Understanding Actual Consumption of Organic Food. *International Journal of Sustainable Economies Management*, 7(2), 1–13. 10.4018/IJSEM.2018040101

Chen, Y. (2017). Sustainable Supply Chains and International Soft Landings: A Case of Wetland Entrepreneurship. In Christiansen, B., & Kasarcı, F. (Eds.), *Corporate Espionage, Geopolitics, and Diplomacy Issues in International Business* (pp. 232–247). Hershey, PA: IGI Global. 10.4018/978-1-5225-1031-4.ch013

Çıtak, L., Akel, V., & Ersoy, E. (2018). Investors' Reactions to the Announcement of New Constituents of BIST Sustainability Index: An Analysis by Event Study and Mean-Median Tests. In Risso, M., & Testarmata, S. (Eds.), *Value Sharing for Sustainable and Inclusive Development* (pp. 270–289). Hershey, PA: IGI Global. 10.4018/978-1-5225-3147-0.ch012

D'Aleo, V., D'Aleo, F., & Bonanno, R. (2018). New Food Industries Toward a New Level of Sustainable Supply: Success Stories, Business Models, and Strategies. In Erokhin, V. (Ed.), *Establishing Food Security and Alternatives to International Trade in Emerging Economies* (pp. 74–97). Hershey, PA: IGI Global. 10.4018/978-1-5225-2733-6.ch004

Dagevos, H., & Reinders, M. J. (2018). Flexitarianism and Social Marketing: Reflections on Eating Meat in Moderation. In Bogueva, D., Marinova, D., & Raphaely, T. (Eds.), *Handbook of Research on Social Marketing and Its Influence on Animal Origin Food Product Consumption* (pp. 105–120). Hershey, PA: IGI Global. 10.4018/978-1-5225-4757-0.ch007

Danisman, G. O. (2022). What Drives Eco-Design Innovations in European SMEs? In Akkucuk, U. (Ed.), *Disruptive Technologies and Eco-Innovation for Sustainable Development* (pp. 191–206). IGI Global. https://doi.org/10.4018/978-1-7998-8900-7.ch011

Deenapanray, P. N., & Ramma, I. (2017). Adaptations to Climate Change and Climate Variability in the Agriculture Sector in Mauritius: Lessons from a Technical Needs Assessment. In I. Management Association (Ed.), *Natural Resources Management: Concepts, Methodologies, Tools, and Applications* (pp. 655-680). Hershey, PA: IGI Global. https://doi.org/10.4018/978-1-5225-0803-8.ch030

Deenapanray, P. N., & Ramma, I. (2017). Adaptations to Climate Change and Climate Variability in the Agriculture Sector in Mauritius: Lessons from a Technical Needs Assessment. In I. Management Association (Ed.), *Natural Resources Management: Concepts, Methodologies, Tools, and Applications* (pp. 655-680). Hershey, PA: IGI Global. https://doi.org/10.4018/978-1-5225-0803-8.ch030

Deshpande, S., Basu, S. K., Li, X., & Chen, X. (2017). Smart, Innovative and Intelligent Technologies Used in Drug Designing. In I. Management Association (Ed.), *Pharmaceutical Sciences: Breakthroughs in Research and Practice* (pp. 1175-1191). Hershey, PA: IGI Global. https://doi.org/10.4018/978-1-5225-1762-7.ch045

Dlamini, P. N. (2017). Use of Information and Communication Technologies Tools to Capture, Store, and Disseminate Indigenous Knowledge: A Literature Review. In Ngulube, P. (Ed.), *Handbook of Research on Theoretical Perspectives on Indigenous Knowledge Systems in Developing Countries* (pp. 225–247). Hershey, PA: IGI Global. 10.4018/978-1-5225-0833-5.ch010

Dolejsova, M., & Kera, D. (2017). The Fermentation GutHub Project and the Internet of Microbes. In Konomi, S., & Roussos, G. (Eds.), *Enriching Urban Spaces with Ambient Computing, the Internet of Things, and Smart City Design* (pp. 25–46). Hershey, PA: IGI Global. 10.4018/978-1-5225-0827-4.ch002

Dolunay, O. (2018). A Paradigm Shift: Empowering Farmers to Eliminate the Waste in the Form of Fresh Water and Energy Through the Implementation of 4R+T. In I. Management Association (Ed.), *Sustainable Development: Concepts, Methodologies, Tools, and Applications* (pp. 882-892). Hershey, PA: IGI Global. https://doi.org/10.4018/978-1-5225-3817-2.ch039

Dube, P., Heijman, W. J., Ihle, R., & Ochieng, J. (2018). The Potential of Traditional Leafy Vegetables for Improving Food Security in Africa. In Erokhin, V. (Ed.), *Establishing Food Security and Alternatives to International Trade in Emerging Economies* (pp. 220–243). Hershey, PA: IGI Global. 10.4018/978-1-5225-2733-6.ch011

Duruji, M. M., & Urenma, D. F. (2017). The Environmentalism and Politics of Climate Change: A Study of the Process of Global Convergence through UNFCCC Conferences. In I. Management Association (Ed.), *Natural Resources Management: Concepts, Methodologies, Tools, and Applications* (pp. 77-108). Hershey, PA: IGI Global. 10.4018/978-1-5225-0803-8.ch004

Dutta, U. (2017). Agro-Geoinformatics, Potato Cultivation, and Climate Change. In Londhe, S. (Ed.), *Sustainable Potato Production and the Impact of Climate Change* (pp. 247–271). Hershey, PA: IGI Global. 10.4018/978-1-5225-1715-3.ch012

Edirisinghe, R., Stranieri, A., & Wickramasinghe, N. (2017). A Taxonomy for mHealth. In Wickramasinghe, N. (Ed.), *Handbook of Research on Healthcare Administration and Management* (pp. 596–615). Hershey, PA: IGI Global. 10.4018/978-1-5225-0920-2.ch036

Ekpeni, N. M., & Ayeni, A. O. (2018). Global Natural Hazard and Disaster Vulnerability Management. In Eneanya, A. (Ed.), *Handbook of Research on Environmental Policies for Emergency Management and Public Safety* (pp. 83–104). Hershey, PA: IGI Global. 10.4018/978-1-5225-3194-4.ch005

Ene, C., Voica, M. C., & Panait, M. (2017). Green Investments and Food Security: Opportunities and Future Directions in the Context of Sustainable Development. In Mieila, M. (Ed.), *Measuring Sustainable Development and Green Investments in Contemporary Economies* (pp. 163–200). Hershey, PA: IGI Global. 10.4018/978-1-5225-2081-8.ch007

Escamilla, I., Ruíz, M. T., Ibarra, M. M., Soto, V. L., Quintero, R., & Guzmán, G. (2018). Geocoding Tweets Based on Semantic Web and Ontologies. In Lytras, M., Aljohani, N., Damiani, E., & Chui, K. (Eds.), *Innovations, Developments, and Applications of Semantic Web and Information Systems* (pp. 372–392). Hershey, PA: IGI Global. 10.4018/978-1-5225-5042-6.ch014

Escribano, A. J. (2018). Marketing Strategies for Trendy Animal Products: Sustainability as a Core. In Quoquab, F., Thurasamy, R., & Mohammad, J. (Eds.), *Driving Green Consumerism Through Strategic Sustainability Marketing* (pp. 169–203). Hershey, PA: IGI Global. 10.4018/978-1-5225-2912-5.ch010

Eudoxie, G., & Roopnarine, R. (2017). Climate Change Adaptation and Disaster Risk Management in the Caribbean. In Ganpat, W., & Isaac, W. (Eds.), *Environmental Sustainability and Climate Change Adaptation Strategies* (pp. 97–125). Hershey, PA: IGI Global. 10.4018/978-1-5225-1607-1.ch004

Farmer, L. S. (2017). Data Analytics for Strategic Management: Getting the Right Data. In Wang, V. (Ed.), *Encyclopedia of Strategic Leadership and Management* (pp. 810–822). Hershey, PA: IGI Global. 10.4018/978-1-5225-1049-9.ch056

Fattal, L. R. (2017). Catastrophe: An Uncanny Catalyst for Creativity. In Shin, R. (Ed.), *Convergence of Contemporary Art, Visual Culture, and Global Civic Engagement* (pp. 244–262). Hershey, PA: IGI Global. 10.4018/978-1-5225-1665-1.ch014

Forti, I. (2017). A Cross Reading of Landscape through Digital Landscape Models: The Case of Southern Garda. In Ippolito, A. (Ed.), *Handbook of Research on Emerging Technologies for Architectural and Archaeological Heritage* (pp. 532–561). Hershey, PA: IGI Global. 10.4018/978-1-5225-0675-1.ch018

Gharbi, A., De Runz, C., & Akdag, H. (2017). Urban Development Modelling: A Survey. In Faiz, S., & Mahmoudi, K. (Eds.), *Handbook of Research on Geographic Information Systems Applications and Advancements* (pp. 96–124). Hershey, PA: IGI Global. 10.4018/978-1-5225-0937-0.ch004

Ghosh, I., & Ghoshal, I. (2018). Implications of Trade Liberalization for Food Security Under the ASEAN-India Strategic Partnership: A Gravity Model Approach. In Erokhin, V. (Ed.), *Establishing Food Security and Alternatives to International Trade in Emerging Economies* (pp. 98–118). Hershey, PA: IGI Global. 10.4018/978-1-5225-2733-6.ch005

Glaves, H. M. (2017). Developing a Common Global Framework for Marine Data Management. In Diviacco, P., Leadbetter, A., & Glaves, H. (Eds.), *Oceanographic and Marine Cross-Domain Data Management for Sustainable Development* (pp. 47–68). Hershey, PA: IGI Global. 10.4018/978-1-5225-0700-0.ch003

Godulla, A., & Wolf, C. (2018). Future of Food: Transmedia Strategies of National Geographic. In Gambarato, R., & Alzamora, G. (Eds.), *Exploring Transmedia Journalism in the Digital Age* (pp. 162–182). Hershey, PA: IGI Global. 10.4018/978-1-5225-3781-6.ch010

Gomes, P. P. (2018). Food and Environment: A Review on the Sustainability of Six Different Dietary Patterns. In Obayelu, A. (Ed.), *Food Systems Sustainability and Environmental Policies in Modern Economies* (pp. 15–31). Hershey, PA: IGI Global. 10.4018/978-1-5225-3631-4.ch002

Gonzalez-Feliu, J. (2018). Sustainability Evaluation of Green Urban Logistics Systems: Literature Overview and Proposed Framework. In Paul, A., Bhattacharyya, D., & Anand, S. (Eds.), *Green Initiatives for Business Sustainability and Value Creation* (pp. 103–134). Hershey, PA: IGI Global. 10.4018/978-1-5225-2662-9.ch005

Goodland, R. (2017). A Fresh Look at Livestock Greenhouse Gas Emissions and Mitigation. In I. Management Association (Ed.), *Natural Resources Management: Concepts, Methodologies, Tools, and Applications* (pp. 124-139). Hershey, PA: IGI Global. 10.4018/978-1-5225-0803-8.ch006

Goundar, S., & Appana, S. (2018). Mainstreaming Development Policies for Climate Change in Fiji: A Policy Gap Analysis and the Role of ICTs. In I. Management Association (Ed.), *Sustainable Development: Concepts, Methodologies, Tools, and Applications* (pp. 402-432). Hershey, PA: IGI Global. 10.4018/978-1-5225-3817-2.ch020

Granell-Canut, C., & Aguilar-Moreno, E. (2018). Geospatial Influence in Science Mapping. In M. Khosrow-Pour, D.B.A. (Ed.), *Encyclopedia of Information Science and Technology, Fourth Edition* (pp. 3473-3483). Hershey, PA: IGI Global. 10.4018/978-1-5225-2255-3.ch302

Grigelis, A., Blažauskas, N., Gelumbauskaitė, L. Ž., Gulbinskas, S., Suzdalev, S., & Ferrarin, C. (2017). Marine Environment Data Management Related to the Human Activity in the South-Eastern Baltic Sea (The Lithuanian Sector). In Diviacco, P., Leadbetter, A., & Glaves, H. (Eds.), *Oceanographic and Marine Cross-Domain Data Management for Sustainable Development* (pp. 282–302). Hershey, PA: IGI Global. 10.4018/978-1-5225-0700-0.ch012

Guma, I. P., Rwashana, A. S., & Oyo, B. (2018). Food Security Policy Analysis Using System Dynamics: The Case of Uganda. *International Journal of Information Technologies and Systems Approach*, 11(1), 72–90. 10.4018/IJITSA.2018010104

Guma, I. P., Rwashana, A. S., & Oyo, B. (2018). Food Security Indicators for Subsistence Farmers Sustainability: A System Dynamics Approach. *International Journal of System Dynamics Applications*, 7(1), 45–64. 10.4018/IJSDA.2018010103

Gupta, P., & Goyal, S. (2017). Wildlife Habitat Evaluation. In Santra, A., & Mitra, S. (Eds.), *Remote Sensing Techniques and GIS Applications in Earth and Environmental Studies* (pp. 258–264). Hershey, PA: IGI Global. 10.4018/978-1-5225-1814-3.ch013

Hanson, T., & Hildebrand, E. (2018). GPS Travel Diaries in Rural Transportation Research: A Focus on Older Drivers. In I. Management Association (Ed.), *Intelligent Transportation and Planning: Breakthroughs in Research and Practice* (pp. 609-625). Hershey, PA: IGI Global. 10.4018/978-1-5225-5210-9.ch027

Hartman, M. B. (2017). Research-Based Climate Change Public Education Programs. In I. Management Association (Ed.), *Natural Resources Management: Concepts, Methodologies, Tools, and Applications* (pp. 992-1003). Hershey, PA: IGI Global. 10.4018/978-1-5225-0803-8.ch046

Hashim, N. (2017). Zanzibari Seaweed: Global Climate Change and the Promise of Adaptation. In I. Management Association (Ed.), *Natural Resources Management: Concepts, Methodologies, Tools, and Applications* (pp. 365-391). Hershey, PA: IGI Global. https://doi.org/10.4018/978-1-5225-0803-8.ch019

Herrera, J. E., Argüello, L. V., Gonzalez-Feliu, J., & Jaimes, W. A. (2017). Decision Support System Design Requirements, Information Management, and Urban Logistics Efficiency: Case Study of Bogotá, Colombia. In Jamil, G., Soares, A., & Pessoa, C. (Eds.), *Handbook of Research on Information Management for Effective Logistics and Supply Chains* (pp. 223–238). Hershey, PA: IGI Global. 10.4018/978-1-5225-0973-8.ch012

Huizinga, T., Ayanso, A., Smoor, M., & Wronski, T. (2017). Exploring Insurance and Natural Disaster Tweets Using Text Analytics. *International Journal of Business Analytics*, 4(1), 1–17. 10.4018/IJBAN.2017010101

Hung, K., Kalantari, M., & Rajabifard, A. (2017). An Integrated Method for Assessing the Text Content Quality of Volunteered Geographic Information in Disaster Management. *International Journal of Information Systems for Crisis Response and Management*, 9(2), 1–17. 10.4018/IJISCRAM.2017040101

Husnain, A., & Avdic, A. (2018). Identifying the Contemporary Status of E-Service Sustainability Research. In I. Management Association (Ed.), *Sustainable Development: Concepts, Methodologies, Tools, and Applications* (pp. 467-485). Hershey, PA: IGI Global. https://doi.org/10.4018/978-1-5225-3817-2.ch022

Iarossi, M. P., & Ferro, L. (2017). "The Past is Never Dead. It's Not Even Past": Virtual Archaeological Promenade. In Ippolito, A., & Cigola, M. (Eds.), *Handbook of Research on Emerging Technologies for Digital Preservation and Information Modeling* (pp. 228–255). Hershey, PA: IGI Global. 10.4018/978-1-5225-0680-5. ch010

Ignjatijević, S., & Cvijanović, D. (2018). Analysis of Serbian Honey Production and Exports. In *Exploring the Global Competitiveness of Agri-Food Sectors and Serbia's Dominant Presence: Emerging Research and Opportunities* (pp. 109–139). Hershey, PA: IGI Global. 10.4018/978-1-5225-2762-6.ch005

Jana, S. K., & Karmakar, A. K. (2017). Globalization, Governance, and Food Security: The Case of BRICS. In I. Management Association (Ed.), *Natural Resources Management: Concepts, Methodologies, Tools, and Applications* (pp. 692-712). Hershey, PA: IGI Global. https://doi.org/10.4018/978-1-5225-0803-8.ch032

Jana, S. K., & Karmakar, A. K. (2017). Food Security in Asia: Is There Convergence? In I. Management Association (Ed.), *Natural Resources Management: Concepts, Methodologies, Tools, and Applications* (pp. 109-123). Hershey, PA: IGI Global. https://doi.org/10.4018/978-1-5225-0803-8.ch005

John, J., & Kumar, S. (2018). A Locational Decision Making Framework for Ship-breaking Under Multiple Criteria. In I. Management Association (Ed.), *Operations and Service Management: Concepts, Methodologies, Tools, and Applications* (pp. 504-527). Hershey, PA: IGI Global. 10.4018/978-1-5225-3909-4.ch024

John, J., & Srivastava, R. K. (2018). Decision Insights for Shipbreaking using Environmental Impact Assessment: Review and Perspectives. *International Journal of Strategic Decision Sciences*, 9(1), 45–62. 10.4018/IJSDS.2018010104

Joshi, Y., & Rahman, Z. (2018). Determinants of Sustainable Consumption Behaviour: Review and Conceptual Framework. In Paul, A., Bhattacharyya, D., & Anand, S. (Eds.), *Green Initiatives for Business Sustainability and Value Creation* (pp. 239–262). Hershey, PA: IGI Global. 10.4018/978-1-5225-2662-9.ch011

Juma, D. W., Reuben, M., Wang, H., & Li, F. (2018). Adaptive Coevolution: Realigning the Water Governance Regime to the Changing Climate. In I. Management Association (Ed.), *Hydrology and Water Resource Management: Breakthroughs in Research and Practice* (pp. 346-357). Hershey, PA: IGI Global. https://doi.org/10.4018/978-1-5225-3427-3.ch014

K., S., & Tripathy, B. K. (2018). Neighborhood Rough-Sets-Based Spatial Data Analytics. In M. Khosrow-Pour, D.B.A. (Ed.), *Encyclopedia of Information Science and Technology, Fourth Edition* (pp. 1835-1844). Hershey, PA: IGI Global. https://doi.org/10.4018/978-1-5225-2255-3.ch160

Kabir, F. (2018). Towards a More Gender-Inclusive Climate Change Policy. In Mahtab, N., Haque, T., Khan, I., Islam, M., & Wahid, I. (Eds.), *Handbook of Research on Women's Issues and Rights in the Developing World* (pp. 354–369). Hershey, PA: IGI Global. 10.4018/978-1-5225-3018-3.ch022

Kabir, F. (2018). Towards a More Gender-Inclusive Climate Change Policy. In I. Management Association (Ed.), *Climate Change and Environmental Concerns: Breakthroughs in Research and Practice* (pp. 525-540). Hershey, PA: IGI Global. https://doi.org/10.4018/978-1-5225-5487-5.ch027

Kanyamuka, J. S., Jumbe, C. B., & Ricker-Gilbert, J. (2018). Making Agricultural Input Subsidies More Effective and Profitable in Africa: The Role of Complementary Interventions. In Obayelu, A. (Ed.), *Food Systems Sustainability and Environmental Policies in Modern Economies* (pp. 172–187). Hershey, PA: IGI Global. 10.4018/978-1-5225-3631-4.ch008

Karimi, H., & Gholamrezafahimi, F. (2017). Study of Integrated Coastal Zone Management and Its Environmental Effects: A Case of Iran. In Singh, R., Singh, A., & Srivastava, V. (Eds.), *Environmental Issues Surrounding Human Overpopulation* (pp. 64–88). Hershey, PA: IGI Global. 10.4018/978-1-5225-1683-5.ch004

Kaya, I. R., Hutabarat, J., & Bambang, A. N. (2018). "Sasi": A New Path to Sustain Seaweed Farming From Up-Stream to Down-Stream in Kotania Bay, Molucass. *International Journal of Social Ecology and Sustainable Development*, 9(2), 28–36. 10.4018/IJSESD.2018040103

Khader, V. (2018). Technologies for Food, Health, Livelihood, and Nutrition Security. In I. Management Association (Ed.), *Food Science and Nutrition: Breakthroughs in Research and Practice* (pp. 94-112). Hershey, PA: IGI Global. https://doi.org/10 .4018/978-1-5225-5207-9.ch005

Kocadağlı, A. Y. (2017). The Temporal and Spatial Development of Organic Agriculture in Turkey. In Ganpat, W., Dyer, R., & Isaac, W. (Eds.), *Agricultural Development and Food Security in Developing Nations* (pp. 130–156). Hershey, PA: IGI Global. 10.4018/978-1-5225-0942-4.ch006

Koundouri, P., Giannouli, A., & Souliotis, I. (2017). An Integrated Approach for Sustainable Environmental and Socio-Economic Development Using Offshore Infrastructure. In I. Management Association (Ed.), *Renewable and Alternative Energy: Concepts, Methodologies, Tools, and Applications* (pp. 1581-1601). Hershey, PA: IGI Global. 10.4018/978-1-5225-1671-2.ch056

Kumar, A., & Dash, M. K. (2017). Sustainability and Future Generation Infrastructure on Digital Platform: A Study of Generation Y. In Ray, N. (Ed.), *Business Infrastructure for Sustainability in Developing Economies* (pp. 124–142). Hershey, PA: IGI Global. 10.4018/978-1-5225-2041-2.ch007

Kumar, A., Mukherjee, A. B., & Krishna, A. P. (2017). Application of Conventional Data Mining Techniques and Web Mining to Aid Disaster Management. In Kumar, A. (Ed.), *Web Usage Mining Techniques and Applications Across Industries* (pp. 138–167). Hershey, PA: IGI Global. 10.4018/978-1-5225-0613-3.ch006

Kumar, C. P. (2017). Impact of Climate Change on Groundwater Resources. In I. Management Association (Ed.), *Natural Resources Management: Concepts, Methodologies, Tools, and Applications* (pp. 1094-1120). Hershey, PA: IGI Global. 10.4018/978-1-5225-0803-8.ch052

Kumari, S., & Patil, Y. (2017). Achieving Climate Smart Agriculture with a Sustainable Use of Water: A Conceptual Framework for Sustaining the Use of Water for Agriculture in the Era of Climate Change. In Rao, P., & Patil, Y. (Eds.), *Reconsidering the Impact of Climate Change on Global Water Supply, Use, and Management* (pp. 122–143). Hershey, PA: IGI Global. 10.4018/978-1-5225-1046-8.ch008

Kumari, S., & Patil, Y. (2018). Achieving Climate Smart Agriculture With a Sustainable Use of Water: A Conceptual Framework for Sustaining the Use of Water for Agriculture in the Era of Climate Change. In I. Management Association (Ed.), *Climate Change and Environmental Concerns: Breakthroughs in Research and Practice* (pp. 111-133). Hershey, PA: IGI Global. 10.4018/978-1-5225-5487-5.ch006

Kursah, M. B. (2017). Least-Cost Pipeline using Geographic Information System: The Limit to Technicalities. *International Journal of Applied Geospatial Research*, 8(3), 1–15. 10.4018/ijagr.2017070101

Lahiri, S., Ghosh, D., & Bhakta, J. N. (2017). Role of Microbes in Eco-Remediation of Perturbed Aquatic Ecosystem. In Bhakta, J. (Ed.), *Handbook of Research on Inventive Bioremediation Techniques* (pp. 70–107). Hershey, PA: IGI Global. 10.4018/978-1-5225-2325-3.ch004

Lallo, C. H., Smalling, S., Facey, A., & Hughes, M. (2018). The Impact of Climate Change on Small Ruminant Performance in Caribbean Communities. In I. Management Association (Ed.), *Climate Change and Environmental Concerns: Breakthroughs in Research and Practice* (pp. 193-218). Hershey, PA: IGI Global. 10.4018/978-1-5225-5487-5.ch010

Laurini, R. (2017). Nature of Geographic Knowledge Bases. In Faiz, S., & Mahmoudi, K. (Eds.), *Handbook of Research on Geographic Information Systems Applications and Advancements* (pp. 29–60). Hershey, PA: IGI Global. 10.4018/978-1-5225-0937-0.ch002

Lawrence, J., Simpson, L., & Piggott, A. (2017). Protected Agriculture: A Climate Change Adaptation for Food and Nutrition Security. In I. Management Association (Ed.), *Natural Resources Management: Concepts, Methodologies, Tools, and Applications* (pp. 140-158). Hershey, PA: IGI Global. 10.4018/978-1-5225-0803-8.ch007

Leadbetter, A., Cheatham, M., Shepherd, A., & Thomas, R. (2017). Linked Ocean Data 2.0. In Diviacco, P., Leadbetter, A., & Glaves, H. (Eds.), *Oceanographic and Marine Cross-Domain Data Management for Sustainable Development* (pp. 69–99). Hershey, PA: IGI Global. 10.4018/978-1-5225-0700-0.ch004

Lucas, M. R., Rego, C., Vieira, C., & Vieira, I. (2017). Proximity and Cooperation for Innovative Regional Development: The Case of the Science and Technology Park of Alentejo. In Carvalho, L. (Ed.), *Handbook of Research on Entrepreneurial Development and Innovation Within Smart Cities* (pp. 199–228). Hershey, PA: IGI Global. 10.4018/978-1-5225-1978-2.ch010

Ma, X., Beaulieu, S. E., Fu, L., Fox, P., Di Stefano, M., & West, P. (2017). Documenting Provenance for Reproducible Marine Ecosystem Assessment in Open Science. In Diviacco, P., Leadbetter, A., & Glaves, H. (Eds.), *Oceanographic and Marine Cross-Domain Data Management for Sustainable Development* (pp. 100–126). Hershey, PA: IGI Global. 10.4018/978-1-5225-0700-0.ch005

Mabe, L. K., & Oladele, O. I. (2017). Application of Information Communication Technologies for Agricultural Development through Extension Services: A Review. In Tossy, T. (Ed.), *Information Technology Integration for Socio-Economic Development* (pp. 52–101). Hershey, PA: IGI Global. 10.4018/978-1-5225-0539-6.ch003

Malomo, B. I. (2018). A Review of Psychological Resilience as a Response to Natural Hazards in Nigeria. In Eneanya, A. (Ed.), *Handbook of Research on Environmental Policies for Emergency Management and Public Safety* (pp. 147–165). Hershey, PA: IGI Global. 10.4018/978-1-5225-3194-4.ch008

Manchiraju, S. (2018). Predicting Behavioral Intentions Toward Sustainable Fashion Consumption: A Comparison of Attitude-Behavior and Value-Behavior Consistency Models. In I. Management Association (Ed.), *Fashion and Textiles: Breakthroughs in Research and Practice* (pp. 1-21). Hershey, PA: IGI Global. 10.4018/978-1-5225-3432-7.ch001

Manzella, G. M., Bartolini, R., Bustaffa, F., D'Angelo, P., De Mattei, M., Frontini, F., & Spada, A. (2017). Semantic Search Engine for Data Management and Sustainable Development: Marine Planning Service Platform. In Diviacco, P., Leadbetter, A., & Glaves, H. (Eds.), *Oceanographic and Marine Cross-Domain Data Management for Sustainable Development* (pp. 127–154). Hershey, PA: IGI Global. 10.4018/978-1-5225-0700-0.ch006

Mbonigaba, J. (2018). Comparing the Effects of Unsustainable Production and Consumption of Food on Health and Policy Across Developed and Less Developed Countries. In Obayelu, A. (Ed.), *Food Systems Sustainability and Environmental Policies in Modern Economies* (pp. 124–158). Hershey, PA: IGI Global. 10.4018/978-1-5225-3631-4.ch006

McKeown, A. E. (2017). Nurses, Healthcare, and Environmental Pollution and Solutions: Breaking the Cycle of Harm. In I. Management Association (Ed.), *Natural Resources Management: Concepts, Methodologies, Tools, and Applications* (pp. 392-415). Hershey, PA: IGI Global. https://doi.org/10.4018/978-1-5225-0803 -8.ch020

Mili, B., Barua, A., & Katyaini, S. (2017). Climate Change and Adaptation through the Lens of Capability Approach: A Case Study from Darjeeling, Eastern Himalaya. In I. Management Association (Ed.), *Natural Resources Management: Concepts, Methodologies, Tools, and Applications* (pp. 1351-1365). Hershey, PA: IGI Global. https://doi.org/10.4018/978-1-5225-0803-8.ch064

Mir, S. A., Shah, M. A., Mir, M. M., & Iqbal, U. (2017). New Horizons of Nanotechnology in Agriculture and Food Processing Industry. In Nayak, B., Nanda, A., & Bhat, M. (Eds.), *Integrating Biologically-Inspired Nanotechnology into Medical Practice* (pp. 230–258). Hershey, PA: IGI Global. 10.4018/978-1-5225-0610-2.ch009

Moallem, M., Sterrett, W. L., Gordon, C. R., Sukhera, S. M., Mahmood, A., & Bashir, A. (2021). An Investigation of the Effects of Integrating Computing and Project- or Problem-Based Learning in the Context of Environmental Sciences: A Case of Pakistani STEM Teachers. In Schroth, S., & Daniels, J. (Eds.), *Building STEM Skills Through Environmental Education* (pp. 49–89). IGI Global. https://doi.org/10.4018/978-1-7998-2711-5.ch003

Mujere, N., & Moyce, W. (2017). Climate Change Impacts on Surface Water Quality. In Ganpat, W., & Isaac, W. (Eds.), *Environmental Sustainability and Climate Change Adaptation Strategies* (pp. 322–340). Hershey, PA: IGI Global. 10.4018/978-1-5225-1607-1.ch012

Mukherjee, A. B., Krishna, A. P., & Patel, N. (2018). Geospatial Technology for Urban Sciences. In *Geospatial Technologies in Urban System Development: Emerging Research and Opportunities* (pp. 99–120). Hershey, PA: IGI Global. 10.4018/978-1-5225-3683-3.ch005

Nagarajan, S. K., & Sangaiah, A. K. (2017). Vegetation Index: Ideas, Methods, Influences, and Trends. In Kumar, N., Sangaiah, A., Arun, M., & Anand, S. (Eds.), *Advanced Image Processing Techniques and Applications* (pp. 347–386). Hershey, PA: IGI Global. 10.4018/978-1-5225-2053-5.ch016

Naraine, L., & Meehan, K. (2017). Strengthening Food Security with Sustainable Practices by Smallholder Farmers in Lesser Developed Economies. In Ganpat, W., Dyer, R., & Isaac, W. (Eds.), *Agricultural Development and Food Security in Developing Nations* (pp. 57–81). Hershey, PA: IGI Global. 10.4018/978-1-5225-0942-4.ch003

Naraine, L., & Meehan, K. (2017). Strengthening Food Security with Sustainable Practices by Smallholder Farmers in Lesser Developed Economies. In Ganpat, W., Dyer, R., & Isaac, W. (Eds.), *Agricultural Development and Food Security in Developing Nations* (pp. 57–81). Hershey, PA: IGI Global. 10.4018/978-1-5225-0942-4.ch003

Naraine, L., & Meehan, K. (2017). Strengthening Food Security with Sustainable Practices by Smallholder Farmers in Lesser Developed Economies. In Ganpat, W., Dyer, R., & Isaac, W. (Eds.), *Agricultural Development and Food Security in Developing Nations* (pp. 57–81). Hershey, PA: IGI Global. 10.4018/978-1-5225-0942-4.ch003

Nikolaou, K., Tsakiridou, E., Anastasiadis, F., & Mattas, K. (2017). Exploring Alternative Distribution Channels of Agricultural Products. *International Journal of Food and Beverage Manufacturing and Business Models*, 2(2), 36–66. 10.4018/IJFBMBM.2017070103

Nishat, K. J., & Rahman, M. S. (2018). Disaster, Vulnerability, and Violence Against Women: Global Findings and a Research Agenda for Bangladesh. In Mahtab, N., Haque, T., Khan, I., Islam, M., & Wahid, I. (Eds.), *Handbook of Research on Women's Issues and Rights in the Developing World* (pp. 235–250). Hershey, PA: IGI Global. 10.4018/978-1-5225-3018-3.ch014

O'Hara, S., Jones, D., & Trobman, H. B. (2018). Building an Urban Food System Through UDC Food Hubs. In Burtin, A., Fleming, J., & Hampton-Garland, P. (Eds.), *Changing Urban Landscapes Through Public Higher Education* (pp. 116–143). Hershey, PA: IGI Global. 10.4018/978-1-5225-3454-9.ch006

Obayelu, A. E. (2018). Integrating Environment, Food Systems, and Sustainability in Feeding the Growing Population in Developing Countries. In Obayelu, A. (Ed.), *Food Systems Sustainability and Environmental Policies in Modern Economies* (pp. 1–14). Hershey, PA: IGI Global. 10.4018/978-1-5225-3631-4.ch001

Othman, R., Nath, N., & Laswad, F. (2018). Environmental Reporting and Accounting: Sustainability Hybridisation. In Azevedo, G., da Silva Oliveira, J., Marques, R., & Ferreira, A. (Eds.), *Handbook of Research on Modernization and Accountability in Public Sector Management* (pp. 130–158). Hershey, PA: IGI Global. 10.4018/978-1-5225-3731-1.ch007

Ouadi, A., & Zitouni, A. (2021). Phasor Measurement Improvement Using Digital Filter in a Smart Grid. In Recioui, A., & Bentarzi, H. (Eds.), *Optimizing and Measuring Smart Grid Operation and Control* (pp. 100–117). IGI Global. https://doi .org/10.4018/978-1-7998-4027-5.ch005

Padigala, B. S. (2018). Traditional Water Management System for Climate Change Adaptation in Mountain Ecosystems. In I. Management Association (Ed.), *Climate Change and Environmental Concerns: Breakthroughs in Research and Practice* (pp. 630-655). Hershey, PA: IGI Global. 10.4018/978-1-5225-5487-5.ch033

Panda, C. K. (2018). Mobile Phone Usage in Agricultural Extension in India: The Current and Future Perspective. In Mtenzi, F., Oreku, G., Lupiana, D., & Yonazi, J. (Eds.), *Mobile Technologies and Socio-Economic Development in Emerging Nations* (pp. 1–21). Hershey, PA: IGI Global. 10.4018/978-1-5225-4029-8.ch001

Pandian, S. L., Yarrakula, K., & Chaudhury, P. (2018). GIS-Based Decision Support System for Village Level: A Case Study in Andhra Pradesh. In Chaudhuri, S., & Ray, N. (Eds.), *GIS Applications in the Tourism and Hospitality Industry* (pp. 275–295). Hershey, PA: IGI Global. 10.4018/978-1-5225-5088-4.ch012

Quaranta, G., & Salvia, R. (2017). Social-Based Product Innovation and Governance in The Milk Sector: The Case of Carciocacio and Innonatura. In Tarnanidis, T., Vlachopoulou, M., & Papathanasiou, J. (Eds.), *Driving Agribusiness With Technology Innovations* (pp. 293–310). Hershey, PA: IGI Global. 10.4018/978-1-5225-2107-5. ch015

Rahman, M. K., Schmidlin, T. W., Munro-Stasiuk, M. J., & Curtis, A. (2017). Geospatial Analysis of Land Loss, Land Cover Change, and Landuse Patterns of Kutubdia Island, Bangladesh. *International Journal of Applied Geospatial Research*, 8(2), 45–60. 10.4018/IJAGR.2017040104

Rajack-Talley, T. A. (2017). Agriculture, Trade Liberalization and Poverty in the ACP Countries. In Ganpat, W., Dyer, R., & Isaac, W. (Eds.), *Agricultural Development and Food Security in Developing Nations* (pp. 1–20). Hershey, PA: IGI Global. 10.4018/978-1-5225-0942-4.ch001

Rajamanickam, S. (2018). Exploring Landscapes in Regional Convergence: Environment and Sustainable Development in South Asia. In I. Management Association (Ed.), *Sustainable Development: Concepts, Methodologies, Tools, and Applications* (pp. 1051-1087). Hershey, PA: IGI Global. https://doi.org/10.4018/978-1-5225-3817-2.ch047

Rizvi, S. M., & Dearden, A. (2018). KHETI: ICT Solution for Agriculture Extension and Its Replication in Open and Distance Learning. In Pandey, U., & Indrakanti, V. (Eds.), *Open and Distance Learning Initiatives for Sustainable Development* (pp. 163–174). Hershey, PA: IGI Global. 10.4018/978-1-5225-2621-6.ch008

Roşu, L., & Macarov, L. I. (2017). Management of Drought and Floods in the Dobrogea Region. In I. Management Association (Ed.), *Agri-Food Supply Chain Management: Breakthroughs in Research and Practice* (pp. 372-403). Hershey, PA: IGI Global. https://doi.org/10.4018/978-1-5225-1629-3.ch016

Rouzbehani, K., & Rouzbehani, S. (2018). Mapping Women's World: GIS and the Case of Breast Cancer in the US. *International Journal of Public Health Management and Ethics*, 3(1), 14–25. 10.4018/IJPHME.2018010102

Roy, D. (2018). Success Factors of Adoption of Mobile Applications in Rural India: Effect of Service Characteristics on Conceptual Model. In M. Khosrow-Pour, D.B.A. (Ed.), *Green Computing Strategies for Competitive Advantage and Business Sustainability* (pp. 211-238). Hershey, PA: IGI Global. https://doi.org/10.4018/978-1-5225-5017-4.ch010

Sajeva, M., Lemon, M., & Sahota, P. S. (2017). Governance for Food Security: A Framework for Social Learning and Scenario Building. *International Journal of Food and Beverage Manufacturing and Business Models*, 2(2), 67–84. 10.4018/IJFBMBM.2017070104

Sambhanthan, A., & Potdar, V. (2017). A Study of the Parameters Impacting Sustainability in Information Technology Organizations. *International Journal of Knowledge-Based Organizations*, 7(3), 27–39. 10.4018/IJKBO.2017070103

Sanga, C., Kalungwizi, V. J., & Msuya, C. P. (2017). Bridging Gender Gaps in Provision of Agricultural Extension Service Using ICT: Experiences from Sokoine University of Agriculture (SUA) Farmer Voice Radio (FVR) Project in Tanzania. In I. Management Association (Ed.), *Discrimination and Diversity: Concepts, Methodologies, Tools, and Applications* (pp. 682-697). Hershey, PA: IGI Global. https://doi.org/10.4018/978-1-5225-1933-1.ch031

Santra, A., & Mitra, D. (2017). Role of Remote Sensing in Potential Fishing Zone Forecast. In Santra, A., & Mitra, S. (Eds.), *Remote Sensing Techniques and GIS Applications in Earth and Environmental Studies* (pp. 243–257). Hershey, PA: IGI Global. 10.4018/978-1-5225-1814-3.ch012

Schaap, D. (2017). SeaDataNet: Towards a Pan-European Infrastructure for Marine and Ocean Data Management. In Diviacco, P., Leadbetter, A., & Glaves, H. (Eds.), *Oceanographic and Marine Cross-Domain Data Management for Sustainable Development* (pp. 155–177). Hershey, PA: IGI Global. 10.4018/978-1-5225-0700-0.ch007

Seckin-Celik, T. (2017). Sustainability Reporting and Sustainability in the Turkish Business Context. In Akkucuk, U. (Ed.), *Ethics and Sustainability in Global Supply Chain Management* (pp. 115–132). Hershey, PA: IGI Global. 10.4018/978-1-5225-2036-8.ch006

Segbefia, A. Y., Barnes, V. R., Akpalu, L. A., & Mensah, M. (2018). Environmental Location Assessment for Seaweed Cultivation in Ghana: A Spatial Multi-Criteria Approach. *International Journal of Applied Geospatial Research*, 9(1), 51–64. 10.4018/IJAGR.2018010104

Sen, Y. (2018). How to Manage Sustainability: A Framework for Corporate Sustainability Tools. In I. Management Association (Ed.), *Sustainable Development: Concepts, Methodologies, Tools, and Applications* (pp. 568-589). Hershey, PA: IGI Global. 10.4018/978-1-5225-3817-2.ch026

Shamshiry, E., Abdulai, A. M., Mokhtar, M. B., & Komoo, I. (2015). Regional Landfill Site Selection with GIS and Analytical Hierarchy Process Techniques: A Case Study of Langkawi Island, Malaysia. In Thomas, P., Srihari, M., & Kaur, S. (Eds.), *Handbook of Research on Cultural and Economic Impacts of the Information Society* (pp. 248–282). Hershey, PA: IGI Global. 10.4018/978-1-4666-8598-7.ch011

Sharma, Y. K., Mangla, S. K., Patil, P. P., & Uniyal, S. (2018). Analyzing Sustainable Food Supply Chain Management Challenges in India. In Ram, M., & Davim, J. (Eds.), *Soft Computing Techniques and Applications in Mechanical Engineering* (pp. 162–180). Hershey, PA: IGI Global. 10.4018/978-1-5225-3035-0.ch008

Silvestrelli, P. (2018). The Impact of Events: To Which Extent Are They Sustainable for Tourist Destinations? Some Evidences From Expo Milano 2015. In Risso, M., & Testarmata, S. (Eds.), *Value Sharing for Sustainable and Inclusive Development* (pp. 185–204). Hershey, PA: IGI Global. 10.4018/978-1-5225-3147-0.ch008

Silvius, G. (2018). Sustainability Evaluation of IT/IS Projects. In I. Management Association (Ed.), *Sustainable Development: Concepts, Methodologies, Tools, and Applications* (pp. 26-40). Hershey, PA: IGI Global. https://doi.org/10.4018/978-1 -5225-3817-2.ch002

Singh, R., Srivastava, P., Singh, P., Upadhyay, S., & Raghubanshi, A. S. (2017). Human Overpopulation and Food Security: Challenges for the Agriculture Sustainability. In Singh, R., Singh, A., & Srivastava, V. (Eds.), *Environmental Issues Surrounding Human Overpopulation* (pp. 12–39). Hershey, PA: IGI Global. 10.4018/978-1-5225-1683-5.ch002

Singh, R., Srivastava, P., Singh, P., Upadhyay, S., & Raghubanshi, A. S. (2017). Human Overpopulation and Food Security: Challenges for the Agriculture Sustainability. In Singh, R., Singh, A., & Srivastava, V. (Eds.), *Environmental Issues Surrounding Human Overpopulation* (pp. 12–39). Hershey, PA: IGI Global. 10.4018/978-1-5225-1683-5.ch002

Srivastava, N. (2017). Climate Change Mitigation: Collective Efforts and Responsibly. In I. Management Association (Ed.), *Natural Resources Management: Concepts, Methodologies, Tools, and Applications* (pp. 64-76). Hershey, PA: IGI Global. https://doi.org/10.4018/978-1-5225-0803-8.ch003

Stanganelli, M., & Gerundo, C. (2017). Understanding the Role of Urban Morphology and Green Areas Configuration During Heat Waves. *International Journal of Agricultural and Environmental Information Systems*, 8(2), 50–64. 10.4018/ IJAEIS.2017040104

Stewart, M. K., Hagood, D., & Ching, C. C. (2017). Virtual Games and Real-World Communities: Environments that Constrain and Enable Physical Activity in Games for Health. *International Journal of Game-Based Learning*, 7(1), 1–19. 10.4018/ IJGBL.2017010101

Stone, R. J. (2017). Modelling the Frequency of Tropical Cyclones in the Lower Caribbean Region. In Ganpat, W., & Isaac, W. (Eds.), *Environmental Sustainability and Climate Change Adaptation Strategies* (pp. 341–349). Hershey, PA: IGI Global. 10.4018/978-1-5225-1607-1.ch013

Syed, A., & Jabeen, U. A. (2018). Climate Change Impact on Agriculture and Food Security. In Eneanya, A. (Ed.), *Handbook of Research on Environmental Policies for Emergency Management and Public Safety* (pp. 223–237). Hershey, PA: IGI Global. 10.4018/978-1-5225-3194-4.ch012

Tam, G. C. (2017). The Global View of Sustainability. In *Managerial Strategies and Green Solutions for Project Sustainability* (pp. 1–24). Hershey, PA: IGI Global. 10.4018/978-1-5225-2371-0.ch001

Tam, G. C. (2017). Understanding Project Sustainability. In *Managerial Strategies and Green Solutions for Project Sustainability* (pp. 110–139). Hershey, PA: IGI Global. 10.4018/978-1-5225-2371-0.ch005

Tam, G. C. (2017). Perspectives on Sustainability. In *Managerial Strategies and Green Solutions for Project Sustainability* (pp. 53–76). Hershey, PA: IGI Global. 10.4018/978-1-5225-2371-0.ch003

Tang, M., & Karunanithi, A. T. (2018). Visual Logic Maps (vLms). In *Advanced Concept Maps in STEM Education: Emerging Research and Opportunities* (pp. 108–149). Hershey, PA: IGI Global. 10.4018/978-1-5225-2184-6.ch005

Taşçıoğlu, M., & Yener, D. (2018). The Value and Scope of GIS in Marketing and Tourism Management. In Chaudhuri, S., & Ray, N. (Eds.), *GIS Applications in the Tourism and Hospitality Industry* (pp. 189–211). Hershey, PA: IGI Global. 10.4018/978-1-5225-5088-4.ch009

Thanh Tung, B. (2021). Pharmacology and Therapeutic Applications of Resveratrol. In Hussain, A., & Behl, S. (Eds.), *Treating Endocrine and Metabolic Disorders With Herbal Medicines* (pp. 321–333). IGI Global. https://doi.org/10.4018/978-1-7998-4808-0.ch014

Tianming, G. (2018). Food Security and Rural Development on Emerging Markets of Northeast Asia: Cases of Chinese North and Russian Far East. In Erokhin, V. (Ed.), *Establishing Food Security and Alternatives to International Trade in Emerging Economies* (pp. 155–176). Hershey, PA: IGI Global. 10.4018/978-1-5225-2733-6.ch008

Tiftikçigil, B. Y., Yaşgül, Y. S., & Güriş, B. (2017). Sustainability of Foreign Trade Deficit in Energy: The Case of Turkey. In Ray, N. (Ed.), *Business Infrastructure for Sustainability in Developing Economies* (pp. 94–109). Hershey, PA: IGI Global. 10.4018/978-1-5225-2041-2.ch005

Tiwari, S., Vaish, B., & Singh, P. (2018). Population and Global Food Security: Issues Related to Climate Change. In I. Management Association (Ed.), *Climate Change and Environmental Concerns: Breakthroughs in Research and Practice* (pp. 41-64). Hershey, PA: IGI Global. https://doi.org/10.4018/978-1-5225-5487-5.ch003

Toujani, A., & Achour, H. (2018). A Data Mining Framework for Forest Fire Mapping. In I. Management Association (Ed.), *Information Retrieval and Management: Concepts, Methodologies, Tools, and Applications* (pp. 771-794). Hershey, PA: IGI Global. 10.4018/978-1-5225-5191-1.ch033

Tripathy, B., & K., S. B. (2017). Rough Fuzzy Set Theory and Neighbourhood Approximation Based Modelling for Spatial Epidemiology. In I. Management Association (Ed.), *Public Health and Welfare: Concepts, Methodologies, Tools, and Applications* (pp. 1257-1268). Hershey, PA: IGI Global. https://doi.org/10.4018/978-1-5225-1674-3.ch058

Trukhachev, A. (2017). New Approaches to Regional Branding through Green Production and Utilization of Existing Natural Advantages. In I. Management Association (Ed.), *Advertising and Branding: Concepts, Methodologies, Tools, and Applications* (pp. 1758-1778). Hershey, PA: IGI Global. 10.4018/978-1-5225-1793-1.ch081

Tsobanoglou, G. O., & Vlachopoulou, E. I. (2017). Social-Ecological Systems in Local Fisheries Communities. In Korres, G., Kourliouros, E., & Michailidis, M. (Eds.), *Handbook of Research on Policies and Practices for Sustainable Economic Growth and Regional Development* (pp. 306–316). Hershey, PA: IGI Global. 10.4018/978-1-5225-2458-8.ch026

Tuydes-Yaman, H., & Karatas, P. (2018). Evaluation of Walkability and Pedestrian Level of Service. In I. Management Association (Ed.), *Intelligent Transportation and Planning: Breakthroughs in Research and Practice* (pp. 264-291). Hershey, PA: IGI Global. 10.4018/978-1-5225-5210-9.ch012

Uddin, S., Chakravorty, S., Ray, A., & Sherpa, K. S. (2018). Optimal Location of Sub-Station Using Q-GIS and Multi-Criteria Decision Making Approach. *International Journal of Decision Support System Technology*, 10(2), 65–79. 10.4018/IJDSST.2018040104

Uddin, S., Chakravorty, S., Sherpa, K. S., & Ray, A. (2018). Power Distribution System Planning Using Q-GIS. *International Journal of Energy Optimization and Engineering*, 7(2), 61–75. 10.4018/IJEOE.2018040103

Uzun, F. V. (2018). Natural Resources Management. In Eneanya, A. (Ed.), *Handbook of Research on Environmental Policies for Emergency Management and Public Safety* (pp. 1–21). Hershey, PA: IGI Global. 10.4018/978-1-5225-3194-4.ch001

V, M., Agrawal, R., Sharma, V., & T.N., K. (2018). Supply Chain Social Sustainability and Manufacturing. In I. Management Association (Ed.), *Technology Adoption and Social Issues: Concepts, Methodologies, Tools, and Applications* (pp. 226-252). Hershey, PA: IGI Global. https://doi.org/10.4018/978-1-5225-5201-7.ch011

van der Vliet-Bakker, J. M. (2017). Environmentally Forced Migration and Human Rights. In Akrivopoulou, C. (Ed.), *Defending Human Rights and Democracy in the Era of Globalization* (pp. 146–180). Hershey, PA: IGI Global. 10.4018/978-1-5225-0723-9.ch007

Vargas-Hernández, J. G. (2022). A Comprehensive Entrepreneurship Model for the Internationalization of Green Innovation Businesses. In Akkucuk, U. (Ed.), *Disruptive Technologies and Eco-Innovation for Sustainable Development* (pp. 131–149). IGI Global. https://doi.org/10.4018/978-1-7998-8900-7.ch008

Vaskov, A. G., Lin, Z. Y., Tyagunov, M. G., Shestopalova, T. A., & Deryugina, G. V. (2018). Design of Renewable Sources GIS for ASEAN Countries. In Kharchenko, V., & Vasant, P. (Eds.), *Handbook of Research on Renewable Energy and Electric Resources for Sustainable Rural Development* (pp. 1–25). Hershey, PA: IGI Global. 10.4018/978-1-5225-3867-7.ch001

Vázquez, D. G., & Gil, M. T. (2017). Sustainability in Smart Cities: The Case of Vitoria-Gasteiz (Spain) – A Commitment to a New Urban Paradigm. In Carvalho, L. (Ed.), *Handbook of Research on Entrepreneurial Development and Innovation Within Smart Cities* (pp. 248–268). Hershey, PA: IGI Global. 10.4018/978-1-5225-1978-2.ch012

Wahab, I. N., & Soonthodu, S. (2018). Geographical Information System in Eco-Tourism. In Chaudhuri, S., & Ray, N. (Eds.), *GIS Applications in the Tourism and Hospitality Industry* (pp. 61–75). Hershey, PA: IGI Global. 10.4018/978-1-5225-5088-4.ch003

Weiss-Randall, D. (2018). Cultivating Environmental Justice. In *Utilizing Innovative Technologies to Address the Public Health Impact of Climate Change: Emerging Research and Opportunities* (pp. 110–143). Hershey, PA: IGI Global. 10.4018/978-1-5225-3414-3.ch004

Weiss-Randall, D. (2018). Cultivating Resilience. In *Utilizing Innovative Technologies to Address the Public Health Impact of Climate Change: Emerging Research and Opportunities* (pp. 204–235). Hershey, PA: IGI Global. 10.4018/978-1-5225-3414-3.ch007

Weiss-Randall, D. (2018). Climate Change Solutions: Where Do We Go From Here? In *Utilizing Innovative Technologies to Address the Public Health Impact of Climate Change: Emerging Research and Opportunities* (pp. 236–268). Hershey, PA: IGI Global. 10.4018/978-1-5225-3414-3.ch008

Whyte, K. P., List, M., Stone, J. V., Grooms, D., Gasteyer, S., Thompson, P. B., . . . Bouri, H. (2018). Uberveillance, Standards, and Anticipation: A Case Study on Nanobiosensors in U.S. Cattle. In I. Management Association (Ed.), *Biomedical Engineering: Concepts, Methodologies, Tools, and Applications* (pp. 577-596). Hershey, PA: IGI Global. https://doi.org/10.4018/978-1-5225-3158-6.ch025

Wulff, E. (2017). Data and Operational Oceanography: A Review in Support of Responsible Fisheries and Aquaculture. In Diviacco, P., Leadbetter, A., & Glaves, H. (Eds.), *Oceanographic and Marine Cross-Domain Data Management for Sustainable Development* (pp. 303–324). Hershey, PA: IGI Global. 10.4018/978-1-5225-0700-0.ch013

Yener, D. (2017). Geographic Information Systems and Its Applications in Marketing Literature. In Faiz, S., & Mahmoudi, K. (Eds.), *Handbook of Research on Geographic Information Systems Applications and Advancements* (pp. 158–172). Hershey, PA: IGI Global. 10.4018/978-1-5225-0937-0.ch006

Yu, K., Liu, Y., & Sharma, A. (2021). Analyze the Effectiveness of the Algorithm for Agricultural Product Delivery Vehicle Routing Problem Based on Mathematical Model. *International Journal of Agricultural and Environmental Information Systems*, 12(3), 26–38. https://doi.org/10.4018/IJAEIS.2021070103

Zhou, X., Sharma, A., & Mohindru, V. (2021). Research on Linear Programming Algorithm for Mathematical Model of Agricultural Machinery Allocation. *International Journal of Agricultural and Environmental Information Systems*, 12(3), 1–12. https://doi.org/10.4018/IJAEIS.2021070101

About the Contributors

Idris Ganiyu is an interdisciplinary researcher with experience that spans the banking and educational sectors. He has a PhD in Human Resource Management from the prestigious University of KwaZulu-Natal in South Africa. Dr Ganiyu has published widely in indexed Journals and has presented his research papers at various Universities across the world including Harvard University. As an avid researcher, Dr Ganiyu is an external examiner for many Universities. He is regularly invited to facilitate workshops on primary data analysis and the various tools for analysing quantitative and qualitative data. He has supervised postgraduate students in Nigeria and South Africa. He is currently supervising Doctoral Students at the Management College of South Africa. Dr Ganiyu is currently a Senior Lecturer in Human Resource Management at York St John University in the United Kingdom.

Odunayo Magret Olarewaju is a distinguished Author, an award-winning researcher and an Ass. P in Accounting, with special focus on financial management, financial econometrics, Environmental Accounting, Managerial Accounting and finance.

Adejoke Ige-Olaobaju obtained her PhD in Work and Employment Relations from the university of Leeds, United Kingdom (UK). She is currently a Senior Lecturer in Learning and Teaching within the Human Resource Management & Organisational Behaviour subject group at the University of Northampton. She is a Module leader, Personal Academic Tutor and Dissertation supervisor on Postgraduate (PG) and Undergraduate (UG) programmes at the university. Ade is a Fellow of the Higher Education Academy, UK. She is also an Academic Member, Chartered Institute of Personnel and Development, Member, British University Industrial Relations Association and Member, British Academy of Management. Ade presented a guest lecture on the topic: Revitalizing Talent Management: The way Forward at Kyrgyz Economic University, Kyrgyzstan in 2021. She provided academic consultancy support during the development of HRMI (UG) programme for one of the universities of Northampton partners outside the UK.

Sulaiman Olusegun Atiku is a Professor of Human Resources and Director of Research at Harold Pupkewitz Graduate School of Business (HP-GSB), Namibia University of Science and Technology, Namibia. He is a pragmatic researcher specializing in Strategic Human Resource Management. His current research area of interest includes Human Capital Formation for Industry 5.0, and People Management Practices for Promoting Sustainability. He has over 16 years' experience in Higher Education. A native of Lagos, Sulaiman graduated from the University of KwaZulu-Natal with a PhD degree in Human Resource Management. His Master of Science degree was awarded in Human Resources and Industrial Relations at Lagos State University, Nigeria. His Bachelor of Science (Honours) degree was also awarded at Lagos State University in the field of Industrial Relations and Personnel Management. He has lectured several courses in his field home/abroad and published many scholarly articles in international journals. He is a member of International Labour and Employment Relations Association (ILERA), Nigerian Institute of Management (NIM), and Institute of People Management (IPM) South Africa.

***

About the Contributors

Omolola A. Arise is a distinguished academic within the Faculty of Accounting, Finance, and Tax Administration at MANCOSA, boasting over nine years of experience in academia. She has recently earned her Ph.D. in Accounting from KwaZulu-Natal University, specializing in Environmental Management Accounting and Sustainability, Climate Change, and Financial Accounting. Dr. Arise's prior academic appointment was as a lecturer in the Management Accounting Department at the Durban University of Technology (DUT), where she also received her Master of Accounting degree, graduating Cum laude and receiving the Dean's Merit Award. Her professional journey includes serving in the Grants Department of the Research Directorate at DUT, transitioning from administrative to academic roles. Dr. Arise is a prolific researcher, with several publications in accredited journals and participation in conferences at both the national and international levels, underscoring her commitment to the field of accounting and its intersection with environmental sustainability.

Elizabeth Babafemi is a senior lecturer in Human Resource Management (HRM) and have been actively teaching since 2017. Elizabeth completed my PhD in 2022 with a qualitative study focusing on women finance directors in the UK, shedding light on the challenges and experiences unique to this demographic in a corporate setting. Before her academic career, she established and operated an HR consultancy business, catering to Small and Medium Enterprises (SMEs) in Hertfordshire and London, providing assistance on various HRM-related issues and fostering people development. Her research endeavours centre around several key areas within the field, encompassing HRM, diversity management, equality, diversity, and inclusion practices. She is particularly interested in topics such as women leaders, women on boards, and women's careers, addressing issues like the gender pay gap and income inequalities. Furthermore, my research extends to areas like reward systems, performance management, strategic leadership, and employment relations.

Ana Batlles-delaFuente holds a Ph.D. from the Doctoral Program in Economic, Business, and Legal Sciences at the University of Almería. She graduated in Business Administration and Management and completed two Master's degrees in Business Economics and Teaching. Currently, she is actively engaged in research in the field of Applied Economics, with a particular focus on aspects related to the circular economy and sustainable production.

Malvern Chiboiwa is a university lecturer and holds a PhD in Human Resource Management. He has published research on different aspects of Human Resource Management. He has published research on HRM concepts such as job satisfaction, organisational citizenship behaviour, employee turnover, employee engagement, and employee innovation. Malvern has also reviewed published work on Green Marketing. His present focus is on how human resource management can contribute to environmental sustainability.

Vanessa Gaffar is professor in Management Study Program, Faculty of Economics and Business Education, Universitas Pendidikan Indonesia and currently a vice dean for academic affairs in the faculty. She had her bachelor degree in Accounting from Universitas Padjadjaran, Bandung Indonesia and hold her MBA degree form Wright State University, Dayton Ohio, USA. She completed her doctoral degree form Universitas Padjadjaran, Bandung Indonesia. Her research interests are in the field of customer relationship management, social media, tourism marketing, sports tourism, and entrepreneurship. Her works have been published in national and reputable international journals, conference proceedings as well as books. She is also an active reviewer for reputable international journals and national accredited journals. She has been doing international collaborations with research partners from Malaysia, Thailand, The Philippines, Spain, The Netherlands, USA, and United Kingdom. She spent her two-year time as a research fellow at the Centre for Trust, Peace and Social Relations, Coventry University, UK.

Saman Hassanzadeh Amin is an associate professor at the department of Mechanical, Industrial and, Mechatronics Engineering, Toronto Metropolitan University.

Manohar Kapse is presently associated with Symbiosis International University, SCMHRD Pune as Assistant Professor in Business Analytics. He had been awarded PhD in Management from Institute of Management Studies, Devi Ahilya Vishwavidyalaya Indore in 2013. He had also done Master Degree in Statistics and M. Phil in Statistics.He teaches Machine Learning, Statistics, Multivariate Analysis, R, SPSS, Research Methodology and Quantitative Techniques. He had been part of many workshops on Machine Learning, SPSS and R as Resource Person in many renowned Institutes, Such as AIIMS Bhopal, Medi-caps University, Jagran Lakecity University, SVIM Indore, APEX Institute of Management, etc. Other than Teaching he is also involved in statistical consultancy to researchers working in different Sciences. He had published more than 20 research papers, many cases and one books.

Tika Koeswandi is a lecturer in Entrepreneurship Study Program and a doctoral candidate in Doctor of Management Study Program, Faculty of Economics and Business Education, Universitas Pendidikan Indonesia. Currently, she is receiving doctoral scholarship from Beasiswa Pendidikan Indonesia under Ministry of Education and Culture Republic of Indonesia. Her research in interests are in the field of Entrepreneurial Marketing, especially Micro, Small and Medium Enterprises (MSMEs), including global marketing, strategic marketing, and marketing communications. Since 2019, she has published seven ISBN books and a scopus-indexed book chapter. She also owns 15 intellectual properties. In 2023, in collaboration with Prof. Vanessa Gaffar & Annisa Ciptagustia, she created an Android-based application called AVANTI 1.0 that aims to help MSMEs in creating Social CRM. She is also a trainer and mentor for MSMEs in the Digital Entrepreneurship Academy and Talents Scouting Academy program under Ministry of Communications and Informatics of Indonesia since 2021. She is very interested in holding international collaborations with global partners in teaching, publications and other fields.

Nyikiwa Mavunda is a part-time Lecturer teaching Human Resource Management at the University of KwaZulu-Natal. She earned her PhD in Human Resource Management from the University of KwaZulu-Natal, where her research study was based on digital transformation and HR practices. She has over a decade of diverse experience in the Human Resources (HR) field and has developed a comprehensive skill set that spans various facets of HR management. Her career includes working for the Department of Education and the University, allowing her to adapt to different organisational cultures and needs. The key areas of her expertise include Recruitment, Employee Engagement, Performance Management, Training and Development, HR Policy and Compliance, and Organisational Development. Her primary research interests include HR Digital Transformation, Change Management, Human Resource Information Systems, Organisational behaviour, Talent Management, Training and Development, and Strategic Human Resource Management. Her teaching is aimed at creating an inclusive and dynamic learning environment. She believes in the importance of adapting teaching methods to meet diverse student needs and incorporating innovative technologies to enhance learning experiences. Her diverse experiences across different educational settings have equipped her with a broad skill set and a deep understanding of effective teaching practices.

Sybert Mutereko is the director of the Center for Postgraduate Studies. He holds master's and doctoral degrees in Policy and Development Studies. Sybert's research interests are in public administration, local government, public policy, policy implementation and evaluation, education policy, education management and vocational education. His current research focuses on the supply and demand dynamics of the public sector labour market.

About the Contributors

Meshel Muzuva holds a PhD in Management Sciences from Durban University of Technology. She also holds a Master of Commerce in Finance from the University of KwaZulu Natal and a Bachelor of Commerce Honours Degree in Banking from the National University of Science and Technology. Currently, she works as a Lecturer at MANCOSA Department of Accounting, Tax, and Finance since 2019. Dr. Muzuva has a strong passion for economics and finance. Her academic contributions include teaching undergraduate and postgraduate courses such as microeconomics, macroeconomics, finance for managers, corporate finance, and international trade theory. She has also supervised postgraduate students' dissertations. Beyond teaching, Dr. Muzuva is dedicated to advancing the field through rigorous research and collaboration. She has significantly contributed to academic discourse by publishing articles in economics and finance. Additionally, she serves as a reviewer for the African Journal of Business and Economic Research, contributing to scholarly discussions in the field. And she is an Internal Moderator for Doctor of Business Administration (DBA). Dr. Muzuva actively participates in webinars and international conferences, sharing insights and findings with a global audience. Her journey is characterized by a relentless pursuit of knowledge dissemination and academic excellence, shaping minds, and contributing to the growth of the academic community. Through her dedication to teaching, research, and collaboration, Dr. Muzuva continues to inspire and drive progress in the field of economics and finance.

Precious Okunhon (Ph.D) is a Lecturer at the University of Northampton, she teaches and supervises various courses within the Department of Human Resource Management & Leadership. Her research area encompasses areas which includes Human Resource Management (HRM), organisational culture, organisational climate, Business Ethics, Green HRM, and Environmental sustainability. Dr. Precious characterizes herself as an inclusive educator, employing a teaching methodology that fosters active student participation in acquiring the necessary knowledge for their professional growth. Her holistic approach to teaching extends beyond imparting educational content, emphasizing mentorship and assisting students in cultivating transferable skills to navigate future challenges and opportunities.

Odunayo Olarewaju obtained her first and second degree in Accounting from Obafemi Awolowo University, Nigeria with cum laude and she obtained her PhD in Accounting from University of KwaZulu-Natal, South Africa.She has held faculty positions in colleges in South Africa and University in Nigeria before relocating to South Africa. As an accomplished researcher with several articles published in DHET accredited peer-reviewed journals to her credit, she is a Board Member and Ed. She is a freelance training facilitator at SASH Business Management for training and consulting, Durban, South Africa (http://sashbm.com/#facilitators). She has also presented academic papers at accredited international conferences in United States of America, Canada, United Kingdom, South Africa, Nigeria and United Arab Emirate amongst many others. Odunayo is a distinguished scholar holding the record of the Fastest Doctorate Degree Completion in Africa (source https://temmybalogun.com/how-nigerias -olarewaju-odunayo-completed-her-phd-in23-months/) alongside other extraordinary academic laurels such as; One of the 16th Powerful women making Impact at DUT published in Mail and Guardian and The Mercury on the 28th and 31st August, 2020 respectively. Durban University of Technology, 14th position out of the top 30 Researchers of the year. November 22, 2019. Faculty Top Researcher of the year, Faculty of Accounting and Informatics, Durban University of Technology. November 22, 2019. Best Presenter at the second International Conference on Humanities, Accounting and Social sciences, University of Toronto, Canada. July 13th Best full paper presented at Southern African Accounting Association (SAAA), KwaZulu-Natal Regional Teaching and Learning Conference, Riverside Hotel, Durban, South Africa. June 29th Best Graduating Female Accounting Student, 2009/2010 Session, Obafemi Awolowo University, Ile-Ife, Nigeria by Society of Women Accountants of Nigeria (SWAN). March, 2011. Dr. Olarewaju has mentored many masters and honours degree students to successful graduation and still has several masters and doctoral students in Accounting and Finance under her supervision. Dr. Olarewaju is an Accounting Technician and a Chartered Accountant under the Institute of Chartered Accountants of Nigeria (ICAN), Business Accountant in Practice under the Southern African Institute of Business Accountants (BAP(SA), an Associate Chartered Management Accountants (ACMA) and Chartered Global Management Accountant (CGMA). Dr. Olarewaju is a Leading editorial board member of London Jounal Press, United Kingdon. https://journalspress.com/ljp?layout=profile&user=4827&view =profile. An Editorial Board member for International Journal of Business and Economics Research (IJBER), reviewer to nine accredited Journals, reviewer to many local and international conferences and post-graduate examiner for various universities in South Africa and Namibia.

Oladejo Olufemi Michael earned his PhD in Management Studies from the University of KwaZulu-Natal. His doctoral research delved into "Training and Development for Academic Staff in Public Universities in South-West Nigeria". His research interests cut across the fields of Management Science and Public Administration, focusing on Organisational Management and Sustainability, Training and Development, Policy Development, Implementation, and Evaluation. He is proficient in both qualitative and quantitative research methodologies. Currently, he is a Postdoctoral Fellow at the University of KwaZulu-Natal. With a strong commitment to teamwork, goal achievement, and lifelong learning, Dr. Oladejo is devoted to making meaningful contributions to the academic field,

Felicia Oseghale is an administrative officer at the Alvan Ikoku University of Education. She holds an MA in Leadership and Management in Education from the University of South Wales, Newport. Her research interest centres on global talent management in oil and gas sector, green leadership and work-life balance in the academic sector.

Raphael Oseghale is a Senior Lecturer in Human Resource Management at Hertfordshire Business School, University of Hertfordshire. Raphael received his PhD in Business studies (Talent management) from Swansea University. He is a Fellow of Higher Education Academy (FHEA). His research interests centre on applying artificial intelligence to personnel selection and other key aspects of HRM such as leadership/talent development, career management, knowledge transfer and employee wellbeing management. Raphael has secured and participated in over £2.5 million worth of funded research projects. He has published in journals such as Business Strategy and the Environment, Human Resource Development International, Employee Relations. His scholarly work has also been published as chapters in peer-reviewed books.

Naghmeh Rabiei is a PhD student at the department of Mechanical, Industrial, and Mechatronics Engineering, Toronto Metropolitan University.

Sunil Sharma has completed his PhD in Mechanical Engineering from Indian Institute of Technology Ropar, Punjab, INDIA. Currently, he is working as an Associate Professor at Lovely Professional University, Phagwara. He is a graduate in the discipline of Mechanical Engineering from Punjab Technical University (currently I.K. Gujral Punjab Technical University), Kapurthala, Punjab, INDIA and postgraduate in Manufacturing and Materials from Oklahoma State University, Stillwater, USA. His research interests are primarily in the field of biomimicry, sustainability, creativity and design process.

Vinod Sharma, an Associate Professor of Management at Symbiosis Centre for Management and Human Resource Development (SCMHRD) in Pune, boasts 23 years of rich experience spanning academia and industry. His diverse management roles equip him to excel as both a researcher and instructor. Dr. Sharma's expertise lies in Marketing Research & Analytics, Marketing Strategy, and Consumer Behaviour. With over 85 authored papers and 3 completed international projects, he's a prolific academician. Additionally, he has facilitated numerous management development programs for prestigious organizations such as IOC, NIBSCOM, MSME, and FIEO.

Amrik Singh is working as Professor in the School of Hotel Management and Tourism at Lovely Professional University, Punjab, India. He obtained his Ph.D. degree in Hotel Management from Kurukshetra University, Kurukshetra. He started his academic career at Lovely Professional University, Punjab, India in the year 2007. He has published more than 40 research papers in UGC and peer-reviewed and Scopus/Web of Science) journals. He has published 12 patents and 01 patent has been granted in the inter-disciplinary domain. Dr. Amrik Singh participated and acted as a resource person in various national and international conferences, seminars, research workshops, and industry talks. His area of research interest is accommodation management, ergonomics, green practices, human resource management in hospitality, waste management, AR VR in hospitality, etc. He is currently guiding 8Ph.D. scholars and 2 Ph.D. scholars have been awarded Ph.D.

Saeed Zolfaghari is a professor in the department of Mechanical, Industrial, and Mechatronics Engineering at Toronto Metropolitan University

Index

B

Biodiversity 47, 58, 67, 68, 96, 160, 165, 167, 190
biomimicry 100, 104, 110

C

circular economy 4, 5, 6, 17, 47, 48, 49, 50, 54, 66, 67, 68, 69, 70, 81, 83, 88, 89, 90, 97, 102, 104, 105, 152, 167, 172, 178, 183, 184, 188, 198, 214, 222, 232, 233, 235
Climate Change 22, 24, 41, 52, 54, 57, 58, 62, 66, 68, 82, 83, 86, 88, 89, 92, 96, 114, 120, 132, 136, 137, 145, 147, 148, 149, 152, 164, 167, 173, 184, 190, 194, 196, 198, 199, 220, 222, 223, 227
Corporate Sustainability 216
cradle to grave 97, 102, 103, 104

E

ecodesign 100
economic analysis 65, 68, 69, 82, 83, 84
economic perspective 80
Employee 23, 28, 91, 111, 112, 113, 114, 115, 116, 117, 118, 119, 120, 121, 122, 124, 125, 126, 128, 138, 140, 142, 143, 146, 148, 151, 152, 153, 154, 156, 157, 158, 192, 199, 201, 202, 213
Environmental concerns 21, 55, 57, 61, 66, 67
Environmental Management 14, 18, 21, 22, 26, 33, 40, 41, 64, 82, 104, 108, 113, 124, 125, 126, 127, 138, 140, 141, 151, 154, 156, 157, 192, 196, 209, 210
Environmental Responsibility 1, 92, 99, 107, 112, 115, 189, 190, 199, 201, 202, 204, 207, 208
Environmental sustainability 13, 20, 24, 43, 66, 70, 109, 111, 112, 113, 114, 115, 116, 117, 118, 119, 121, 122, 125, 127, 128, 132, 137, 138, 139, 140, 142, 146, 155, 158, 160, 162, 163, 164, 174, 190, 191, 192, 193, 194, 196, 198, 200, 212

G

Ghana 20, 21, 29, 39, 40, 62, 64, 127
green economy 140, 153, 168, 217, 218, 219, 220, 223, 226, 228, 229, 230, 232, 233, 234, 235
Green HRM 115, 124, 125, 126, 127, 128, 129, 130, 139, 153, 154, 155, 157
Green Human Resource Management 111, 112, 113, 121, 124, 125, 126, 127, 128, 129, 130, 131, 137, 147, 151, 152, 153, 154, 155, 156, 157, 158, 213
Green Marketing 110, 159, 161, 162, 163, 168, 169, 170, 171, 172, 173, 179, 180, 181, 182, 183, 184, 185, 186, 187, 188
Green organisational culture 117, 138, 140, 143, 213
Green Organizational Culture 123, 152, 189, 190, 191, 192, 193, 194, 195, 196, 197, 198, 199, 204, 205, 206, 207, 209, 211, 212
Green practices 21, 24, 25, 26, 27, 30, 33, 34, 37, 38, 40, 41, 91, 112, 114, 115, 118, 119, 121, 122, 132, 137, 143, 200, 207
green product 86, 87, 88, 89, 90, 91, 92, 93, 94, 95, 96, 97, 98, 99, 100, 101, 102, 103, 104, 105, 106, 107, 108, 109, 110, 163, 164, 166, 169, 184, 185, 187
Green Product design 86, 87, 88, 89, 90, 91, 92, 93, 94, 95, 96, 97, 98, 99, 100, 101, 102, 103, 104, 106, 107, 108, 109, 110

H

Hotel Industry 20, 21, 31, 41, 184

I

Indonesia 155, 159, 160, 162, 163, 166, 168, 169, 170, 171, 172, 173, 177, 180,

182, 183, 185, 186, 187, 188

L

lca 69, 71, 73, 74, 75, 95, 101, 104, 198
lcca 69, 70, 71, 72, 73, 75, 80, 81, 102
Lifecycle assessment 99
local government 177, 217, 219, 223, 224, 225, 227, 230, 232, 233, 234, 235

M

Management 2, 4, 5, 6, 7, 11, 13, 14, 15, 16, 17, 18, 21, 22, 23, 24, 25, 26, 33, 34, 35, 36, 37, 38, 40, 41, 42, 43, 44, 45, 46, 47, 48, 49, 50, 51, 52, 54, 55, 56, 57, 58, 59, 60, 61, 62, 63, 64, 65, 66, 67, 68, 69, 70, 71, 82, 83, 84, 86, 89, 94, 99, 100, 104, 108, 110, 111, 112, 113, 114, 115, 116, 118, 119, 120, 121, 122, 123, 124, 125, 126, 127, 128, 129, 130, 131, 132, 133, 134, 135, 137, 138, 139, 140, 141, 142, 143, 144, 145, 146, 147, 148, 149, 150, 151, 152, 153, 154, 155, 156, 157, 158, 159, 161, 162, 167, 171, 172, 174, 175, 176, 177, 183, 185, 187, 188, 192, 196, 198, 200, 205, 207, 209, 210, 211, 213, 215, 216, 217, 218, 219, 220, 221, 222, 223, 224, 225, 226, 227, 228, 229, 230, 231, 232, 233, 234, 235, 236
monetary evaluation 69, 80

O

Operations research techniques 1, 3, 8, 9, 15

R

Recycling 2, 5, 6, 8, 9, 12, 13, 15, 18, 19, 21, 23, 24, 25, 33, 34, 35, 36, 43, 44, 45, 46, 47, 48, 49, 50, 52, 53, 54, 55, 57, 58, 59, 60, 61, 63, 83, 85, 87, 90, 92, 99, 100, 102, 104, 105, 112, 114, 122, 142, 152, 163, 170, 175, 186, 198, 200, 217, 219, 220, 221, 222,

223, 224, 226, 228, 229, 230, 234, 235
Returned products 1, 8, 11, 16
Reverse logistics 1, 2, 3, 4, 6, 15, 17, 18, 19, 102, 104

S

SDGs 67, 68, 69, 80, 81, 132, 133, 134, 135, 136, 138, 139, 140, 141, 144, 146, 147, 148, 149, 157, 197, 198, 221
Social responsibility 88, 114, 124, 151, 155, 157, 166, 174, 175, 183, 195, 198, 209, 212, 213, 215, 232
Sustainability 1, 4, 7, 13, 15, 20, 21, 22, 23, 24, 25, 26, 27, 38, 43, 46, 58, 59, 63, 64, 66, 67, 68, 69, 70, 71, 82, 83, 85, 88, 90, 91, 92, 95, 96, 97, 103, 104, 105, 106, 108, 109, 110, 111, 112, 113, 114, 115, 116, 117, 118, 119, 120, 121, 122, 124, 125, 126, 127, 128, 129, 130, 131, 132, 137, 138, 139, 140, 142, 144, 145, 146, 147, 148, 149, 152, 153, 154, 155, 156, 157, 158, 160, 161, 162, 163, 164, 165, 166, 170, 171, 173, 174, 175, 182, 183, 184, 185, 186, 188, 189, 190, 191, 192, 193, 194, 195, 196, 197, 198, 199, 200, 201, 202, 203, 204, 205, 206, 207, 208, 210, 211, 212, 213, 214, 215, 216, 221, 223, 226, 230, 232, 233, 234, 235, 236
sustainability development goals 131
sustainable 6, 12, 14, 17, 18, 20, 21, 22, 23, 24, 25, 26, 27, 28, 33, 35, 37, 38, 39, 40, 42, 43, 44, 45, 46, 47, 48, 49, 50, 51, 52, 54, 55, 58, 59, 62, 63, 64, 65, 66, 67, 68, 69, 70, 71, 72, 79, 80, 81, 82, 83, 84, 85, 86, 87, 88, 89, 90, 91, 92, 95, 96, 98, 99, 101, 102, 103, 104, 105, 106, 107, 110, 111, 112, 113, 114, 115, 117, 118, 121, 122, 126, 127, 129, 130, 131, 132, 133, 134, 135, 136, 137, 138, 139, 140, 141, 142, 143, 144, 145, 146, 147, 148, 149, 150, 152, 153, 154, 155, 156, 157, 158, 159, 162, 163, 164, 165, 166, 167, 168, 170, 171, 172,

173, 174, 175, 181, 182, 183, 184,
186, 187, 188, 189, 190, 191, 192,
193, 194, 195, 196, 197, 198, 199,
200, 201, 202, 203, 205, 206, 207,
208, 209, 210, 211, 212, 213, 214,
215, 217, 218, 219, 220, 221, 223,
225, 226, 228, 229, 230, 231, 232,
233, 234, 235, 236

Sustainable Development 44, 45, 46, 62,
65, 67, 68, 69, 80, 81, 82, 83, 85, 110,
114, 132, 133, 135, 141, 142, 143,
144, 148, 149, 152, 154, 155, 156,
157, 158, 167, 168, 171, 172, 182,
187, 189, 190, 191, 193, 194, 196,
197, 198, 200, 207, 208, 209, 210,
211, 212, 213, 215, 218, 221, 229,
232, 233, 234, 236

sustainable development goals 65, 67, 68,
80, 81, 83, 85, 132, 133, 142, 144, 148,
149, 152, 154, 155, 157, 158, 168, 171,
190, 197, 198, 215, 221

Sustainable Performance Management
131, 132, 133, 134, 135, 137, 143, 144,
145, 146, 147, 148, 149, 150, 155, 158

Sustainable practices 20, 21, 22, 24, 25,
26, 27, 28, 33, 35, 37, 52, 54, 55, 58,
59, 71, 90, 95, 103, 111, 112, 113,
115, 121, 122, 137, 141, 146, 166,
168, 173, 183, 192, 194, 195, 196,
200, 201, 202, 206, 230

Sustainable Waste management 43, 44,
45, 46, 47, 48, 54, 175, 217, 218, 219,

220, 223, 225, 226, 228, 229, 230, 231,
233, 234, 235

W

Waste 2, 4, 5, 6, 7, 11, 12, 13, 14, 16, 17,
18, 19, 20, 21, 22, 23, 24, 25, 28, 33,
34, 35, 36, 37, 38, 40, 41, 42, 43, 44,
45, 46, 47, 48, 49, 50, 51, 52, 53, 54,
55, 56, 57, 58, 59, 60, 61, 62, 63, 64,
67, 68, 87, 88, 89, 90, 92, 98, 99, 100,
101, 102, 104, 122, 132, 136, 139,
142, 159, 162, 164, 166, 167, 168,
169, 171, 172, 174, 175, 176, 177,
178, 198, 200, 201, 202, 203, 204,
205, 206, 217, 218, 219, 220, 221,
222, 223, 224, 225, 226, 227, 228,
229, 230, 231, 232, 233, 234, 235, 236

Waste Disposal 44, 45, 50, 54, 62, 174,
222, 224, 226, 227, 228, 229

Waste Management 7, 13, 14, 17, 18, 21,
22, 25, 33, 34, 35, 36, 37, 38, 40, 41,
42, 43, 44, 45, 46, 47, 48, 49, 50, 51,
52, 54, 55, 56, 57, 58, 59, 60, 61, 62,
63, 64, 159, 162, 167, 171, 172, 175,
176, 198, 217, 218, 219, 220, 221,
222, 223, 224, 225, 226, 227, 228,
229, 230, 231, 232, 233, 234, 235, 236

Waste management Hierarchy 44, 46, 52,
54, 55, 221

wastewater 65, 66, 67, 68, 69, 70, 71, 72,
74, 76, 80, 81, 82, 83, 84, 85

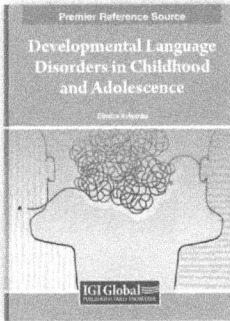

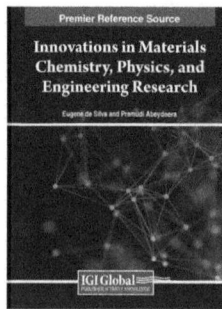

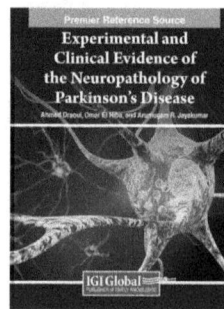

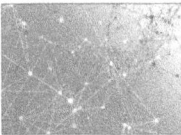

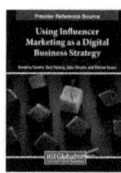

9 798369 325957